# Lecture Notes in Computer Science 8612

*Commenced Publication in 1973*
Founding and Former Series Editors:
Gerhard Goos, Juris Hartmanis, and Jan van Leeuwen



# FoLLI Publications on Logic, Language and Information

## Subline of Lectures Notes in Computer Science

Glyn Morrill Reinhard Muskens
Rainer Osswald Frank Richter (Eds.)

# Formal Grammar

19th International Conference, FG 2014
Tübingen, Germany, August 16-17, 2014
Proceedings

Springer

Volume Editors

Glyn Morrill
Universitat Politècnica de Catalunya
Departament de Llenguatges i Sistemes Informàtics
Barcelona, Spain
E-mail: morrill@lsi.upc.edu

Reinhard Muskens
Tilburg University, Department of Philosophy
Tilburg, The Netherlands
E-mail: r.a.muskens@gmail.com

Rainer Osswald
Heinrich-Heine-University Düsseldorf
Department of Linguistics and Information Science
Düsseldorf, Germany
E-mail: osswald@phil.hhu.de

Frank Richter
Goethe University Frankfurt am Main
Institute for English and American Studies
Department of Linguistics
Frankfurt, Germany
E-mail: frankmrichter@gmail.com

ISSN 0302-9743 e-ISSN 1611-3349
ISBN 978-3-662-44120-6 e-ISBN 978-3-662-44121-3
DOI 10.1007/978-3-662-44121-3
Springer Heidelberg New York Dordrecht London

Library of Congress Control Number: 2014943008

LNCS Sublibrary: SL 1 – Theoretical Computer Science and General Issues



*Typesetting:* Camera-ready by author, data conversion by Scientific Publishing Services, Chennai, India

Printed on acid-free paper

Springer is part of Springer Science+Business Media (www.springer.com)

# Preface

FG provides a forum for the presentation of new and original research on formal grammar, mathematical linguistics and the application of formal and mathematical methods to the study of natural language. Themes of interest include, but are not limited to:

- Formal and computational phonology, morphology, syntax, semantics and pragmatics
- Model-theoretic and proof-theoretic methods in linguistics
- Logical aspects of linguistic structure
- Constraint-based and resource-sensitive approaches to grammar
- Learnability of formal grammar
- Integration of stochastic and symbolic models of grammar
- Foundational, methodological and architectural issues in grammar and linguistics
- Mathematical foundations of statistical approaches to linguistic analysis

Previous Formal Grammar meetings were held in Barcelona (1995), Prague (1996), Aix-en-Provence (1997), Saarbrücken (1998), Utrecht (1999), Helsinki (2001), Trento (2002), Vienna (2003), Nancy (2004), Edinburgh (2005), Malaga (2006), Dublin (2007), Hamburg (2008), Bordeaux (2009), Copenhagen (2010), Ljubljana (2011), Opole (2012), and Düsseldorf (2013).

The present volume collects the papers from the 19th Conference on Formal Grammar celebrated in Tübingen in 2014. The conference comprised 10 contributed papers selected from 19 high quality submissions, and two invited contributions, by Thomas Icard and Christian Retoré.

We thank for their support the local organisers of ESSLLI 2014, with which the conference was colocated.

May 2014

Glyn Morrill
Reinhard Muskens
Rainer Osswald
Frank Richter

# Organization

## Program Committee

| | |
|---|---|
| Alexander Clark | King's College London, UK |
| Benoit Crabbé | Université Paris 7, France |
| Berthold Crysmann | CNRS - LLF, France |
| Denys Duchier | Université d'Orleans, France |
| Annie Foret | IRISA - IFSIC, France |
| Nissim Francez | Technion, Israel |
| Philippe de Groote | Inria Nancy, France |
| Laura Kallmeyer | University of Düsseldorf, Germany |
| Makoto Kanazawa | National Institute of Informatics, Japan |
| Greg Kobele | University of Chicago, USA |
| Valia Kordoni | Humboldt-Universität Berlin, Germany |
| Wolfgang Maier | University of Düsseldorf, Germany |
| Stefan Müller | Freie Universität Berlin, Germany |
| Gerald Penn | University of Toronto, Canada |
| Christian Retoré | Université Bordeaux 1, France |
| Manfred Sailer | Goethe University Frankfurt, Germany |
| Ed Stabler | UCLA, USA |
| Jesse Tseng | CNRS - CLLE-ERSS, France |
| Oriol Valentín | Universitat Politècnica de Catalunya, Spain |

## Program Chairs

| | |
|---|---|
| Glyn Morrill | Universitat Politècnica de Catalunya, Spain |
| Reinhard Muskens | Tilburg University, The Netherlands |
| Rainer Osswald | Heinrich-Heine-Universität Düsseldorf, Germany |

## Standing Committee

| | |
|---|---|
| Glyn Morrill | Universitat Politècnica de Catalunya, Spain |
| Reinhard Muskens | Tilburg University, The Netherlands |
| Rainer Osswald | Heinrich-Heine-Universität Düsseldorf, Germany |
| Frank Richter | Goethe Universität, Frankfurt am Main, Germany |

# Table of Contents

## Invited Contributions



## Contributed Papers

# Higher-Order Syllogistics

Thomas F. Icard, III

Department of Philosophy
Stanford University, California

**Abstract.** We propose a distinction between bottom-up and top-down systems of natural logic, with the classical syllogism epitomizing the first and the Monotonicity Calculus the second. We furthermore suggest it useful to view top-down systems as higher-order generalizations of broadly syllogistic systems. We illustrate this view by proving a result of independent interest: we axiomatize the first-order/single-type fragment of a higher-order calculus for reasoning about inclusion and exclusion (MacCartney and Manning, 2009; Icard, 2012). We show this logic is equivalent to a syllogistic logic with *All* and nominal complementation, in fact a fragment of a system recently studied (Moss, 2010b).

**Keywords:** syllogistics, natural logic, exclusion, surface reasoning.

## 1 Motivation

Systems of syllogistic logic long held the distinction of being the centerpiece of the study of logic and language, so much so that Kant famously declared that the subject seemed "to all appearances to be finished and complete" (1997, Bxviii-xix). With the development of relational logic in the late 19$^{\text{th}}$ century and quantification theory in the early 20$^{\text{th}}$ century, spurred mostly by problems in foundations of mathematics, logical interest in the traditional systems waned. Recently, however, work on syllogistic[1] logics and extensions thereof has seen something of a renaissance, due in large part to concern with computational issues. It has been shown that restricting one's logical language to match some controlled fragment of *natural* language can result in tractable, even polynomial time, satisfiability testing (McAllester and Givan, 1992); indeed, the traditional syllogism provides one such example (Pratt-Hartmann, 2004). Much recent work has explored the space of logical systems and their decision problems, which result from restricting the syntax of first or even higher-order logic in some linguistically motivated way (see, e.g., Moss 2010a for a comprehensive overview).

This research program has sometimes gone under the heading of 'natural logic', since it concerns logical systems inspired by natural language. There is a second tradition within so called natural logic, related but distinct, which allows the syntax of the language to be arbitrary—e.g., admitting all expressions of some (syntactically parsed, recursively defined) natural language—but restricts

[1] In this paper, I will be using the terms 'syllogism' and 'syllogistic' loosely, following the recent literature (see, e.g., Pratt-Hartmann and Moss 2009; Moss 2010a, etc.).

G. Morrill et al. (Eds.): Formal Grammar 2014, LNCS 8612, pp. 1–14, 2014.

what semantic features of expressions are inferentially visible. The classic example of such a system is the Monotonicity Calculus of van Benthem (1991) and Sánchez-Valencia (1991), which defines proof systems based on *monotonicity* and *antitonicity* of functions over expressions parsed, e.g., in some version of categorial grammar (see van Benthem 2008 or Icard and Moss 2014 for recent overviews). This tradition has also generated a long line of interesting logical work, expanding and extending the original calculus (e.g., Dowty 1994; Bernardi 2002; MacCartney and Manning 2009, among many others), and has found its way into computational applications (MacCartney and Manning, 2007).

While this difference in approach—bottom-up controlled syntax vs. top-down controlled semantics—is significant, I want to demonstrate in this note that, from a certain formal point of view, at least some top-down systems such as the Monotonicity Calculus should be seen as *generalizations* of bottom-up syllogistic fragments. Specifically, in many cases of interest, the higher-order systems can be seen as lifting the syllogistic systems from reasoning within a single type (e.g., the type $e \to t$ of predicates) to reasoning with arbitrary types, as well as reasoning between types. In the next section §2, I will illustrate this simple idea with the Monotonicity Calculus on the one hand, and a syllogistic system involving only the quantifier *All* on the other. Next, in §3 I will use the idea to distill a very simple axiomatization of the first-order (single-type) fragment of a higher-order calculus for reasoning about exclusion relations (MacCartney and Manning, 2009; Icard, 2012), a result of independent interest. In §4 I will explain what we can learn from this about the full higher-order Exclusion Calculus and questions about its axiomatization. Finally, §5 will conclude with some speculative remarks and pointers to further questions.

## 2 Monotonicity and the *All* Fragment

The Monotonicity Calculus captures monotonicity reasoning at arbitrary types in the function hierarchy. Here we follow the presentation in Icard and Moss (2013, 2014). Starting with basic types $e$ and $t$, we add new function types $\sigma \xrightarrow{+} \tau$, $\sigma \xrightarrow{-} \tau$, and $\sigma \xrightarrow{\cdot} \tau$, corresponding to monotone, antitione, and non-monotone functions, respectively. Type domains are preordered sets $\mathbb{D}_\tau = (D_\tau, \leq_\tau)$, antecedently given for basic types, and obtained by taking all functions preordered pointwise for functional types. That is, $\mathbb{D}_{\sigma \xrightarrow{m} \tau} = (D_{\sigma \xrightarrow{m} \tau}, \leq_{\sigma \xrightarrow{m} \tau})$, where $f \leq_{\sigma \xrightarrow{m} \tau} g$ iff $f(a) \leq_\tau g(a)$ for all $a \in D_\sigma$. If $m = \cdot$, then $D_{\sigma \xrightarrow{m} \tau}$ is the set of all functions from $D_\sigma$ to $D_\tau$; if $m = +$, it is the set of monotone functions; if $m = -$, it is the set of antitone functions. Given a set of typed constants, we can define a set of typed terms by allowing any type-compatible function application. One can then reason about inequalities $s \preceq t$ between terms $s$ and $t$ of compatible type.

A system of type domains $\{\mathbb{D}_\tau\}_\tau$, together with an interpretation function $[\![\cdot]\!]$ sending terms to elements of the appropriate type domain, provides a model $\mathcal{M} = \langle \{\mathbb{D}_\tau\}_\tau, [\![\cdot]\!] \rangle$ of such a typed language, whereby $\mathcal{M} \vDash s \preceq t$, iff $[\![s]\!] \leq_\tau [\![t]\!]$. There thus arises an obvious question of completeness for any given class of models, and proof system for deriving inequality statements. The question is answered for several cases in Icard and Moss (2013), where the main two rules of inference, relating terms of different types, are those found already in early work on the Monotonicity Calculus (van Benthem, 1991; Sánchez-Valencia, 1991):[2]

$$\text{(Monotone)}\quad \frac{u \preceq v}{t_{\uparrow}(u) \preceq t_{\uparrow}(v)} \qquad\qquad \text{(Antitone)}\quad \frac{u \preceq v}{t_{\downarrow}(v) \preceq t_{\downarrow}(u)}$$

with $t_{\uparrow}$ of monotone functional type and $t_{\downarrow}$ antitone. Together with a rule capturing the pointwise order on functions, and rules for partial order (see Fig. 1 below), this system is complete with respect to the general class of applicative (or Henkin) models. Adding two further rules gives completeness with respect to the class of standard structures in which each domain comes with upper and lower bounds for all pairs of elements (called *weakly complete* in Icard and Moss 2013), a property which standard boolean domains certainly satisfy.

When applied to natural language, the Monotonicity Calculus captures entailments between sentences that might otherwise be difficult to translate accurately into a logical form. To take an example inspired by a comment in Geurts and van der Slik (2005), while first-order renderings of so called donkey-sentences are notoriously controversial, the following entailment is patent:

$$\frac{\text{Everyone who knows a foreign language speaks it at home}}{\text{Everyone who knows a foreign language speaks it at home or at work}}$$

For examples like this, the top-down inference strategy characteristic of the Monotonicity Calculus is particularly appropriate. If we replace *Everyone* with the non-first-order-definable *Most people*, the advantage is even more pronounced.

A natural question, even if it admits an obvious answer in this case, is what fragment of first-order logic this kind of reasoning captures, for example, if we restrict the language to only variables of type $e \rightarrow t$ and relations between them. Let us explicitly introduce a language of variables and *All* statements:

$$\mathsf{Var} \quad ::= \quad X \mid Y \mid Z \mid \ldots \qquad\qquad L_{all} \quad ::= \quad \mathsf{Var} \Rightarrow \mathsf{Var}$$

Thus, instead of writing *All X are Y*, as is typical in work on syllogistics, we will simply write $X \Rightarrow Y$. We can again interpret such statements in preorders $\mathbb{D} = \langle D, \leq_D \rangle$ together with an interpretation function $[\![\cdot]\!]$ mapping each variable to an element of $D$, so that $\langle \mathbb{D}, [\![\cdot]\!] \rangle \models X \Rightarrow Y$, just in case $[\![X]\!] \leq_D [\![Y]\!]$.

In what might be the simplest of all completeness proofs, Moss (2010a) shows that this language can be axiomatized over the class of boolean lattices by the two basic rules of calculus $\mathcal{C}_0$ in Figure 1.[3] These are essentially just the two axioms for preorders. That is, the logic of *All* by itself, as a relation between sets or predicates, coincides with the logic of preorders.

[2] Usually these rules are stated for arbitrary complex terms $t$ with $u$ some *subterm occurrence* of $t$, shown to be in monotone (or antitone) position. The two versions are equivalent in the presence of the other rules (Icard and Moss, 2013, 2014).

[3] Strictly speaking, the proof shows completeness with respect to set-based semantics, with $\Rightarrow$ interpreted in terms of subset. But of course, since every powerset algebra can be viewed as a boolean lattice, the result as stated is an obvious consequence.

$$\text{(I)}\quad \frac{}{X \Rightarrow X} \qquad\qquad \text{(II)}\quad \frac{X \Rightarrow Y \qquad Y \Rightarrow Z}{X \Rightarrow Z}$$

**Fig. 1.** Calculus $\mathcal{C}_0$: The Logic of *All* (from Moss 2010a)

This logic also coincides with that of the Monotonicity Calculus when restricting to a single type. After all, the type domain for each type is just a preorder (or perhaps a boolean lattice, in which case Moss' result applies). It is in this sense that the *All* fragment can be said to coincide with the single-type fragment of the Monotonicity Calculus. The main two inter-type rules—one for functions preserving the order from type to type (Monotone), the other for functions reversing the order (Antitione)—give a very natural extension of the *All* fragment to the higher-order setting. If all we have is a preorder, preserving and reversing the order arguably exhaust the reasonable properties of such functions.

From one point of view, the rules of $\mathcal{C}_0$ capture everything there is to know about *All*. By results of van Benthem (1984) and Westerståhl (1984), if $Q$ is a standard quantifier (i.e., conservative, quantitative, having extension, and showing variety), and satisfies the rules of $\mathcal{C}_0$, then $Q$ must be the quantifier *All*.

But from another point of view, $\mathcal{C}_0$ falls short of what we can use the Monotonicity Calculus to tell us about *All*; we are not even using the monotonicity rules when working within a single type. As a fragment of the traditional syllogism, this essentially just gives us the *Barbara* rule. As a fragment of first-order logic, $\mathcal{C}_0$ is also quite poor. If $P_1, P_2, \ldots$ are a collection of one-place predicates, we essentially have just one axiom—$\forall x(P_i x \supset P_i x)$—and one rule—from $\forall x(P_i x \supset P_j x)$ and $\forall x(P_j x \supset P_k x)$, infer $\forall x(P_i x \supset P_k x)$.

We can invoke the full Monotonicity Calculus and add typed constants:

$$\mathsf{all} : p \xrightarrow{-} (p \xrightarrow{+} t) \qquad \mathsf{some} : p \xrightarrow{+} (p \xrightarrow{+} t) \qquad \mathsf{no} : p \xrightarrow{-} (p \xrightarrow{-} t)\ ,$$

where $p \equiv (e \to t)$, in order to obtain a slightly more powerful logical fragment. We can now derive many more of the traditional syllogisms, e.g., *Darii*:

$$\frac{\mathit{No}(Z,Y) \qquad \mathit{All}(X,Z)}{\mathit{No}(X,Y)} \qquad \equiv \qquad \frac{X \Rightarrow Z}{\mathsf{no}(Z,Y) \Rightarrow \mathsf{no}(X,Y)}$$

provided that we translate $\Rightarrow$ (when between predicates) and the constant all both in terms of the universal quantifier. As van Eijck (2005) has demonstrated, every valid syllogism can be derived with exactly one instance of a monotonicity/antitonicity rule, and at most one application of symmetry or existential import. Still, as a fragment of first-order logic, the resulting system is rather small. Monotonicity of all, e.g., gives us distribution: from $\forall x(P_i x \supset P_j x)$, infer $\forall x(P_i x) \supset \forall x(P_j x)$, with the same result for some. One attraction of the Monotonicity Calculus is that it can be seamlessly extended with higher-order expressions, e.g., with a constant most or any other monotonic/antitonic expression, which elude both first-order logic and the traditional syllogism.[4] There is no

[4] Though see Endrullis and Moss (2014) for an elegant syllogistic system with *Most*.

reason not to add such expressions. As far as the inference system is concerned, these are all just more function constants with monotonicity-marked types.

To summarize this warm-up discussion, the simple syllogistic system characterizing *All* by itself coincides with the Monotonicity Calculus when restricted to a single type, in fact any single type with a preordered domain. The full version of the calculus with higher-order types puts these systems together and relates them *via* the monotonicity rules, which is perhaps the most natural such extension in the setting of preorders. It is in this sense that it is a generalization of the *All* fragment. Finally, once we have the full system, we can go back and add constants for quantifiers, in effect adding further rules of inference. Once we do that, however, there is no reason to stop at first-order-definable expressions.

## 3 Exclusion and *All* with Complements

MacCartney and Manning (2009) introduced an informal extension of the Monotonicity Calculus, to deal with relations not only of inclusion, but also *exclusion*. Here we follow Icard (2012), which showed how to formalize the system as a genuine extension of the Monotonicity Calculus. Let us begin with an exclusion language for a single type, before moving to the full higher-order language in the next section. Here we will present our main result.

For a given set $\mathsf{Var}$ of variables (or constants; it does not matter here), we add to $\preceq$ several further relations.[5] Call the language $\mathcal{L}_x^1$:

$$\mathcal{L}_x^1 \quad ::= \quad \mathsf{Var} \preceq \mathsf{Var} \quad | \quad \mathsf{Var} \succeq \mathsf{Var} \quad | \quad \mathsf{Var} \pitchfork \mathsf{Var} \quad | \quad \mathsf{Var} \smile \mathsf{Var}$$

$\mathcal{L}_x^1$ is naturally interpreted in bounded distributive lattices. For a structure $\mathcal{M} = \langle \mathbb{D}, [\![\cdot]\!] \rangle$, where $\mathbb{D} = (D, \leq_D)$ comes with $\wedge$ and $\vee$ operations, a 0 element, and a 1 element, and $[\![\cdot]\!] : \mathsf{Var} \to D$ interprets the variables, our truth clauses will be:

$$\begin{array}{lll}
\mathcal{M} \models X \preceq Y & \text{iff} & [\![X]\!] \wedge [\![Y]\!] = [\![X]\!]; \\
\mathcal{M} \models X \succeq Y & \text{iff} & [\![X]\!] \vee [\![Y]\!] = [\![X]\!]; \\
\mathcal{M} \models X \pitchfork Y & \text{iff} & [\![X]\!] \wedge [\![Y]\!] = 0; \\
\mathcal{M} \models X \smile Y & \text{iff} & [\![X]\!] \vee [\![Y]\!] = 1;
\end{array}$$

Just as the Monotonicity Calculus on a single type corresponds to the *All* syllogistic fragment, it turns out $\mathcal{L}_x^1$ corresponds to a natural syllogistic fragment, adding only nominal complementation to the *All* fragment. Following Moss

[5] We are ignoring here the three additional relations from Icard (2012) and MacCartney and Manning (2009): $\#$, $\equiv$, and $\curlywedge$. Any reasoning system for these four connectives can easily be extended to incorporate the rest.

Note that for truth-value types, $\pitchfork$ is the relation of *contrary*, and $\smile$ is that of *subcontrary*. Thus, with all of these relations together, we can express the traditional square of opposition. Also, it is worth pointing out that the quantified sentences stand in these relations only if we assume non-vacuity of all predicates.

(2010b) (see also Pratt-Hartmann and Moss 2009), who considered a slightly richer language with *Some* as well, we extend the *All* language as follows:

$$\mathsf{CVar} \quad ::= \quad \mathsf{Var} \mid \mathsf{Var}' \qquad\qquad L^1_{all,'} \quad ::= \quad \mathsf{CVar} \Rightarrow \mathsf{CVar}$$

Here $X'$ is intuitively the complement of $X$. We do not allow variables of the form $X''$. A bijective translation between $\mathcal{L}^1_x$ and $L^1_{all,'}$ can be given as follows:[6]

$$\begin{array}{ccc} X \preceq Y & \equiv & X \Rightarrow Y \\ X \succeq Y & \equiv & X' \Rightarrow Y' \\ X \pitchfork Y & \equiv & X \Rightarrow Y' \\ X \smile Y & \equiv & X' \Rightarrow Y \end{array}$$

**Fig. 2.** Correspondence between $\mathcal{L}^1_x$ and $L^1_{all,'}$

If $S$ is an $L^1_{all,'}$-sentence, let $S^*$ be the translation of $S$ into $\mathcal{L}^1_x$. These two languages are not mere notational variants. As described above, algebraically, $\mathcal{L}^1_x$-expressions are naturally interpreted in bounded distributive lattices. The existence of complements is not necessary. $L^1_{all,'}$-expressions, on the other hand, are naturally interpreted in orthoposets (see Moss 2010b and the Appendix), which have complements but need not in general be lattices. Both of these generalize boolean lattices, and in fact, in boolean lattices, their interpretations coincide. In this case we interpret $X'$ as the boolean complement of $X$.

**Lemma 1.** Suppose $\mathbb{D} = (D, \leq_D)$ is a boolean lattice and $[\![\cdot]\!] : \mathsf{Var} \to D$. Then $\langle \mathbb{D}, [\![\cdot]\!] \rangle \models S$, iff $\langle \mathbb{D}, [\![\cdot]\!] \rangle \models S^*$.

*Proof.* Simply check all four cases. For example,

$$\begin{array}{rcl} \langle \mathbb{D}, [\![\cdot]\!] \rangle \models X' \Rightarrow Y & \Longleftrightarrow & [\![X']\!] \leq_D [\![Y]\!] \\ & \Longleftrightarrow & [\![X]\!]' \leq_D [\![Y]\!] \\ & \Longleftrightarrow & [\![X]\!] \vee [\![Y]\!] = 1 \\ & \Longleftrightarrow & \langle \mathbb{D}, [\![\cdot]\!] \rangle \models X \smile Y \\ & \Longleftrightarrow & \langle \mathbb{D}, [\![\cdot]\!] \rangle \models (X' \Rightarrow Y)^* . \end{array}$$

The other cases are just as easy.

Lemma 1 will allow us to transfer a completeness theorem for $L^1_{all,'}$ directly to $\mathcal{L}^1_x$. Our axiomatization of $L^1_{all,'}$ with respect to boolean domains follows a recent result from Moss (2010b), which in our setting can be quite simplified. The rules of the logic $\mathcal{C}_1$ are given in Figure 3, where $A, B, C$ range over $\mathsf{CVar}$ expressions $X, X', Y, Y', \ldots$; and if $A = X'$, then $A' = X$.

[6] In fact, relating exclusion to complement was already suggested by van Benthem (2008) and Moss (p.c., 2010).

$$\text{(I)}\ \frac{}{A \Rightarrow A} \qquad \text{(II)}\ \frac{A \Rightarrow B \qquad B \Rightarrow C}{A \Rightarrow C}$$

$$\text{(III)}\ \frac{A \Rightarrow A'}{A \Rightarrow B} \qquad \text{(IV)}\ \frac{A \Rightarrow B'}{B \Rightarrow A'}$$

**Fig. 3.** Calculus $\mathcal{C}_1$: The logic of *All* with nominal complementation

If $\Gamma$ is a set of statements in $L^1_{all,'}$, and $S$ is a statement in $L^1_{all,'}$, then let us write $\Gamma \vdash_c S$ if there is a finite tree with $S$ as the root and each node either an element of $\Gamma$ or the result of an application of one of the rules in Figure 3 from its parents. Call the resulting complement calculus $\mathcal{C}_1$.

**Theorem 1 (Completeness of $\mathcal{C}_1$).** $\Gamma \vdash_c S$ if and only if $\Gamma \vDash S$.

For a proof, mostly following but also simplifying Moss (2010b) (because this is a smaller language), see the technical Appendix.

This simple axiomatization belies the amount of information it captures about exclusion reasoning. First, notice that there are two instances of (I), capturing the reflexivity of $\preceq$ and $\succeq$. Two instances of (IV) capture the fact that $\preceq$ and $\succeq$ are inverses; the other two instances derive the symmetry of $\pitchfork$ and $\smile$. For example, one instance of (IV) gives:

$$\frac{X \Rightarrow Y'}{Y \Rightarrow X'} \quad \equiv \quad \frac{X \pitchfork Y}{Y \pitchfork X}$$

Rule (III) captures the interaction between between $\pitchfork$ and $\preceq$, and that between $\smile$ and $\succeq$. For example, we have as an instance of (III):

$$\frac{X' \Rightarrow X}{X' \Rightarrow Y'} \quad \equiv \quad \frac{X \smile X}{X \succeq Y}$$

Most strikingly, there are eight instances of (II), reproduced here in Fig. 4:

$$\frac{X \preceq Y \qquad Y \preceq Z}{X \preceq Z} \qquad \frac{X \pitchfork Y \qquad Y \smile Z}{X \preceq Z}$$

$$\frac{X \succeq Y \qquad Y \succeq Z}{X \succeq Z} \qquad \frac{X \smile Y \qquad Y \pitchfork Z}{X \succeq Z}$$

$$\frac{X \preceq Y \qquad Y \pitchfork Z}{X \pitchfork Z} \qquad \frac{X \pitchfork Y \qquad Y \succeq Z}{X \pitchfork Z}$$

$$\frac{X \succeq Y \qquad Y \smile Z}{X \smile Z} \qquad \frac{X \smile Y \qquad Y \preceq Z}{X \smile Z}$$

**Fig. 4.** The eight instances of rule (II) translated into $\mathcal{L}^1_x$

Indeed, these rules subsume all of the rules proposed in Icard (2012) that concern a single type.[7] Lemma 1 and Theorem 1 in fact show:

**Corollary 1.** The translation of the calculus $\mathcal{C}_1$ gives a sound and complete proof system for $\mathcal{L}^1_x$ with respect to boolean lattices, and also with respect to bounded distributive lattices.

*Proof.* Theorem 1 and Lemma 1 show these rules are sound and complete with respect to boolean domains. Since the rules are sound in general bounded distributive lattices as well (see Icard 2012), the second result also follows.

Satisfyingly, the translation in Fig. 2 allows us to express eighteen rules (none of which seem to be redundant) in terms of four simple schemata. This also makes apparent that the single-type fragment of the Exclusion Calculus can be seen as extending that of the Monotonicity Calculus by just two further rules, (III) and (IV), in a language with an added complement operator.

## 4 Projecting Exclusion and Complements

As with the Monotonicity Calculus, most of the interest in the Exclusion Calculus comes from the higher-order setting. Instead of considering the full functional hierarchy, in this section we examine only a small extension of the language $L^1_{all,'}$, with variables $\mathsf{Var} = \{X, Y, Z, \dots\}$ interpretated in a boolean lattice $\mathbb{D} = (D, \leq_D)$, as before, and a set of function symbols $\Phi = \{\phi, \psi, \dots\}$ interpreted as functions from $\mathbb{D}$ to some other boolean lattice $\mathbb{E} = (E, \leq_E)$. Our simple second-level 'syllogistic' language $L^2_{all,'}$ will be generated by the following grammar:

$$\Phi\mathsf{exp} \quad ::= \quad \Phi(\mathsf{Var}) \mid \Phi(\mathsf{Var})' \qquad L^2_{all,'} \quad ::= \quad L^1_{all,'} \mid \Phi\mathsf{exp} \Rightarrow \Phi\mathsf{exp}$$

That is, we can make statements about pairs of (possibly complemented) variables, or about pairs of functions applied to variables (possibly complemented).

By analogy with the Monotonicity Calculus, and following Icard (2012), functions in $\Phi$ are marked as being from one of six semantically distinguished classes: (merely) monotone, (merely) antitione, and four more refining these two classes.[8]

1. A function $f : \mathbb{D} \to \mathbb{E}$ is (completely) **additive** if for all $a, b \in D$, $f(a \vee b) = f(a) \vee f(b)$, and moreover $f(1_D) = 1_E$.
2. A function $f : \mathbb{D} \to \mathbb{E}$ is (completely) **multiplicative** if for all $a, b \in D$, $f(a \wedge b) = f(a) \wedge f(b)$, and moreover $f(0_D) = 0_E$.

---

[7] Notice also that five instances of (II) correspond to the five valid syllogisms using only the $a$ and $e$ forms: *Barbara, Cesare, Celarent, Camestres, Calemes.*

[8] As claimed in Icard (2012), the connection between these classes of functions—in particular the anti-additive, and the anti-additive, anti-multiplicative (sometimes called anti-morphic) functions—raises the interest of this logical system from the point of view of grammar. These classes are correlated with the distribution of negative polarity items (Zwarts, 1981). Thus, a combined system for parsing and inference, in the style of Dowty (1994) or Bernardi (2002), for example, would be attractive here.

3. A function $f : \mathbb{D} \to \mathbb{E}$ is (completely) **anti-additive** if for all $a, b \in D$, $f(a \vee b) = f(a) \wedge f(b)$, and moreover $f(1_D) = 0_E$.
4. A function $f : \mathbb{D} \to \mathbb{E}$ is (completely) **anti-multiplicative** if for all $a, b \in D$, $f(a \wedge b) = f(a) \vee f(b)$, and moreover $f(0_D) = 1_E$.

Also as with the Monotonicity Calculus, these semantic classes are manifested in valid rules, giving rise to a full-fledged Exclusion Calculus. In the following Fig. 5, $\phi_\uparrow$ is monotone, $\phi_+$ is additive, $\phi_\times$ is multiplicative, $\phi_\downarrow$ is antitone, $\phi_{\overline{+}}$ is anti-additive, and $\phi_{\overline{\times}}$ is anti-multiplicative.

$$\frac{X \Rightarrow Y}{\phi_\uparrow(X) \Rightarrow \phi_\uparrow(Y)} \qquad \frac{X' \Rightarrow Y'}{\phi_\uparrow(X)' \Rightarrow \phi_\uparrow(Y)'}$$

$$\frac{X' \Rightarrow Y}{\phi_+(X)' \Rightarrow \phi_+(Y)} \qquad \frac{X \Rightarrow Y'}{\phi_\times(X) \Rightarrow \phi_\times(Y)'}$$

$$\frac{X \Rightarrow Y}{\phi_\downarrow(X)' \Rightarrow \phi_\downarrow(Y)'} \qquad \frac{X' \Rightarrow Y'}{\phi_\downarrow(X) \Rightarrow \phi_\downarrow(Y)}$$

$$\frac{X' \Rightarrow Y}{\phi_{\overline{+}}(X) \Rightarrow \phi_{\overline{+}}(Y)'} \qquad \frac{X \Rightarrow Y'}{\phi_{\overline{\times}}(X)' \Rightarrow \phi_{\overline{\times}}(Y)}$$

**Fig. 5.** Calculus $\mathcal{C}_2$: Projecting exclusion relations from one type to another

Let us call $\mathcal{C}_2$ the calculus that results from adding the rules above to those from $\mathcal{C}_1$, with appropriate restrictions given by formula types. It seems very likely that the methods described in the Appendix can be combined with those used in Icard and Moss (2013) to prove the following:

**Conjecture:** $\mathcal{C}_2$ is complete for $L^2_{all,'}$ w.r.t. bounded distributive lattices.

These eight rule schemata summarize the twenty-four informative entries in the so called projection table (Icard 2012, see also MacCartney and Manning 2009), characterizing how functions 'project' the different relations from one type to another. The translation into this 'syllogistic' language now allows summarizing forty-two axioms in terms of only twelve perspicuous schemata. In particular, the rules above in Fig. 5 make transparent why we have chosen the function classes as we have: each one reflects a distinct pattern of projection, and together they represent all possible projection patterns.

As in the case of monotonicity, we can still go beyond what we have primitively in the calculus, adding constants for quantifiers all, some, no, etc., together with their marked types. Modulo the above conjecture, completeness for this fragment would be obtained by adding the specific rules from Fig. 5 appropriate for these constants; after all, their projectivity behavior is all that the inferential calculus can see. But, again, there is no reason to stop at first-order-definable quantifiers. For instance, if we let most be interpreted as a function from a boolean 'powerset' domain $\wp(E)$ to the truth value domain $\mathbf{2}$, such that $[\![\mathsf{most}]\!](S) = 1$ iff $\frac{|S|}{|E|} > \theta$ for some determined threshold $\theta \in [0.5, 1)$, then it is easy to check that most is multiplicative. An instance of the multiplicative rule above in Fig. 5 in this case would give:

$$\frac{\mathsf{aardvark} \Rightarrow \mathsf{wombat}'}{\mathsf{most}(\mathsf{aardvark}) \Rightarrow \mathsf{most}(\mathsf{wombat})'} \equiv \frac{\mathsf{aardvark} \pitchfork \mathsf{wombat}}{\mathsf{most}(\mathsf{aardvark}) \pitchfork \mathsf{most}(\mathsf{wombat})}$$

This captures good arguments like the following: Nothing is both an aardvark and a wombat; if most of my friends are aardvarks, this precludes the possibility that most of my friends are wombats. And so on.

It is also natural to consider extending this translation to the full higher-order setting of the Exclusion Calculus. It is expected that the same methods applied to the Monotonicity Calculus in Icard and Moss (2013) should be applicable in this case, with similar completeness results for special cases of model classes. What we have seen here is that the question of completeness for this richer system, which at first appears daunting, can be reduced to the question for a 'higher-order' version of the *All* fragment with complementation.

## 5 A Broader View

The two cases we have examined in this note hardly scratch the surface of what can be explored once we view top-down natural logical systems as higher-order versions of bottom-up natural logics. Starting with *All*, arguably the simplest syllogistic fragment, if we generalize this in a natural way to the higher-order setting, the Monotonicity Calculus is the result. Adding to *All* a complement operator and generalizing again in a natural way gives us the Exclusion Calculus. These are of course very basic syllogistic fragments. A natural question now is whether there are attractive higher-order versions of more complex syllogistic fragments. To take just one example, Pratt-Hartmann and Moss (2009) consider extensions of the full syllogism (with and without complements) to languages with relational statements, capturing arguments like:

$$\frac{\text{All wombats are marsupials}}{\text{All who respect a wombat respect a marsupial}}$$

which go beyond the traditional syllogistic. Arguments like this can be formalized in the Monotonicity Calculus, provided we take as a backdrop the full boolean $\lambda$-calculus (van Benthem, 1991), and mark $\exists$/some as being monotone:

$$\frac{\mathsf{wombat} \preceq \mathsf{marsupial}}{\lambda x.\exists(\lambda y.\mathsf{respect}(x,y) \wedge \mathsf{wombat}(y)) \preceq \lambda x.\exists(\lambda y.\mathsf{respect}(x,y) \wedge \mathsf{marsupial}(y))}$$

But perhaps there is a simple extension of the system described in §2 that does not make use of $\lambda$-abstraction, for instance adding primitive predicate types $\forall(m,r)$ and $\exists(m,r)$ as in Pratt-Hartmann and Moss (2009).

Another obvious question is why it seems more natural in top-down systems to adopt a main connective corresponding to *All*.[9] To be sure, the $\preceq$ relation we have considered here just is the *All* relation between type $e \to t$ objects. Between truth-values it is entailment, and for other, higher types it is a more abstract 'inclusion' relation. In fact, it is one of the fundamental insights of early work in logical semantics and natural logic that these relations make sense all the way up the (boolean) type hierarchy (Keenan and Faltz, 1984; van Benthem, 1991). But of course, to match the traditional syllogism more closely, we might also consider a primitive relation between terms corresponding to *Some*. A connective $\Cap$—with $t \Cap s$ meaning $[\![t]\!] \wedge [\![s]\!] \neq 0$—would make sense between terms of any type associated with a lower semi-lattice domain, thus certainly any boolean domain. Such a connective would not be unnatural; it would give us conjunction for truth-value-type expressions, and assertions of non-empty intersection for properties, for example. Moreover, it is clear that the projectivity behavior of functional expressions would be quite different from anything we have seen. For instance, single-argument $\exists/\mathsf{some}$ would exhibit the following pattern, distinguishing it from other additive quantifiers like $\exists^{\geq n}/\mathsf{at\text{-}least\text{-}}n$:

$$\frac{X \Cap Y}{F \preceq \mathsf{some}(X)}$$

for arbitrary $F$ of type $t$. It would be worthwhile exploring what the resulting system in the style of the Monotonicity or Exclusion Calculus would look like, since this would give a higher-order analogue of the full classical syllogism.

In the background here, of course, are also questions of complexity. While the complexity landscape is well studied for bottom-up natural logics (McAllester and Givan, 1992; Pratt-Hartmann, 2004; Moss, 2010a), including where the boundary lies between decidability and undecidability, these questions are relatively unexplored for top-down natural logics. How much more complex is a higher-order system than its 'flat' syllogistic fragment? Is there some useful correspondence between the complexity hierarchy of systems on the two sides?

From a modern point of view, traditional bottom-up syllogistic logics can be seen as capturing a limited range of inferences by restricting the language on which inference is based. This typically gives rise to an abstract structure: in the case of the *All* fragment, a preorder; for *All* with complements, an algebraic structure with complements. Where this structure makes sense and promises to be useful in more general types, and especially where there is interaction between the structures of different types, it is reasonable to look at higher-order versions of these systems. I have explained how we can view both the Monotonicity Calculus and the Exclusion Calculus through this lens, in such a way as to facilitate new results of independent interest. I submit that there is much more to be explored and gained in this connection.

[9] Incidentally, van Eijck (1985) has proven every valid syllogism can be obtained by a simple transformation of *Barbara*, thus showing *All* is in some sense all one needs for the classical syllogism, provided we add enough other machinery.

## References


van Benthem, J.: Questions about quantifiers. Journal of Symbolic Logic 49(2), 443–466 (1984)

van Benthem, J.: Language in Action: Categories, Lambdas, and Dynamic Logic. Studies in Logic, vol. 130. Elsevier, Amsterdam (1991)

van Benthem, J.: A brief history of natural logic. In: Chakraborty, M., Löwe, B., Mitra, M.N., Sarukkai, S. (eds.) Logic, Navya-Nyaya and Applications, Homage to Bimal Krishna Matilal, College Publications, London (2008)

Bernardi, R.: Reasoning with Polarity in Categorial Type Logic. PhD thesis, University of Utrecht (2002)

Dowty, D.: The role of negative polarity and concord marking in natural language reasoning. In: Proceedings of Semantics and Linguistic Theory (SALT) IV (1994)

van Eijck, J.: Generalized quantifiers and traditional logic. In: van Benthem, J., ter Meulen, A. (eds.) Generalized Quantifiers, Theory, and Applications. Foris, Dordrecht (1985)

van Eijck, J.: Syllogistics = monotonicity + symmetry + existential import. Technical Report SEN-R0512, CWI, Amsterdam (2005)

Endrullis, J., Moss, L.S.: Syllogistic logic with "Most". Unpublished ms (2014)

Geurts, B., van der Slik, F.: Monotonicity and processing load. Journal of Semantics 22 (2005)

Icard, T.F.: Inclusion and exclusion in natural language. Studia Logica 100(4), 705–725 (2012)

Icard, T.F., Moss, L.S.: A complete calculus of monotone and antitone higher-order functions. Unpublished ms (2013)

Icard, T.F., Moss, L.S.: Recent progress on monotonicity. Linguistic Issues in Language Technology 9 (2014)

Kant, I.: Critique of Pure Reason. Cambridge University Press (1997)

Keenan, E.L., Faltz, L.M.: Boolean Semantics for Natural Language. Springer (1984)

MacCartney, B., Manning, C.D.: Natural logic for textual inference. In: Proceedings of the ACL Workshop on Textual Entailment and Paraphrasing (2007)

MacCartney, B., Manning, C.D.: An extended model of natural logic. In: Proceedings of the Eighth International Conference on Computational Semantics, IWCS-8 (2009)

McAllester, D.A., Givan, R.: Natural language syntax and first-order inference. Artificial Intelligence 56, 1–20 (1992)

Moss, L.S.: Logics for natural language inference. ESSLLI 2010 Course Notes (2010a)

Moss, L.S.: Syllogistic logic with complements. In: van Benthem, J., Gupta, A., Pacuit, E. (eds.) Games, Norms, and Reasons: Logic at the Crossroads, pp. 185–203. Springer (2010b)

Pratt-Hartmann, I.: Fragments of language. Journal of Logic, Language, and Information 13, 207–223 (2004)

Pratt-Hartmann, I., Moss, L.S.: Logics for the relational syllogistic. The Review of Symbolic Logic 2(4), 647–683 (2009)

Sánchez-Valencia, V.: Studies on Natural Logic and Categorial Grammar. PhD thesis, Universiteit van Amsterdam (1991)

Westerståhl, D.: Some results on quantifiers. Notre Dame Journal of Formal Logic 25(2), 152–170 (1984)

Zwarts, F.: Negatief polaire uitdrukkingen I. GLOT 4, 35–132 (1981)

# Appendix

In this Appendix we sketch the proof of the following theorem.

**Theorem 1 (Completeness of $\mathcal{C}_1$).** $\Gamma \vdash_c S$ if and only if $\Gamma \models S$.

The soundness direction is by induction on length of proofs. Note, however, that there are many cases to check for each rule schema, since $A, B, C$ can be a variable or a complemented variable in each case. In all, there are eighteen distinct rule instances to verify.

We prove completeness of $\mathcal{C}_1$ *via* a kind of canonical model construction. The proof here is a variant on one from Moss (2010b) for a richer language including the quantifier *Some* as well. The proof in our case is simpler, but does not obviously follow from Moss' result, so we give it here. The strategy closely follows Moss' exposition otherwise. First note the following fact about $\mathcal{C}_1$:

$$\{A \Rightarrow B, A \Rightarrow B'\} \vdash_c A \Rightarrow A' \tag{1}$$

The proof is simple:

$$\text{(II)}\ \dfrac{A \Rightarrow B \qquad \dfrac{A \Rightarrow B'}{B \Rightarrow A'}\,\text{(IV)}}{A \Rightarrow A'}$$

As it happens, the standard canonical term model built from the syntax of $L^1_{all,'}$ is not in general a boolean lattice. But it can be represented as one, in an appropriate sense for completeness. In that direction, following Moss (2010b), we define a new type of algebraic structure, familiar also from quantum logic:

**Definition 1.** An **orthoposet** is an ordered structure $\mathbb{P} = (P, \leq_P, 0_P, ')$:

1. $(P, \leq_P)$ is a partial order ;
2. $0 \leq_P p$ for all $p \in P$ ;
3. The operation $'$ defines a function from $P$ to $P$, such that:
   (a) $p \leq_P q$ iff $q' \leq_P p'$,
   (b) $p'' = p$, and
   (c) if $p \leq_P q$ and $p \leq_P q'$, then $p = 0$.

The following lemma shows that every orthoposet can be represented by a boolean lattice in an appropriate sense (see Moss 2010b for a proof and a number of citations to other proofs):

**Lemma 2.** For any orthoposet $\mathbb{P}$, there is a boolean lattice $\mathbb{D}$ and a map $\mu$ : $\mathbb{P} \to \mathbb{D}$ such that $\mu(p') = \mu(p)'$, and $p \leq_P q$ iff $\mu(p) \leq_D \mu(q)$.

*Proof (Sketch).* Given $P$, define a *point* $S \subseteq P$ to be an $\leq_P$-upward closed set containing exactly one of $p$ or $p'$ for each $p \in P$. Let $\mathcal{S}$ be the set of all points, and define $\mathbb{D} = (\wp(\mathcal{S}), \subseteq)$. This defines an orthoposet, and in fact a boolean lattice. The map is defined: $\mu(p) = \{S \in \mathcal{S} : p \in S\}$. Checking that $\mu$ has the requisite properties is then routine.

We now proceed to define a kind of 'canonical orthoposet' $\mathbb{V}$ from $\Gamma$, which *via* Lemma 2 we will map to a canonical boolean model.

We first define the set $V$ of points. For each $A \in \mathsf{CVar}$, let

$$[A] = \{B : \Gamma \vdash_c A \Rightarrow B \text{ and } \Gamma \vdash_c B \Rightarrow A\}\ .$$

If there is some $A$ such that $\Gamma \vdash_c A \Rightarrow A'$, then let $0 = [A]$ and $1 = [A']$. Otherwise, add new elements 0 and 1. Let

$$V = \{[A] : A \in \mathsf{CVal}\} \cup \{0, 1\}\ .$$

Define a relation $\leq_V$ such that $[A] \leq_V [B]$ iff $\Gamma \vdash_c A \Rightarrow B$, and $0 \leq_V v \leq_V 1$ for all $v \in V$. Note that if $\Gamma \vdash_c A \Rightarrow A'$, then by rule (III), $\Gamma \vdash_c A \Rightarrow B$ for any $B$; hence $[A] \leq_V v$ for any $v$. The relation $\leq_V$ is thus well-defined.

Define $[A]'$ to be $[X']$ if $A = X$, and $[A'] = [X]$ if $A = X'$. Stipulate that $0' = 1$ and $1' = 0$. Again, if $\Gamma \vdash_c A \Rightarrow A'$, then since $\Gamma \vdash_c A \Rightarrow B$ for any $B$, also $\Gamma \vdash_c B' \Rightarrow A$ for any $B$, by rule (IV). Hence, $'$ is well defined.

Name this structure $\mathbb{V} = (V, \leq_V, 0, ')$. We claim $\mathbb{V}$ is an orthoposet:

1. That $\leq_V$ is a partial order follows from rules (I) and (II) ;
2. We already verified that $0 \leq_V [B]$ for any $B$ ;
3. For the operation $'$, we have:

$$\begin{aligned} [A] \leq_V [B]' &\iff \Gamma \vdash_c A \Rightarrow B' \\ &\overset{\text{(IV)}}{\iff} \Gamma \vdash_c B \Rightarrow A' \\ &\iff [B] \leq_V [A]' \end{aligned}$$

which verifies property (a) of Def. 1. Property (b) follows by definition, and property (c) follows by Eq. (1) above:

$$\begin{aligned} [A] \leq_V [B] \text{ and } [A] \leq_V [B]' &\Rightarrow \Gamma \vdash_c A \Rightarrow B \text{ and } \Gamma \vdash_c A \Rightarrow B' \\ &\overset{(1)}{\Rightarrow} \Gamma \vdash_c A \Rightarrow A' \\ &\Rightarrow [A] = 0\ . \end{aligned}$$

By Lemma 2, there is some boolean lattice $\mathbb{D}$ and map $\mu : \mathbb{V} \to \mathbb{D}$, s.t.

$$\begin{aligned} \Gamma \vdash_c A \Rightarrow B &\iff [A] \leq_V [B] \\ &\iff \mu([A]) \leq_D \mu([B])\ . \end{aligned}$$

Defining $[\![\cdot]\!]_{\mathbb{D}} : L^1_{all,'} \to D$ so that $[\![X]\!]_{\mathbb{D}} = \mu([X])$ and $[\![X']\!]_{\mathbb{D}} = \mu([X])'$,

$$(\mathbb{D}, [\![\cdot]\!]) \vDash A \Rightarrow B \iff \Gamma \vdash_c A \Rightarrow B\ ,$$

for all $A$ and $B$, from which completeness follows.

# Typed Hilbert Epsilon Operators and the Semantics of Determiner Phrases

Christian Retoré

LaBRI, Université de Bordeaux, France
(& MELODI, IRIT-CNRS, Toulouse, France)

**Abstract.** The semantics of determiner phrases, be they definite descriptions, indefinite descriptions or quantified noun phrases, is often assumed to be a fully solved question: common nouns are properties, and determiners are generalised quantifiers that apply to two predicates: the property corresponding to the common noun and the one corresponding to the verb phrase.

We first present a criticism of this standard view. Firstly, the semantics of determiners does not follow the syntactical structure of the sentence. Secondly the standard interpretation of the indefinite article cannot account for nominal sentences. Thirdly, the standard view misses the linguistic asymmetry between the two properties of a generalised quantifier.

In the sequel, we propose a treatment of determiners and quantifiers as Hilbert terms in a richly typed system that we initially developed for lexical semantics, using a many sorted logic for semantical representations. We present this semantical framework called the Montagovian generative lexicon and show how these terms better match the syntactical structure and avoid the aforementioned problems of the standard approach.

Hilbert terms are rather different from choice functions in that there is one polymorphic operator and not one operator per formula. They also open an intriguing connection between the logic for meaning assembly, the typed lambda calculus handling compositionality and the many-sorted logic for semantical representations. Furthermore epsilon terms naturally introduce type-judgements and confirm the claim that type judgments are a form of presupposition.

## 1 Presentation

Determiners and quantifiers are an important ingredient of (computational) semantics, at least of the part of semantics known as formal semantics or compositional semantics, that is concerned with what is asserted, especially by a sentence: such a semantical analysis tells "*who does what*" in a sentence.

Researchers in formal linguistics, must be aware that semantics also includes other aspects like lexical semantics, distributional semantics, vectors of words for which there exist far more efficient natural language processing tools. These aspects of semantics rather concern *what a text speaks about.*

G. Morrill et al. (Eds.): Formal Grammar 2014, LNCS 8612, pp. 15–33, 2014.

Of course both aspect are needed to understand the meaning, both for our human use of language and for the design of applications in natural language processing, like question answering by web searching. For instance, if one wants to know which guitar(s) played a rock star during a concert, the negation makes it difficult to extract the wanted information:

(1) a. *Question: Which guitars did he play at the concert.*
b. Funny he didn't play a Fender at that concert at least for one song. (web)

The standard treatment of determiners and quantifiers is to view them as generalised quantifiers, i.e. as functions of two predicates. In this paper we argue that although such an account "*works*" it is not really satisfactory mainly because it does not provide determiners with a proper logical form that can be interpreted on its own (as in the nominal phrase 2, or when we just hear the indefinite noun phrase of example 3) that would follow syntax (in example 4 generalised quantifiers require a predicate "*Keith sang _*" which does not correspond to any constituent)— furthermore in the case of indefinite determiners it introduces a misleading symmetry between topic (theme) and comment (rheme) as example 5 shows: these sentences do not speak about the same group.

(2) Cars, cars, cars....[1]

(3) a. Some philosophy students ....
b. *We already have some image(s) in mind.*
c. Some philosophy students are "free spirits" who travel, read, and seek to live a non-traditional life.

(4) Keith sang a song I never heard of.

(5) a. Some professors are smokers.
b. Some smokers are professors.

## 2 The Standard Logical Form of Determiners

The idea of Montague semantics is to map sentences to formulae of higher order logic (their logical forms) in a way which implements the Fregean principle of compositionality: typed functions (lambda terms) associated with words in the lexicon are composed according to the syntax. The glue logic is simply typed lambda calculus, over two types, **e** for entities or individuals and **t** for propositions (that may there after be endowed with a truth value).

These typed lambda terms use two kinds of constants: connectives and quantifiers on the one hand and individual constants and $n$-ary predicates for the precise language to be described — for instance a binary predicate like *delighted* has the type $\mathbf{e} \rightarrow \mathbf{e} \rightarrow \mathbf{t}$.

A small example goes as follows. Assume the syntax says that the structure of the sentence "*Keith sang a song.*" is

[1] Unless otherwise stated examples are from the Web.

| Constant | Type |
|---|---|
| $\exists$ | $(\mathbf{e} \to \mathbf{t}) \to \mathbf{t}$ |
| $\forall$ | $(\mathbf{e} \to \mathbf{t}) \to \mathbf{t}$ |

| Constant | Type |
|---|---|
| not | $\mathbf{t} \to \mathbf{t}$ |
| and | $\mathbf{t} \to (\mathbf{t} \to \mathbf{t})$ |
| or | $\mathbf{t} \to (\mathbf{t} \to \mathbf{t})$ |
| implies | $\mathbf{t} \to (\mathbf{t} \to \mathbf{t})$ |

| Constant | Type |
|---|---|
| *played*, *sang* | $\mathbf{e} \to (\mathbf{e} \to \mathbf{t})$ |
| *song* | $(\mathbf{e} \to \mathbf{t})$ |
| *Keith* | $\mathbf{e}$ |

**Fig. 1.** Logical constants and language constants

$$\text{(a (song))}(\lambda y \text{ Keith sang } y)$$

where the function is always the term on the left. On the semantical side, this means that "*sang*" is applied first to the property of "*being a song*" and to the property "*was sung by Keith*". If the semantical terms are as in the lexicon in Figure 2, placing the semantical terms in place of the words yields a large λ-term that can be reduced:

$$\Big(\big(\lambda P^{\mathbf{e}\to\mathbf{t}}\ \lambda Q^{\mathbf{e}\to\mathbf{t}}\ (\exists^{(\mathbf{e}\to\mathbf{t})\to\mathbf{t}}\ (\lambda z^{\mathbf{e}}(\&(P\ z)(Q\ z))))$$
$$(\lambda u^{\mathbf{e}}.\mathtt{song}(u))\big)(\lambda y^{\mathbf{e}}(\mathtt{sang}^{\mathbf{e}\to\mathbf{t}}\ \mathtt{Keith})y)\Big)$$
$$\downarrow \beta$$
$$\lambda P^{\mathbf{e}\to\mathbf{t}}\ \lambda Q^{\mathbf{e}\to\mathbf{t}}\ (\exists^{(\mathbf{e}\to\mathbf{t})\to\mathbf{t}}\ (\lambda Z^{\mathbf{e}}(\&((\lambda u^{\mathbf{e}}.\mathtt{song}(u))\ z)$$
$$((\lambda y^{\mathbf{e}}(\mathtt{sang}^{\mathbf{e}\to\mathbf{t}}\ \mathtt{Keith})\ y)\ z))))$$
$$\downarrow \beta$$
$$(\exists^{(e\to t)\to t}\ (\lambda y^{e}(\&(\mathtt{song}^{e\to t}\ y)((\mathtt{sang}^{e\to(e\to t)}\ \mathtt{Keith})\ y))))$$

This λ-term of type $\mathbf{t}$ that can be called the *logical form* of the sentence, represents the following formula of predicate calculus (admittedly more pleasant to read):

$$\exists y.\ (\mathtt{song}(y)\ \&\ \mathtt{sang}(\mathtt{Keith}, y))$$

**Fig. 2.** A simple semantical lexicon

| | |
|---|---|
| **word** | ***semantical type*** $u^*$<br>***semantics : λ-term of type*** $u^*$<br>$x^v$ *the variable or constant $x$ is of type $v$* |
| *a* | $(\mathbf{e} \to \mathbf{t}) \to ((\mathbf{e} \to \mathbf{t}) \to \mathbf{t})$<br>$\lambda P^{\mathbf{e}\to\mathbf{t}}\ \lambda Q^{\mathbf{e}\to\mathbf{t}}\ (\exists^{(\mathbf{e}\to\mathbf{t})\to\mathbf{t}}\ (\lambda z^{\mathbf{e}}(\&^{\mathbf{t}\to(\mathbf{t}\to\mathbf{t})}(P\ z)(Q\ z))))$ |
| *song* | $\mathbf{e} \to \mathbf{t}$<br>$\lambda x^{\mathbf{e}}(\mathtt{song}^{\mathbf{e}\to\mathbf{t}}\ x)$ |
| *sang* | $\mathbf{e} \to (\mathbf{e} \to \mathbf{t})$<br>$\lambda y^{\mathbf{e}}\ \lambda x^{\mathbf{e}}\ ((\mathtt{sang}^{\mathbf{e}\to(\mathbf{e}\to\mathbf{t})}\ x)\ y)$ |
| *Keith* | $\mathbf{e}$<br>$\mathtt{Keith}$ |

This algorithm actually works because of the following result:
There is a one to one correspondence between:

- the first order formulae over a first (respectively higher order) order language $\mathcal{L}$
- the closed normal lambda terms of type $\mathbf{t}$ with constants that correspond to connectives, quantifiers and to the constants, functions and predicates in $\mathcal{L}$.

The computation of the semantics of a sentence boils down to complete the following steps (see e.g. [22, Chapter 3]):

1. Parse the sentence, and turn the syntactic structure into a (linear) lambda term of type $\mathbf{t}$ (at least a functor argument structure, that is a binary tree with words as leafs and internal nodes specifying which subtree applies to the other one). This step is much easier when syntax is handled with categorial grammars.
2. Insert at each word's place the corresponding semantical lambda term provided by the lexicon.
3. Beta reduce this lambda term, the normal form being a logical formula, the semantical representation of the sentence.

### 2.1 Some Syntactical Inadequacies of the Standard Semantics of Determiners

As noted in the introduction, the standard approach to determiners that we just recalled, is not fully satisfactory, and there are at least three reasons to be disappointed by the standard semantical analysis.

A first point is that when one hears a determiner phrase, he does not need a complete sentence nor the main clause predicate to interpret the determiner phrase. This is easily observed from introspection: the simple utterance of a determiner phrase already suggests some interpretations, and possible referents, and references as individuals (sets of individuals, generic individual). It can also be observed in corpora: novels do include sentences without verbs. This can be observed in examples 2, 3 above or in the following examples: when one reads "*some students*", he has an idea, an image in mind, as well as when he reads "*What a thrill*" or "*an onion*".

(6) Some students do not participate in group experiments or projects.

(7) What a thrill — My thumb instead of an onion. (Sylvia Plath)

A second point is that this formalisation misses the asymmetry between the noun and the main clause predicate in existential statements. This asymmetry is the asymmetry between theme (or topic) and rheme (or comment) vanishes because both are assumed to be predicates and the indefinite determiner simply asserts that something has both properties, and this "*and*" is commutative. Even when both statements are felicitous, their meanings do differ: the sentence and its mirror image do not speak about the same class of objects. In the first case 8 one sentence can be said when speaking about universities or education and the next one when speaking about a company. This difference is even more striking in the example 8c: sentences like the first one can be read and heard (our example is from Internet) while the second one or similar sentences cannot be found on the Internet: the reason is probably that "*crooks*" do not really constitute a class one wants to speak about.

(8) a. Some students are employees.
b. Some employees are students.
c. i. Some politicians are crooks.
ii. Some crooks are politicians. (no such examples on Internet)

A third drawback is that the semantical or logical structure of the sentence does not match the syntactical structure (basically the parse tree) of the sentence. In the example we gave, this is patent: no constituent, no phrase does correspond to $\lambda x.(sang(Keith))x^{\mathbf{e}}$. This is related to the fact that the determiner or quantifier does not apply to a single predicate to form some term that can be interpreted.

(9) a. Keith played some Beatles songs.
b. syntax (Keith (played (some (Beatles songs))))
c. semantics: (some (Beatles songs)) ($\lambda x^{\mathbf{e}}$. Keith played $x$)

## 2.2 Quantification and Lexical Semantics Require a Many Sorted Logic

Let us point out that this Fregean view with a single sort prevents a proper treatment of quantification. Frege managed to express universal quantifiers (determiners like "*each*" or "*every*") and existential quantifiers like "*a*" or "*some*" restricted to a sort, set or type $A$ by using the following equivalences:

(10) a. $\forall x \in M \; P(x) \quad \equiv \forall x \; (M(x) \Rightarrow P(x))$
b. $\exists x \in M \; P(x) \quad \equiv \exists x \; (M(x) \& P(x))$

This treatment does not apply to other quantifiers like percentage or vague quantifiers:

(11) a. for a third of the $x \in M \; P(x) \quad \not\equiv \forall x \; (M(x) \Rightarrow P(x))$
b. for few$x \in M \; P(x) \quad \neg \equiv \exists x \; (M(x) \& P(x))$

Furthermore, as said in the first point of the previous subsection, we would like to have a logical form or a reference for determiner phrases, even though the main predicate is still to come.

(12) a. The Brits
b. The Brits love Australia, more than any other country except their own, according to an online survey for London's Daily Telegraph.

(13) a. Most students.
b. Most students will still be paying back loans from their university days in their 40s and 50s.

This question is related to lexical semantics: what classes are natural, what sorts do we quantify over, what can possibly be the comparison classes that have not been uttered, what are the sorts of complement a verb admit, what verbs can apply to a given sort of objects or of subjects? Our treatment of determiner phrases takes place in a framework that we initially designed for lexical semantics. But let us first speak about an alternative view of determiners and quantifiers.

## 3 Hilbert Operators, Quantifiers, and Determiners

After the quantifier *the one and unique individual such that P ...* introduced by Russell for definite descriptions, Hilbert (with Ackerman and Bernays) intensively used *generic* elements for quantification, the study of which culminated in the second volume of *Grundlagen der Mathematik* [11]. It should be stressed that these operators are introduced and described here with natural language examples, which is not so common in Hilbert's writings. We shall first present the $\epsilon$ operator which recently lead to important work in linguistics in particular with von Heusinger's work. [5,9,10]

### 3.1 An Ancestor to Hilbert Operators: Russell's Iota for Definite Descriptions

The first step due to Russell was to denote by $\iota_x.\ F$ the unique individual enjoying the property $F$ in a definite description like the first sentence below and to remain undetermined when existence and uniqueness do not hold. [28]

(14) The present president of France *was born in Rouen.*
(existence and uniqueness hold)

(15) The present king of France *was born in Pau.*
(existence fails)

(16) The present minister *was born in Barcelona.*
(uniqueness fails)

Of course this operator is not handy from a logical or formal point of view since the negation of "*there exists a unique x such that* $P(x)$" is "*either no x or more than two x enjoys* $\neg P$": its negation is clearly inelegant and indeed there are no well behaved deduction rules for such an operator. However, as observed by von Heusinger the uniqueness even when using the *definite* article is not really mandatory: it should refer to a salient element in the speaker's view, and in many examples the definite description is neither unique nor objectively salient, we shall come back to this point at the end of the present paper.

### 3.2 Hilbert Epsilon and Tau

From this idea, Hilbert introduced an individual existential term defined from a formula: given a formula $F(x)$ with a free variable $x$ one defines the term $\epsilon_x.\ F$ in which the occurrences of $x$ in $F$ are bound (this is the original notation, nowadays this term is often written as $\epsilon x.\ F$). Whenever **some** element, say $a$, enjoys $F$, then the epsilon term $\epsilon_x.\ F$ enjoys $F$.

Dually, Hilbert introduced a universal generic element $\tau_x.\ F$, which corresponds to the generic elements used in mathematical proofs: to establish that a property $P$ holds for every integer, the proof usually starts with "*Let n be an integer, ...*" where $n$ has no other property than being an integer. Consquently when this generic integer has the property, so does any integer. The $\tau$-term $\tau_x.\ F$ is the dual of the $\epsilon$-term $\epsilon_x.\ F$ : $\tau_x.\ F$ enjoys the property $F$ when **every** individual does.

More formally, given a first language $\mathcal{L}$ (constants, variables, function symbols, relation symbols, the later two with an arity) here is a precise definition of the epsilon terms and formulae. Terms and formulae are defined by mutual recursion:

- Any constant in $\mathcal{L}$ is a term.
- Any variable in $\mathcal{L}$ is a term.
- $f(t_1, \ldots, t_p)$ is a term provided each $t_i$ is a term and $f$ is a function symbol of arity $p$
- $\epsilon_x A$ is a term if $A$ is a formula and $x$ a variable and any free occurrence of $x$ in $A$ is bound by $\epsilon_x$
- $\tau_x A$ is a term if $A$ is a formula and $x$ a variable and any free occurrence of $x$ in $A$ is bound by $\tau_x$
- $s = t$ is a formula whenever $s$ and $t$ are terms.
- $R(t_1, \ldots, t_n)$ is a formula provided each $t_i$ is a term and $R$ is a relation symbol of arity $n$
- $A\&B$, $A \vee B$, $A \Rightarrow B$ are formulae if $A$ and $B$ are formulae
- $\neg A$ is formula if $A$ is a formula.

As the example below shows, a formula of first order logic can be recursively translated into a formula of the epsilon calculus, without surprise. Admittedly the epsilon translation of a usual formula may look quite complicated — at least we are not used to them:[2]

(17) a. $\forall x\ \exists y\ P(x, y)$
b. $= \exists y\ P(\tau_x P(x, y), y)$
c. $= P(\tau_x P(x, \epsilon_y P(\tau_x P(x, y), y)), \epsilon_y P(\tau_x P(x, y), y))$

The deduction rules for $\tau$ and $\epsilon$ are the usual rules for quantification:

[2] We shall not use such formulae as semantical representations: indeed, they are even further away from the syntactical structure than usual first order formulae.

- From $A(x)$ with $x$ generic in the proof (no free occurrence of $x$ in any hypothesis), infer $A(\tau_x.\ A(x))$
- From $B(c)$ infer $B(\epsilon_x.\ B(x))$.

The other rules can be found by duality:

- From $A(x)$ with $x$ generic in the proof (no free occurrence of $x$ in any hypothesis), infer $A(\epsilon_x.\ \neg A(x))$
- From $B(c)$ infer $B(\tau_x.\ \neg B(x))$

Hence we have:

$$F(\tau_x.\ F(x)) \equiv \forall x. F(x)$$
$$F(\epsilon_x.\ F(x)) \equiv \exists x.\ F(x)$$
$$\tau_x. A(x) = \epsilon_x. \neg A(x)$$

Because of the latest equation due to the classical negation ($\forall x.\ P(x) \equiv \neg\exists x.\ \neg P(x)$), only one of these two operators $\tau$ and $\epsilon$ is needed: commonly people choose the $\epsilon$ operator.

This logic is known as the *epsilon calculus.*

Hilbert turned these symbols into a mathematically satisfying theory, since it allows to fully describe quantification with simple rules. The first and second epsilon theorem basically say that this is an alternative formulation of first order logic.

**First Epsilon Theorem.** When inferring a formula $C$ without the $\epsilon$ symbol nor quantifiers from formulae $\Gamma$ not involving the $\epsilon$ symbol nor quantifiers the derivation can be done within quantifier free predicate calculus.

**Second Epsilon Theorem.** When inferring a formula $C$ without the $\epsilon$ symbol from formulae $\Gamma$ not involving the $\epsilon$ symbol, the derivation can be done within usual predicate calculus.

In this way, Hilbert provided the first correct proof of Herbrand's theorem (much before mistakes where found and solved by Goldfarb) and a way to prove the consistence of Peano's arithmetic at the same time as Gentzen did.

Later on Asser [2] and Leisenring [13] have been working on epsilon calculus in particular for having models and completeness, and for cut-elimination. Nevertheless, as one reads on *Zentralblatt* MATH these results are misleading as well as the posterior corrections — see in particular [4,19] and the related reviews. Only the proof theoretical aspects of the epsilon calculus seem to have been further investigated with some success in particular by Moser and Zach [23] and Mints [20].[3]

---

[3] While correcting these lines before printing, we just learnt that this great logician Grigori (Grisha) Mints passed away; sincere condolences to his family, friends and to the logic community.

### 3.3 Hilbert's Operators in Natural Language

In Hilbert's book the operators $\epsilon$ and $\tau$ are explained with natural language examples, but a very important and obvious linguistic property is not properly stated: the $\epsilon_x F$ has the type (both in the intuitive and in the formal sense) of a noun phrase, and is meant to be the argument of a predicate (for instance the subject of a verb), thus being a *suppositio* in the medieval sense. [14,12]

Nowadays there has been a renewed interest in the epsilon formulation of quantification, in particular by von Heusinger. He uses a variant of the epsilon for definite descriptions, leaving out the uniqueness of the iota operator of Russell, one reason being that the context often determines a unique object, the most salient one. We call it a "variant" because it is not clear whether one still has the equivalence with ordinary existential quantification: von Heusinger constructs an epsilon term whenever there is an expression like *a man* or *the man* but it is not clear how one asserts that $man(\epsilon_x.\ man(x))$. The distinction between $\epsilon$ and $\eta$ is that the former selects the most salient possible referent, while the later selects a new one.

### 3.4 Hilbert's Operators, Beyond Usual Logic

The study of epsilon operators focused on usual logic, typically first order classical logic within this extended language. Epsilon and the epsilon substitution method were part of Hilbert's program to provide finistic consistency proofs for arithmetic (and even analysis, using second order epsilon). Hence, although by that time people were probably aware that it goes beyond usual first order, none spoke about this extension.

Here is an extremely simple example of a formula of the epsilon calculus without an equivalent in first order logic, that von Heusinger and us use for natural language semantics as explained below:

$$F = P(\epsilon_x Q(x))$$

This formula, according to the aforementioned epsilon rules, entails $G = P(\epsilon_x P(x))$ (i.e. $\exists x.\ P(x)$), but it does not entails $H = Q(\epsilon_x Q(x))$ (i.e. $\exists x.\ Q(x)$). Of course, if one further assumes $H$, then the formulae $F$ and $H$ entail, according to epsilon rules, the $P\&Q(\epsilon_x.\ P\&Q)$ that is $\exists x.\ P\&Q(x) = \exists x.P(x)\&Q(x)$. But there is no first order formula equivalent to this simple epsilon formula $F$.

## 4 Determiners in the Montagovian Generative Lexicon

The standard view in Montague semantics is in perfect accordance with Frege's view of entities: a single universe gathers all entities. Hence a definite or indefinite determiner picks one element from this single sorted universe and a quantifier ranges over this single universe. As said in subsections 2.2 and 2.1, this view of quantification does not really match our linguistic competence nor our cognitive abilities.

This question is related to another part of semantics, namely lexical semantics. If one wants to integrate some lexical issues in a compositional framework, one needs sorts or many base types for entities, in order to specify what should be the nature of the arguments of a given word. This question is related to the type of the semantical constants: what should be the domain of a predicate, what are the relations between these logical constants? Observe, for instance that in Montague semantics a verb phrase and a common noun have the very same type $\mathbf{e} \to \mathbf{t}$, that events are standard entities, and that there is no way to have privileged relation between predicates and arguments: for instance a "*book*" can be "*enjoyed, disliked, read, written, printed, bound, burnt, lost,...*"

As the two questions are linked, we here present a compositional framework for semantics that accounts for both lexical issues and for the present question of determiners and quantifiers.

### 4.1 The Montagovian Generative Lexicon

As observed above, it would be more accurate to have many individual base types rather than just $\mathbf{e}$. Thus, the application of a predicate to an argument may only happen when it makes sense. Some sentences should be ruled out like "*The chair barks.*" or "*Their five is running.*", and this is quite easy when there are several types for individuals: the lexicon can specify "*barks*" and "*is running*" only apply to individuals of type "*animal*". Nevertheless, such a type system needs to incorporate some flexibility. Indeed, in the context of a football match, the second sentence makes sense: "*their five*" can be the player wearing the 5 shirt and who, being "*human*", is an "*animal*" that can "*run*".

Our system is called the Montagovian Generative Lexicon or $\Lambda Ty_n$. Its lambda terms extend the simply typed ones of Montague semantics above. Indeed, we use second order lambda terms from Girard's system $\mathsf{F}$ (1971) [8].

The types of $\Lambda\mathsf{Ty}_n$are defined as follows:

- Constants types $\mathbf{e}_i$ and $\mathbf{t}$, as well as type variables $\alpha, \beta, \ldots$ are types.
- $\Pi\alpha.\ T$ is a type whenever $T$ is a type and $\alpha$ a type variable . The type variable may or may not occur in the type $T$.
- $T_1 \to T_2$ is a type whenever $T_1$ and $T_2$ are types.

The terms of $\Lambda\mathsf{Ty}_n$, are defined as follows:

- A variable of type $T$ i.e. $x : T$ or $x^T$ is a *term*, and there are countably many variables of each type.
- In each type, there can be a countable set of constants of this type, and a constant of type $T$ is a term of type $T$. Such constants are needed for logical operations and for the logical language (predicates, individuals, etc.).
- $(f\ t)$ is a term of type $U$ whenever $t$ is a term of type $T$ and $f$ a term of type $T \to U$.
- $\lambda x^T.\ \tau$ is a term of type $T \to U$ whenever $x$ is variable of type $T$, and $t$ a term of type $U$.

- $t\{U\}$ is a term of type $T[U/\alpha]$ whenever $\tau$ is a term of type $\Pi\alpha.\ T$, and $U$ is a type.
- $\Lambda\alpha.t$ is a term of type $\Pi\alpha.T$ whenever $\alpha$ is a type variable, and $t : T$ a term without any free occurrence of the type variable $\alpha$ in the type of a free variable of $t$.

The later restriction is the usual one on the proof rule for quantification in propositional logic: one should not conclude that $F[p]$ holds for any proposition $p$ when assuming $G[p]$ — i.e. having a free hypothesis of type $G[p]$.

The reduction of the terms in system F or its specialised version $\Lambda\mathsf{Ty}_n$ is defined by the two following reduction schemes that resemble each other:

- $(\lambda x.\tau)u$ reduces to $\tau[u/x]$ (usual $\beta$ reduction).
- $(\Lambda\alpha.\tau)\{U\}$ reduces to $\tau[U/\alpha]$ (remember that $\alpha$ and $U$ are types).

As [7,8] showed reduction is strongly normalising and confluent *every term of every type admits a unique normal form which is reached no matter how one proceeds.* This has a good consequence for us, see e.g. [22, Chapter 3]:

> $\Lambda\mathsf{Ty}_n$ ***terms as formulae of a many-sorted logic*** *If the predicates, the constants and the logical connectives and quantifiers are the ones from a many sorted logic of order n (possibly $n = \omega$) then the closed normal terms of $\Lambda\mathsf{Ty}_n$ of type* $\mathbf{t}$ *unambiguously correspond to many sorted formulae of order n.*

Polymorphism allows a factored treatment of phenomena that treat uniformly families of types and terms. An interesting example is the polymorphic conjunction for copredication: *whenever* an object $x$ of type $\xi$ can be viewed both:.

- as an object of type $\alpha$ (via a term $f_0 : \xi \to \alpha$) to which a property $P^{\alpha\to\mathbf{t}}$ applies
- and as an object of type $\beta$ to which a property $Q^{\beta\to\mathbf{t}}$ applies (via a term $g_0 : \xi \to \beta$),

the fact that $x$ enjoys $P\&Q$ can be expressed by the unique polymorphic term (see explanation in figure 3):

$$(18)\quad \&^{\Pi} = \Lambda\alpha\Lambda\beta\lambda P^{\alpha\to\mathbf{t}}\lambda Q^{\beta\to\mathbf{t}}\Lambda\xi\lambda x^{\xi}\lambda f^{\xi\to\alpha}\lambda g^{\xi\to\beta}. \\ (\&^{\mathbf{t}\to\mathbf{t}\to\mathbf{t}}\ (P\ (f\ x))(Q\ (g\ x)))$$

The lexicon provides each word with:

- A main $\lambda$-term of $\Lambda\mathsf{Ty}_n$, the "usual one" specifying the argument structure of the word.
- A finite number of $\lambda$-terms of $\Lambda\mathsf{Ty}_n$ (possibly none) that implement meaning transfers. Each meaning transfer is declared in the lexicon to be *flexible* (F) or *rigid* (R).

Let us see how such a lexicon works. When a predication requires a type $\psi$ (e.g. Place) while its argument is of type $\sigma$ (e.g. Town) the optional terms in the lexicon can be used to "convert" a Town into a Place.

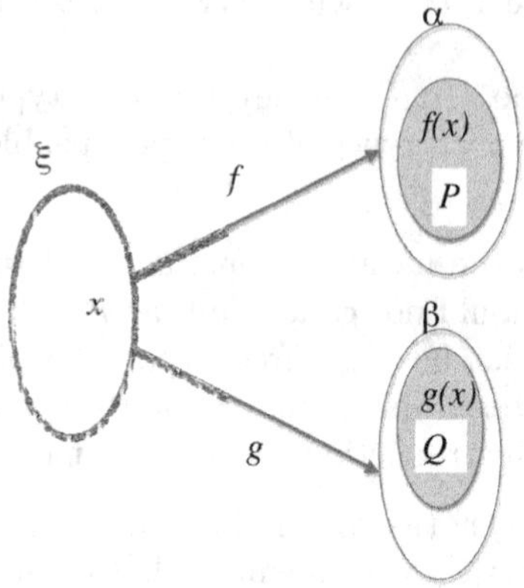


**Fig. 3.** Polymorphic and: $P(f(x))\&Q(g(x))$ $[x{:}\xi,\ f{:}\xi \to \alpha,\ g{:}\xi \to \beta]$

(19) a. Liverpool is spread out.

b. This sentence leads to a type mismatch $spread\_out^{Pl\to\mathbf{t}}(\mathtt{lpl}^T))$, since "*spread_out*" applies to "*places*" (type $Pl$) and not to "*towns*" as "*Liverpool*". This type conflict is solved using the optional term $t_3^{T\to Pl}$ provided by the entry for "*Liverpool*", which turns a town ($T$) into a place ($Pl$) $spread\_out^{Pl\to\mathbf{t}}(t_3^{T\to Pl}\mathtt{lpl}^T))$ — a single optional term is used, the (F)/ (R)difference is useless.

(20) a. Liverpool is spread_out and voted (last Sunday).

b. In this example, the fact that "*Liverpool*" is "*spread_out*" is derived as previously, and the fact "*Liverpool*" "*voted*" is obtained from the transformation of the town into people, which can vote. The two can be conjoined by the polymorphic "*and*" defined above in 18 ($\&^\Pi$) because these transformations are flexible: one can use both of them. We can make this precise using only the rules of our typed calculus. The syntax yields the predicate $(\&^\Pi(is\_spread\_out)^{Pl\to\mathbf{t}}(voted)^{P\to\mathbf{t}})$

**Fig. 4.** A sample lexicon

| word | principal λ-term | optional λ-terms rigid/flexible |
|---|---|---|
| *Liverpool* | $\mathtt{lpl}^T$ | $Id_T : T \to T$ (F)<br>$t_1 : T \to F$ (R)<br>$t_2 : T \to P$ (F)<br>$t_3 : T \to Pl$ (F) |
| *spread_out* | $spread\_out : Pl \to \mathbf{t}$ | |
| *voted* | $voted : P \to \mathbf{t}$ | |
| *won* | $won : F \to \mathbf{t}$ | |

where the base types are defined as follows: $T$ town
$P$ people
$Pl$ place

and consequently the type variables should be instantiated by $\alpha := Pl$ and $\beta := P$ and the exact term is
$\&^{\Pi}\{Pl\}\{P\}(is\_spread\_out)^{Pl\to\mathtt{t}}(voted)^{P\to\mathtt{t}}$
which reduces to:
$\Lambda\xi\lambda x^{\xi}\ \lambda f^{\xi\to\alpha}\lambda g^{\xi\to\beta}(\&^{\mathtt{t}\to\mathtt{t})\to\mathtt{t}}\ (is\_spread\_out\ (f\ x))(voted\ (g\ x)))$.
Syntax also says this term is applied to "*Liverpool*". which forces the instantiation $\xi := T$ and the term corresponding to the sentence is after some reduction steps,
$\lambda f^{T\to Pl}\lambda g^{T\to P}(\&\ (is\_spread\_out\ (f\ \mathtt{lpl}^{T}))(voted\ (g\ \mathtt{lpl}^{T}))))$. Fortunately the optional $\lambda$-terms $t_2 : T \to P$ and $t_3 : T \to Pl$ are provided by the lexicon, and they can both be used, since none of them is rigid. Thus we obtain, as expected
$(\&\ (is\_spread\_out^{Pl\to\mathtt{t}}\ (t_3^{T\to Pl}\ \mathtt{lpl}^{T}))(voted^{Pl\to\mathtt{t}}\ (t_2^{T\to P}\ \mathtt{lpl}^{T})))$

(21) a. # Liverpool voted and won (last Sunday).

b. This third and last example is rejected as expected. Indeed, the transformation of the town into a football club prevents any other transformation (even the identity) to be used with the polymorphic "*and*" ($\&^{\Pi}$) defined above in 18. We obtain the same term as above, with *won* instead of *is_spread_out*:
$\lambda f^{T\to Pl}\lambda g^{T\to P}(\&\ (won\ (f\ \mathtt{lpl}^{T}))(voted\ (g\ \mathtt{lpl}^{T}))))$ and the lexicon provides the two morphisms that would solve the type conflict, but one of them is *rigid*, i.e. we can solely use this one. Consequently no semantics can be derived from this sentence, which is semantically invalid.

The difference between our system and those of [15,1] does not come down to the type systems, which are quite similar, but in the *architecture* which is, in our case, rather *word driven* than type driven. The optional morphisms are anchored in the words, and do not derive from the types. This is supported in our opinion by the fact that some words with the very same ontological type (like French nouns "*classe*" and "*promotion*", that are groups of students in the context of teaching) may undergo different coercions (only the first one can mean a classroom). This rather lexicalist view goes well with the present work that proposes to have specific entries for deverbals, that are derived from the verb entry but not automatically.

This system has been implemented as an extension to the Grail parser [21], with $\lambda$-DRT instead of formulae as $\lambda$-terms. It works fine once the semantical lexicon has been typeset.[4]

We already explored some of the compositional properties (quantifiers, plurals and generic elements,....) of our Montagovian generative lexicon as well as some of the lexical issues (meaning transfers, copredication, fictive motion, deverbals, ... ) [3,27,25,26,17,24].

[4] Syntactical categories are learnt from annotated corpora, but semantical typed $\lambda$-terms cannot yet be learnt, as discussed in the conclusion.

### 4.2 Determiners as Typed Epsilon Operators

As we saw there are many base types that are sorts of the many sorted logic and even more complex types over which one may quantify, a fairly natural semantics for determiners is to pick one element in its sort.

For instance, consider the indefinite determiner "*a*". It should be seen as an operator acting on a noun phrase without determiners that outputs some individual. In order to make things correct and precise, consider the noun phrase, "*a cat*" where "*a*" acts upon "*cats*", and think about the possible types of "*a*", which clearly it depend on what "*cat*" is. Is *cat* a type or a property satisfied by "*cats*" among a larger class or type?

1. If "*cat*" is a type the constant for "*a*" should be of type $\Pi\alpha.\ \alpha$.
2. If "*cat*" is a property, say of a larger type "*animal*", then this constant should take a property of animals of type $animal \to \mathbf{t}$ and yield a cat. Now assume that the property is a more complex property $P$ "*cat which lives nearby*", what should "*a*" do? It should apply to a property of animals like $P$ and yields an entity $x$ that enjoys $P$. Because $x$ enjoys $P$ its type should be "*animal*". In this case the type of the constant corresponding to $P$ should be $\Pi\alpha.\ (\alpha \to \mathbf{t}) \to \alpha$, hence the type does not guarantee by itself that $x$ enjoys $P$ and consequently a presupposition $P(a\ cat)$ has to be added.

We deliberately chose to use option 2 and only this one. Firstly, we cannot avoid this case, because not every property that a determiner may apply to can be assumed to be a type, there would be too many of them. Secondly, the first option can be encoded within the second option. Indeed if there is a type *cat* one can consider a predicate "*being a cat*". Indeed, unsurprisingly, the semantics of predicates and the one of quantifiers and determiners are closely related.

Usually, a determiner or a quantifier applies to one ("*everyone*") or two ("*a*") predicates and yields a proposition. A Hilbert operator combines with *one* predicate and yields a term, an entity. In a many sorted and typed system like $\Lambda\mathsf{Ty}_n$ what is the type of a predicate? The standard type for a predicate is $\mathbf{e} \to \mathbf{t}$, but given the many sorts $\mathbf{e}_i$ we could have predicates that apply to other entity type than $\mathbf{e}$. Is "*cat*" a property of individuals of type "*animal*" if such a type exists or is it a property that may apply to any entity, and which is constantly false outside of the type "*animal*"? If the domain of a predicate is $\mathbf{e}_i$ and not $\mathbf{e}$ (the type of all entities), a predicate $P^{\mathbf{e}_i \to \mathbf{t}}$ canonically extends to a predicate $\overline{P}^{\mathbf{e} \to \mathbf{t}}$ by saying it never holds outside of $\mathbf{e}_i$. Conversely a property like *cat* whose domain is some $\mathbf{e}_i$ (e.g "*animal*") can be restricted to any subtype of $\mathbf{e}_i$, but in case the subtype of $\mathbf{e}_i$ does not include all "*cats*" there dis no way to recover the initial predicate "*cat*" that applies to animals.

Now that we have a proper representation of a predicate in the type system, one may wonder how a type can be reflected as a predicate. For instance what should be the type of a predicate associated with a type, like "*being a cat*" if "*cat*" is a type. Natural domains for the such a predicate could be "*animals*", "*mammals*", "*felines*",... As it is difficult to chose, let us decide that the domain of a given predicate associated with a type always is the largest, the collection of all possible entities **e** which can be restricted as indicated above. Hence "*being of type* $\alpha$" that we write $\widehat{\alpha}$ is of type $\mathbf{e} \to \mathbf{t}$

So far we have not said what are the base type which intervenes in representing predicates and quantifiers. We need several of them, to express selectional restrictions . Asher [1] uses a dozen of ontological types (events, physical objects, human beings, information, etc.) Luo [16] suggests using a flat ontology with common nouns (there are thousands of them) as base types. With Mery we suggested to consider classifiers (100–200) as in languages that have classifiers (sign language, Chinese, Japanese) [18].

As said above the lexicon associate the constant $\epsilon$ of type $\Pi\alpha.\ (\alpha \to \mathbf{t}) \to \alpha$ to the indefinite article — that is an Hilbert/von Heusinger $\epsilon$ adapted to the typed case. Hence the indefinite article is a polymorphic $\epsilon$ that specialises to a type/sort $\{\mathbf{e}_i\}$ and applies to a predicate $P$ of type $\mathbf{e}_i \to \mathbf{t}$ yielding an entity of type $\mathbf{e}_i$. Let us consider an extremely simple example: (*ani* stands for the type of animals):

(22) a. A cat sleeps (under your car).
b. term for "*a*": $\epsilon : \Pi\alpha.\ ((\alpha \to \mathbf{t}) \to \alpha)$
c. term for "*sleep*": $(\lambda x.\ sleeps^{ani \to \mathbf{t}}(x))$
d. term for "*cat*": $(\lambda x.\ cat^{ani \to \mathbf{t}}(x))$
e. syntax: $((a \to cat) \leftarrow sleeps)$
f. semantics: $sleeps(a\ cat)$
g. $(\lambda x.\ sleeps^{ani \to \mathbf{t}}(x))(\epsilon^{\Pi\alpha.\ ((\alpha \to \mathbf{t}) \to \alpha)} cat^{ani \to \mathbf{t}})$
h. $(\lambda x.\ sleeps(x))(\epsilon^{\Pi\alpha.\ ((\alpha \to \mathbf{t}) \to \alpha)}\{ani\} cat^{ani \to \mathbf{t}})$
i. $sleeps^{ani \to \mathbf{t}}(\epsilon^{\Pi\alpha.\ ((\alpha \to \mathbf{t}) \to \alpha)}\{ani\} cat^{ani \to \mathbf{t}}) : \mathbf{t}$ Logical Form
j. $cat(\epsilon^{\Pi\alpha.\ ((\alpha \to \mathbf{t}) \to \alpha)}\{ani\} cat^{ani \to \mathbf{t}}) : \mathbf{t}$ Presupposition

In order to apply "*a*" to "*cat*" a predicate of type $ani \to \mathbf{t}$ the $\epsilon$ must be specialised to $\alpha = ani$. The verb "*sleeps*" can apply to result of "*a cat*" which is of type *ani*, and the final term (22h) is of type **t** as expected — as explained in section provided there actually exists a cat this epsilon formula with out any first order equivalent (see subsection 3.4) can be understood as $\exists x : ani \quad sleep(x)$. Our analysis ought to be completed: nothing tells us that $cat(\epsilon cat)$ $(\exists x.\ cat(x))$, i.e. that a "*cat*" actually exists ... and this needs to be added as a presupposition. In fact, such a presupposition is added as soon as a determiner or an existential quantifier appears: when an utterance "*a cat*" appears, the existence of the corresponding entity ought to be asserted.

We use the word "*presupposition*" with the same sense as Asher [1] when he calls "*presupposition*" a selectional restriction: a verb like "*sleeps*" presupposes that its subject is an "*animal*". This really is some sort of presupposition, indeed it is quite difficult to deny a type judgement, both formally and linguistically:

- Formally: To refute $(a{:}A)$ is not easy. Indeed the complement of a type is not a type, i.e. the negation of $a{:}A$ is not $a{:}\neg A$ — as opposed to $\tilde{A}(x)$ whose negation is easily formulated as $\neg\tilde{A}(x)$

- Linguistically: If one says "*Rex is sleeping in the garden.*" the reply: — "*No, Rex is not an animal*", that *refutes a typing judgment* (*Rex*:*ani*) is difficult to utter out of the blue and needs to be better introduced and justified. On the other hand it is easy to utter an answer that *refutes the proposition:* — "*No, Rex is not sleeping, he just left.*"

### 4.3 A Rather Satisfying Account of Determiners

We started with three objections to the standard account of determiners in Montague semantics. We proposed a model that avoids those three problems:

1. Epsilon are individuals that can be interpreted as such (even though their interpretation does not ensure completeness of the epsilon calculus).
2. With epsilon terms, the syntactical structure and the structure of the logical form match.
3. For an indefinite determiner phrase, which corresponds to an existential statement, there is not anymore an irrelevant symmetry between the noun (topic, theme) and the verb phrase (comment, rheme).

As in von Heusinger's work, one can give a similar account of definite descriptions, the main difference being at the interpretation level: the definite description should be interpreted as the most salient entity in the context. This entity is usually introduced by an indefinite description, that is another epsilon term defined from the same property (from the same logical formulae). The difference between a definite description and an indefinite determiner phrase is that the former one refers to an existing discourse referent while the later one introduces a new discourse referent.

This also provides a natural account of Evan's E-type pronoun [6]: the semantics of the pronoun "*he*" in the example below can be copied from its antecedent to obtain the semantics of these two sentences.

(23) A man entered the conference hall. The man sat nearby the window.

(24) A $\text{man}_1$ entered the conference hall and sat nearby the window. A $\text{man}_2$ ($\neq \text{man}_1$) told him that he just missed two slides.

(25) A man entered the conference hall. He sat nearby the window.

Universal quantification can be treated just like indefinite determiners. A universally quantified NP corresponds to the term $\tau_x.P(x) = \epsilon_x.\neg(P(x))$ (c.f. section 3). The $\tau$-terms are actually much easier to interpret than the $\epsilon$-terms: it's a generic entity with respect to property $P$. Furthermore one can introduce operators for generalised and vague quantifiers like "*most*", "*few*", "*a third of*" etc.

The approach to existential quantification is rather similar to choice functions that have been used in formal semantics, especially in Steedman recent book [29], who also enjoy the three properties above. There are nevertheless some differences:

- The syntax, the definition of epsilon terms, is simple. I think different choice functions are needed for all the formulae, while a single epsilon is enough (and possibly already too much).
- Universal quantification can be treated un just the same way with $\tau_x.P(x) = \epsilon_x.\neg(P(x))$ and even generalised and vague quantifiers can be treated that way.

Of course the challenging difficulty of epsilon is to find the proper notion of model which would give a completeness theorem for all the formulae including the one that do not have a first order equivalent.

## 5 Conclusion

This work is an investigation of the outcomes of the Montagovian generative lexicon, which was designed for lexical semantics, in formal semantics. The many sorted compositional framework seems to be a rich setting to explore some new direction like a typed and richer view of epsilon terms as the semantics of determiner phrases.

We did not elaborate on scope issues: using freely the epsilon and tau operators is a form of underspecification. It involves formulae that are *not* part of first order logic, like: $R(\epsilon_x P(x), \tau_z Q(z))$.

As we showed here, this refinement of Montague semantics draws intriguing connections between type theory — say a judgement $a : A$ — and many sorted logic — a formula $\tilde{A}(a)$: we hope to understand better those issues in future work.

As far as quantification is concerned, we would like to better understand formulae of the epsilon calculus that do not have any equivalent in usual logic and any proper notion of model, complete if possible, would help a lot.

We presently are doing psycholinguistic experiments to see how do we naturally interpret determiner phrases, by confronting sentence to pictures in which they can be true or not, measuring reaction time and recording eye tracking. This will possibly confirm or refute the soundness of some cognitive arguments.

The possibly to model with Hilbert operators generalised quantifiers like "*a third of*" and vague quantifiers like "*many*" if of course very appealing, and we already made some advances in this direction. [25] Nevertheless we should not be too ambitious: basic epsilon terms already goes beyond usual first order logic, and although they do have deduction rules they lack proper models. So the situation is probably much more complicated with Hilbert terms for generalised quantifiers, which do not even have proper deductions rules. Hence such terms are a natural and appealing but mathematically difficult approach to quantification related to the semantics of determiner phrases.

**Acknowledgements.** Thanks to the anonymous colleagues who provided some comments on this paper and to Michele Abrusci, Nicholas Asher, Francis Corblin, Ulrich Kohlenbach, Zhaohui Luo, Richard Moot, Fabio Pasquali for helpful discussions.

## References

1. Asher, N.: Lexical Meaning in context – a web of words. Cambridge University Press (2011)
2. Asser, G.: Theorie der logischen auswahlfunktionen. Zeitschrift für Mathematische Logik und Grundlagen der Mathematik (1957)
3. Bassac, C., Mery, B., Retoré, C.: Towards a Type-Theoretical Account of Lexical Semantics. Journal of Logic Language and Information 19(2), 229–245 (2010), http://hal.inria.fr/inria-00408308/
4. Canty, J.T.: Zbl0327.02013: review of "on an extension of Hilbert's second $\epsilon$-theorem" by T. B. Flanagan (jsl 1975)
5. Egli, U., von Heusinger, K.: The epsilon operator and E-type pronouns. In: Egli, U., Pause, P.E., Schwarze, C., von Stechow, A., Wienold, G. (eds.) Lexical Knowledge in the Organization of Language, pp. 121–141. Benjamins (1995)
6. Evans, G.: Pronouns, quantifiers, and relative clauses (i). Canadian Journal of Philosophy 7(3), 467–536 (1977)
7. Girard, J.Y.: Une extension de l'interprétation de Gödel à l'analyse et son application: l'élimination des coupures dans l'analyse et la théorie des types. In: Fenstad, J.E. (ed.) Proceedings of the Second Scandinavian Logic Symposium. Studies in Logic and the Foundations of Mathematics, vol. 63, pp. 63–92. North Holland, Amsterdam (1971)
8. Girard, J.Y.: The blind spot – lectures on logic. European Mathematical Society (2011)
9. von Heusinger, K.: Definite descriptions and choice functions. In: Akama, S. (ed.) Logic, Language and Computation, pp. 61–91. Kluwer (1997)
10. von Heusinger, K.: Choice functions and the anaphoric semantics of definite nps. Research on Language and Computation 2, 309–329 (2004)
11. Hilbert, D., Bernays, P.: Grundlagen der Mathematik, Bd. 2. Springer (1939), traduction française de F. Gaillard, E. Guillaume et M. Guillaume, L'Harmattan (2001)
12. Kneale, W., Kneale, M.: The development of logic, 3rd edn. Oxford University Press (1986)
13. Leisenring, A.C.: Mathematical logic and Hilbert's $\epsilon$ symbol. University Mathematical Series. Mac Donald & Co. (1967)
14. de Libera, A.: La querelle des universaux de Platon à la fin du Moyen Âge. Des travaux, Seuil (1996), http://books.google.com/books?id=64AEAQAAIAAJ
15. Luo, Z.: Contextual analysis of word meanings in type-theoretical semantics. In: Pogodalla, S., Prost, J.-P. (eds.) LACL 2011. LNCS, vol. 6736, pp. 159–174. Springer, Heidelberg (2011)
16. Luo, Z.: Common nouns as types. In: Béchet, D., Dikovsky, A. (eds.) LACL 2012. LNCS, vol. 7351, pp. 173–185. Springer, Heidelberg (2012)
17. Mery, B., Moot, R., Retoré, C.: Plurals: individuals and sets in a richly typed semantics. In: Yatabe, S. (ed.) Logic and Engineering of Natural Language Semantics 10 (LENLS 10), pp. 143–156, Keio University (2013) ISBN 978-4-915905-57-5
18. Mery, B., Retoré, C.: Semantic types, lexical sorts and classifiers. In: Sharp, B., Zock, M. (eds.) 10th International Workshop on Natural Language Processing and Cognitive Science. Marseilles (September 2013), http://hal.inria.fr/hal-00916722
19. Mints, G.: Zbl0381.03042: review of "cut elimination in a Gentzen-style $\epsilon$-calculus without identity" by Linda Wessels (Z. math Logik Grundl. Math (1977))

20. Mints, G.: Cut elimination for a simple formulation of epsilon calculus. Ann. Pure Appl. Logic 152(1-3), 148–160 (2008)
21. Moot, R.: Wide-coverage French syntax and semantics using Grail. In: Proceedings of Traitement Automatique des Langues Naturelles (TALN), Montreal (2010)
22. Moot, R., Retoré, C.: The Logic of Categorial Grammars. LNCS, vol. 6850. Springer, Heidelberg (2012)
23. Moser, G., Zach, R.: The epsilon calculus and herbrand complexity. Studia Logica 82(1), 133–155 (2006)
24. Real, L., Retoré, C.: Deverbal semantics and the Montagovian generative lexicon $\Lambda\mathsf{Ty}_n$. Journal of Logic, Language and Information, 1–20 (2014), `http://dx.doi.org/10.1007/s10849-014-9187-y`
25. Retoré, C.: Variable types for meaning assembly: a logical syntax for generic noun phrases introduced by "most". Recherches Linguistiques de Vincennes 41, 83–102 (2012), `http://hal.archives-ouvertes.fr/hal-00677312`
26. Retoré, C.: Sémantique des déterminants dans un cadre richement typé. In: Morin, E., Estève, Y. (eds.) Traitement Automatique du Langage Naturel, TALN RECITAL 2013, vol. 1, pp. 367–380. ACL Anthology (2013), `http://www.taln2013.org/actes/`
27. Retoré, C.: The Montagovian generative lexicon $\Lambda Ty_n$: a type theoretical framework for natural language semantics. In: Matthes, R., Schubert, A. (eds.) 19th International Conference on Types for Proofs and Programs (TYPES 2013). Leibniz International Proceedings in Informatics LIPIcs, vol. 27, Dagstuhl Publishing, Germany (2014), `http://hal.archives-ouvertes.fr/hal-00779214`
28. Russell, B.: On denoting. Mind 56(14), 479–493 (1905)
29. Steedman, M.: Taking Scope: The Natural Semantics of Quantifiers. MIT Press (2012)

# Adjuncts and Minimalist Grammars*

Meaghan Fowlie

UCLA Linguistics,
3125 Campbell Hall, UCLA, Los Angeles, California, USA
http://mfowlie.bol.ucla.edu

**Abstract.** The behaviour of adverbs and adjectives has qualities of both ordinary selection and something else, something unique to modifiers. This makes them difficult to model. Modifiers are generally optional and transparent to selection while arguments are required and driven by selection. Cinque [4] proposes that adverbs, functional heads, and descriptive adjectives are underlyingly uniformly ordered across languages and models them by ordinary Merge or selection. Such a model captures only the ordering restrictions on these morphemes; it fails to capture their optionality and transparency to selection. I propose a model of adjunction with a separate Adjoin function that allows the derivation to keep track of both the true head of the phrase and the place in the Cinque hierarchy of the modifier, preventing inverted modifier orders in the absence of Move.

**Keywords:** adjoin, minimalist grammars, adjectives, adverbs, functional projections, ordering, optionality.

## 1 Introduction

Adjuncts are optional, meaning the sentence is grammatical without them. For example, in (1-a), *red* is optional. They are transparent to selection in that the selector seems to select for the features of the head, not those of the intervening adjunct. For example, in (1-b), the gender of *boek* 'book' is neuter. The intervening adjective does not have gender agreement, so *het* selects *boek* for its gender, regardless of the intervening adjunct.

(1) a. The (red) rose **Optionality**
    b. Het mooi-e boek
    the.NEU beautiful-DET book
    'The beautiful book' (Dutch) **Transparency**

Many languages have a default order for adjuncts, with unmarked intonation and without special scopal meaning. For example, English has ordered adjectives.

* Many thanks to Ed Stabler, my dissertation committee chair, as well as to the rest of my committee (Ed Keenan, Martin Monti, and Carson Schutze). Thank you also to Thomas Graf for our MG discussions, UCLA syntax/semantics seminar, audiences at MoL13 and NWLC 2013, and of course to three very helpful anonymous reviewers.

G. Morrill et al. (Eds.): Formal Grammar 2014, LNCS 8612, pp. 34–51, 2014.

(2) a. Wear the enormous ugly green hat
*Wear the hat that is enormous, ugly, and green*
b. #Wear the ugly enormous green hat
*Of your enormous green hats, wear the ugly one.*

This paper proposes a minimalist model of ordered adjuncts, using a new function adjoin that has access to sets of adjuncts for each category and hierarchy levels of adjuncts.

## 2 Minimalist Grammars

I formulate my model as a variant of *Minimalist Grammars* (MGs), which are Stabler's [14] formalisation of Chomsky's [3] notion of feature-driven derivations using the functions Merge and Move. MGs are mildly context-sensitive, putting them in the right general class for human language grammars. They are also simple and intuitive to work with.

At the heart of MGs is a function that takes two structures and puts them together. I will give derived structures as strings as Keenan & Stabler's [10] grammar would generate them.[1]

**Definition 1.** *A* ***Minimalist Grammar*** *is a five-tuple* $G = \langle \Sigma, \mathbf{sel}, \mathbf{lic}, Lex, M\rangle$. *$\Sigma$ is the alphabet.* $\mathbf{sel}\cup\mathbf{lic}$ *are the base features. Let* $F = \{\texttt{+f}, -\texttt{f}, \texttt{=X}, \texttt{X} | \texttt{f} \in \mathbf{lic}, \texttt{X} \in \mathbf{sel}\}$ *be the features.* $Lex \subseteq \Sigma \times F^*$, *and $M$ is the set of operations* ***Merge*** *and* ***Move****. The language $L_G$ is the closure of Lex under $M$. A set $C \subseteq F$ of designated features can be added; these are the types of complete sentences.*

Minimalist Grammars are *feature-driven*, meaning features of lexical items determine which operations can occur and when. There are two finite sets of features, ***selectional*** features **sel** which drive the operation **Merge** and ***licensing*** features **lic** which drive **Move**. **Merge** puts two derived structures together; **Move** operates on the already built structure. Each feature has a positive and negative version. Positive **sel** and **lic** features are `=X` and `+f` respectively, and negatives are `X` and `-f`. Intuitively, negative **sel** features are the categories of lexical items. Merge and Move are defined over ***expressions***: sequences of pairs ⟨derived structure, feature stack⟩. The first pair can be thought of as the "main" structure being built; the remaining are waiting to move.

[1] Keenan & Stabler's grammar also incorporates an additional element: lexical items are triples of string, features, and lexical status, which allows derivation of Spec-Head-Complement order. I will leave this out for simplicity, as it is not really relevant here, as our interest is in spec/adjunct placement, which will always be on the left. For convenience of English reading, I will give sentences in head-spec-complement order, but the formal definition I give here always puts the selected on the left and the selector on the right.

An MG essentially works as follows: **Merge** takes two expressions and combines them into one if the first structure displays `=X` and the second `X` for some $\mathtt{X} \in$ **sel**. The `X` features are deleted, after which the second structure may still have features remaining, meaning the second structure is going to move. It is stored separately by the derivation until the matching positive licensing feature comes up later in the derivation, when the moving structure is combined again; this is **Move**. **Move** also carries the requirement that for each $\mathtt{f}\in$**lic** there be at most one structure waiting to move. This is the ***shortest move constraint (SMC)***.[2]

**Definition 2 (*Merge*).** *For $\alpha, \beta$ sequences of negative* **lic** *features, $s, t$ derived structures, $mvrs_{s,t}$ expressions:*[3]

$$\mathbf{Merge}(s : \mathtt{=X}\alpha :: mvrs_s,\ t : \mathtt{X}\beta :: mvrs_t) = \begin{cases} ts : \alpha :: mvrs_s \cdot mvrs_t & \text{if } \beta = \epsilon \\ (s : \alpha) :: (t : \beta) :: mvrs_s \cdot mvrs_t & \text{if } \beta \neq \epsilon \end{cases}$$

**Definition 3 (*Move*).** *For $\alpha, \beta, \gamma$ sequences of negative* **lic** *features, $s, t$ derived structures, $mvrs$ an expression, suppose $\exists! \langle t, \beta \rangle \in mvrs$ such that $\beta = \mathtt{-f}\gamma$. Then:*

$$\mathbf{Move}(s : \mathtt{+f}\alpha :: mvrs) = \begin{cases} ts : \alpha :: mvrs & \text{if } \gamma = \epsilon \\ s : \alpha :: t : \gamma :: mvrs) & \text{if } \gamma \neq \epsilon \end{cases}$$

In this article I will make use of *derivation trees*, which are trees describing the derivation. They may also be annotated: in addition to the name of function, I (redundantly) include for clarity the derived expressions in the form of strings and features. For example, figure 1 shows derivation trees (annotated and unannotated) of *the wolf* with feature `D`.

Merge
the wolf:D
the:=ND wolf:N
Merge
the:=ND wolf:N

**Fig. 1.** Annotated and unannotated derivation trees

[2] The SMC is based on economy arguments in the linguistic literature [3], but it is also crucial for a type of finiteness: the valid derivation trees of an MG form a regular tree language [11]. The number of possible movers must be finite for the automaton to be finite-state. The SMC could also be modified to allow up to a particular (finite) number of movers for each $\mathtt{f}\in$**lic**.

[3] :: adds an element to a list; · appends two lists.

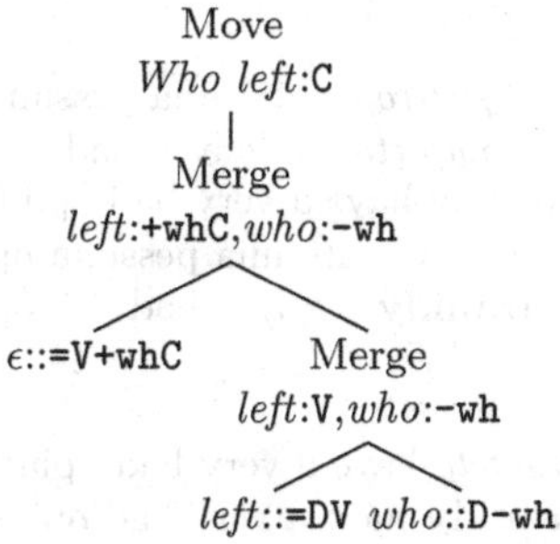


**Fig. 2.** Example: *Who left?*

# 3 Cartography

Despite their optionality, linguists, most famously Cinque [4], argue that certain adjuncts have a default order that is consistent across languages. The phenomena this model is designed to account for are modifiers and other apparently optional projections such as the following.

(3)
a. The small ancient triangular green Irish pagan metal artifact was lost.
b. *The metal green small artifact was lost. **Adjectives**
c. Frankly, John probably once usually arrived early.
d. *Usually, John early frankly once arrived probably. **Adverbs**
e. [Il premio Nobel]$_{top}$, [a chi]$_{wh}$ lo daranno?
[the prize Nobel]$_{top}$, [to whom]$_{wh}$ it give.fut
The Nobel Prize, to whom will they give it? **Left periphery**
f. [$_{DP}$ zhe [$_{NumP}$ yi [$_{ClP}$ zhi [$_{NP}$ bi]]]
[$_{DP}$ this [$_{NumP}$ one [$_{ClP}$ CL [$_{NP}$ pen]]]
'this pen' **Functional projections**

These three phenomena all display *optionality, transparency to selection*, and *strict ordering.* By *transparency* I mean that despite the intervening modifiers, properties of the selected head are relevant to selection. For example, in a classifier language, the correct classifier selects the noun even if adjectives intervene.

The hypothesis that despite their optionality these projections are strictly ordered is part of *syntactic cartography* [12]. Cinque [4], [5] in particular proposes a universal hierarchy of functional heads that select adverbs in their specifiers, yielding an order on both the heads and the adverbs. He proposes a parallel hierarchy of adjectives modifying nouns. These hierarchies are very deep. The adverbs and functional heads incorporate 30 heads and 30 adverbs.

Cinque argues that the surprising universality of adverb order calls for explanation. For example, Italian, English, Bosnian/Serbo-Croatian, Mandarin Chinese, and more show strong preferences for *frankly* to precede *unfortunately.* These arguments continue for a great deal more adverbs.[4]

[4] Data from [4].

(4) Italian

a. **Francamente** ho *purtroppo* una pessima opinione di voi.
**Frankly** have *unfortunately* a bad opinion of you
'Frankly I unfortunately have a very bad opinion of you.'

b. *Purtroppo ho francamente una pessima opinione di voi.
*Unfortuately* have **frankly** a bad opinion of you

(5) English

a. **Frankly**, I *unfortuately* have a very bad opinion of you

b. ?*Unfortunately* I **frankly** have a very bad opinion of you

(6) Bosnian/Serbo-Croatian

a. **lskreno**, ja *naialost* imam jako lose misljenje o vama
**Frankly**, I *unfortunately* have very bad opinion of you.
Frankly, I unfortunately have a very bad opinion of you.'

b. **Naialost*, ja **iskreno** imam jako lose misljenje o varna.
*unfortunately* I **frankly** have very bad opinion of you.

(7) Mandarin Chinese

a. **laoshi-shuo** wo *buxing* dui tamen you pian-jian.
**Frankly**, I *unfortunately* to them have prejudice
'Honestly I unfortunately have prejudice against them.'

b. **buxing* wo **laoshi-shuo** dui tamen you pian-jian.
*unfortunately* I **Frankly** to them have prejudice

Supposing these hierarchies are indeed universal, the grammar should account for it.

## 4 Desiderata

In addition to these three main properties, an account of adjuncts should ideally also account for the following: selectability of adjunct categories, adjuncts of adjuncts, unordered adjuncts, so-called obligatory adjuncts, and adjunct islands.

(8) Mary is **tall** — *tall* is selected by *is*

(9) The **surprisingly short** basketball player — *surprisingly* modifies *short*

(10) a. The alliance officer shot Kaeli **in the cargo hold** *with a gun.*
b. The alliance officer shot Kaeli *with a gun* **in the cargo hold**. English PP adjuncts are unordered

(11) a. He makes a **good** father — *good* is an adjunct but is not optional
b. *He makes a father
c. She worded the letter **carelessly**.
d. ...and Marc did so **carefully**. — *carefully* is an adjunct
e. *She worded the letter. — yet it is not optional

(12) a. He left [because she arrived]$_{\text{adjunct}}$.
b. *Who did he leave because __ arrived? — (some) adjuncts are islands

c. He thinks [she arrived]$_{\text{object}}$.
d. Who does he think _ arrived? Embedded object CPs are not islands; islandhood is a property of adjuncts, not embedded clauses in general.

In sum, an account of adjuncts in minimalist grammars should ideally have the following properties:

1. **Optionality**: sentences should be grammatical with or without adjuncts
2. **Transparency to selection**: If a phrase P is normally selected by head Q, when P has adjuncts Q should still select P.
3. **Order**: there should be a mechanism for forcing an order on adjuncts
4. **Selectability** (8)
   (a) **Efficiency:** All adjectives are possible arguments of the same predicates, so there should be a way to select for any adjective, rather than cross-listing the selector for each adjective category.
5. **Adjuncts of adjuncts** (9)
   (a) **Efficiency:** Similarly to selection, there should be a way to say that, say, an adverbs can adjoin to all adjective, rather than having a homophonous form of the adverb for each adjective category.
6. **Unordered** (10)
7. **Obligatory adjuncts** (11)
8. **Islands** (12)

# 5 Previous Approaches to Adjunction

This section provides a brief overview of three approaches to adjunction. The first two are from a categorial grammar perspective and account for the optionality and, more or less, transparency to selection; however, they are designed to model unordered adjuncts. The last is an MG formalisation of the cartographic approach. Since the cartographic approach takes adjuncts to be regular selectors, unsurprisingly they account for order, but not easily for optionality or transparency to selection.

## 5.1 Traditional MG/CG Solution

To account for the optionality and transparency, a common solution is for a modifier to combine with its modified phrase, and give the result the same category as the original phrase. Traditionally in MGs, an `X`-modifier has features `=XX`: it selects an `X` and the resulting structure has category feature `X`. Similarly, in categorial grammars, an X-modifier has category X/X or X\X. As such, the properties of traditional MG and CG models of adjunction are the same.[5]

[5] This is not the only possible solution using the MG architecture, but rather the traditional solution. Section 5.2 gives a model within MGs that accounts for order.

An anonymous reviewer suggested a different solution, with a set of silent, meaningless heads that turn categories into selectors of their adjuncts, for example $\epsilon$::`=N =Adj =N`. Such a solution does much better on desiderata 4 and 5 than the one given here, but shares with the cartographic solution given in section 5.2 the problem of linguistic undesirability of silent, meaningless elements.

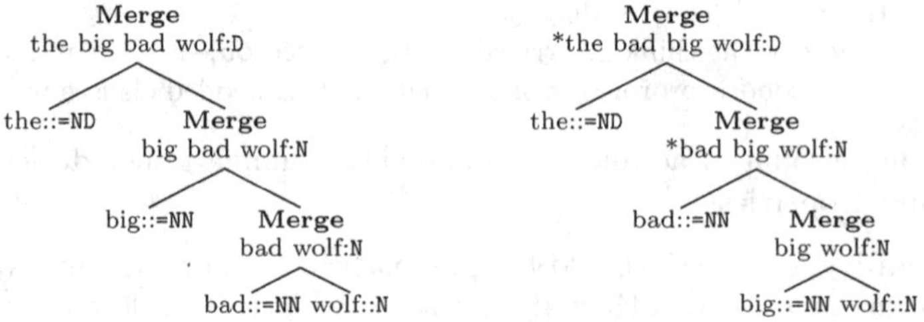


**Fig. 3.** Traditional MG approach

1. **Optionality:** ✓the original category is kept
2. **Transparency to selection: Sort of:** in Fig. 3, *the* selects N, but the N it checks is the one introduced by *bad*, not the one on *wolf.*
3. **Order: No**, the original category is kept so any adjunct may adjoin at any time.
4. **Selectability** Adjuncts need two versions, one for being adjuncts and the other for being selected. For example, *bad*::=NN cannot be selected by anything until it has itself selected an N.
turn an =NN into an N by being selected; however, such a solution predicts the general existence of silent Ns, and zero-derivation of adjectives from nouns; indeed, silent, meaningless versions of any modifiable category and zero-derivation of any modifiable category to its modifiers' categories.

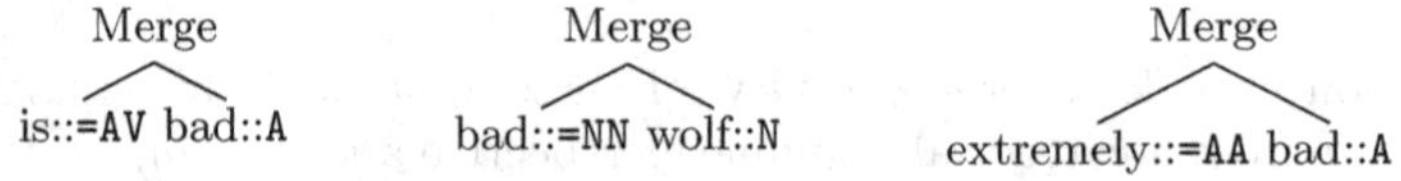


   (a) **Efficiency: No.** Adjuncts have two versions, or else we permit new silent categories and zero-derivation.
5. **Adjuncts of adjuncts** Since adjunction is selection in this model, we have the same problem, but with the same solution: the feature for selection is also the feature for being adjoined to.
   (a) **Efficiency:** The homophony for selection covers adjunction too.
6. **Unordered** ✓ All adjuncts are unordered in this model
7. **Obligatory adjuncts: No**, there is no way to distinguish between an phrase with an adjunct and one without.
8. **Islands** Not without additional constraints. See Graf [8] for an account that will work with the present approach.

**Frey & Gärtner.** Frey & Gärtner [7] propose an improved version of the categorial grammar approach, one which keeps the modified element the head, giving true transparency to selection. They do this by asymmetric feature checking. To the basic MG formalism a third polarity is added for **sel**, ≈X. This polarity drives the added function Adjoin. Adjoin behaves just like Merge except that instead of cancelling both ≈X and X, it cancels only ≈X, leaving the original X in tact. This allows the phrase to be selected or adjoined to again by anything that selects or adjoins to X. This model accounts for optionality and true transparency, but it is not designed to capture ordered adjuncts. Also, since adjuncts don't have categories of their own (just ≈X), it is not clear how to model selection of and adjunction to adjuncts.[6]

## 5.2 Selectional Approach

A third approach is to treat adjuncts just like any other selector. This is the approach implicitly taken by syntactic cartography in mainstream linguistics.[7] Such an approach accounts straightforwardly for order, but not for optionality or transparency; this is unsurprising since the phenomena I am modelling share only ordering restrictions with ordinary selection.

The idea is to take the full hierarchy of modifiers and functional heads, and have each select the one below it; for example, *big* selects *bad* but not vice versa, and *bad* selects *wolf.* However, here we are left with the question of what to do when *bad* is not present, and the phrase is just *the big wolf. big* does not select *wolf.* I will briefly outline one solution, in which the full structure is always present.

We give each modifier and functional head a silent, meaningless version that serves only to tie the higher modifier to the lower, like syntactic glue. For example, we add to the lexicon a silent, meaningless "size" modifier that goes where *big* and *small* and other LIs of category S go.

| | | | |
|---|---|---|---|
| the::=S D | $\epsilon$::=S D | wolf::N | |
| big::=G S | $\epsilon$::=G S | bad::=N G | $\epsilon$::=N G |

This solution doubles substantial portions of the lexicon. Doubling is not computationally significant, but it does indicate a missing generalisation: somehow, it just happens that each of these modifiers has a silent, meaningless doppelganger. Relatedly, the ordering facts are epiphenomenal. There is no universal principle predicting the fairly robust cross-linguistic regularity. Moreover, normally when something silent is in the derivation, we want to say it is contributing something semantically. Here these morphemes are nothing more than a trick to hold the syntax together.

1. **Optionality:** ✓Choose the right version of an LI. Note this is **inefficient**.
2. **Transparency to selection:** No, selection is of the adjunct, not the head. For example, in the lexicon above, *the* selects the (possibly empty) adjunct with features =G S, not the noun.
3. **Order:** ✓This is Merge, so order is determined by the particular lexical items' feature stacks

[6] This is not to say that it cannot be done. [7] has examples with selectional features =≈X, though there is little discussion.

[7] It is not clear whether we should take mainstream syntax approaches to mean that there are always-present, silent, meaningless, functional heads. Another interpretation of their models is that each functional head on the hierarchy has a set of homophones, one for each level down in the hierarchy.

4. **Selectability** ✓This is ordinary Merge, so selection proceeds as usual.
   (a) **Efficiency:** ✓No homophony added for selection
5. **Adjuncts of adjuncts** Adjuncts of adjuncts are simply selectors of adjuncts.
   (a) **Efficiency:** ✓No homophony added for adjuncts of adjuncts
6. **Unordered** Very difficult, but possible with enough homophony.
7. **Obligatory adjuncts: No**, the same thing that allows optionality of adjuncts prevents us from requiring that an adjunct be present.
8. **Islands: No**, not without additional constraints. Again, see [8] for constraints that may work here.

## 6 Proposal

I propose a solution, which I will call Minimalist Grammars with Adjunction (MGAs),[8] which accounts for ordering by indexing phrases according to the hierarchy level of the last adjunct adjoined to them.

A given adjunct phrase $P$ needs four pieces of information: $P$'s category, what $P$ is an adjunct of, what level adjunct $P$ is, and what level the last adjunct that adjoin to $P$ was. We need to know what a category is an adjunct of because that will determine whether, say, an adjective can adjoin to a noun phrase. I include in the grammar a set of adjuncts for each category. The hierarchy level of the adjunct is needed for when it acts as an adjunct. If the phrase it is adjoining to already has a adjunct, we need to check that the new adjunct is higher in the hierarchy. For this purpose, every phrase carries with it an additional number, indexing the level of its last adjunct. The two numbers are kept separate so that adjuncts can have adjuncts, as in *bright blue*. *Bright blue* has an adjunct *bright*, which may affect what further adjuncts can adjoin to it, but which does not affect what the phrase *bright blue* can adjoin to.

To track hierarchy level, each category feature is expanded into a triple consisting of the category feature, the level of the hierarchy of adjuncts the head is at, and the level of hierarchy the whole phrase is at. Hierarchy levels are encoded as natural numbers,[9] and $\leq$ is the usual order on $\mathbb{N}$.[10] These numbers are lexically specified; for example *wolf*:: `[N,0,0]` would be in the lexicon.

By splitting the category into its category and its level as adjunct, we can allow all, say, adjectives, to have the same category. This extends the efficiency gains in [6] to selection of adjuncts and adjuncts of adjuncts.

When adjunct `[Y,n,m]` adjoins to something of category `[X,i,j]`, the resulting phrase is of category `[X,i,m]`, `i,j,n,m` $\in \mathbb{N}$. The second number is what

[8] My earlier paper [6] used this name as well; this model is designed to improve on it.

[9] This is a similar approach to that taken by Adger [1], who proposes a second version of Merge that models the functional heads in a hierarchy. Our approaches differ in that in MGAs the original category is kept under Adjoin. A general discussion of the use of explicit hierarchy in grammars can be found in [2].

[10] $\mathbb{N}$ is simply acting as an index set, and that the maximal depth of hierarchies in a language bounds the actual index set for the grammar.

**Fig. 4.** Adjoin. The category feature of the new phrase is the first two elements of the adjoined-to phrase followed by the second element of the adjunct.

tracks the level of the hierarchy the phrase is at; it is the only thing that can change.

### 6.1 Example

Before I give the full formal definition I will present an example. Suppose we have a grammar in which the adjunct sets are defined as follows:

$$\text{Ad}(\texttt{N})=\{\texttt{Adj, P, C}\}, \text{Ad}(\texttt{Adj})=\{\texttt{Adv, Int}\}, \text{Ad}(\texttt{Adv})=\{\texttt{Int}\}, \text{Ad}(\texttt{V})=\{\texttt{Adv,T}\}$$

We can derive *Apparently, John very often sang* as in figure 5. *very* adjoins to *often* since *often* is at level 0 and *very* is at level 3, and $3 \geq 0$. The whole phrase adjoins to *sang* since it's at level 18 and *sang* is at 0. T Merges to the VP, yielding a phrase at level 25. *Apparently* is at level 26, so it can adjoin.

To get order, we require that the first number of the adjunct be at least as high as the second number of the adjoined-to phrase. For example, in Figure 6, the derivation of *the big bad wolf* works because Adj$\in$**Ad**(N), and $6 > 4 > 0$. The derivation of *the bad big wolf* fails because the category of *big wolf* is [N,0,6]. *bad*::[Adj,4,0] can't adjoin to it because *bad* is a level-4 adjunct, but *big wolf* is already at level 6, and $4 < 6$.

### 6.2 Definition

**Merge** must be trivially redefined for categories as triples. **Merge** only cares about category, so it looks to match the positive selectional feature with the first element of the triple. (**Move** is unchanged.)

**Definition 4 (Merge).** *For $\alpha, \beta \in F^*$; $s, t$ strings:*

$$\textbf{Merge}(\langle s, \texttt{=X}\alpha\rangle :: mvrs_s, \langle t, [\texttt{X},\texttt{i},\texttt{j}]\beta\rangle :: mvrs_t) = \begin{cases} \langle st, \alpha\rangle :: mvrs_s \cdot mvrs_t & \text{if } \beta = \epsilon \\ \langle s, \alpha\rangle :: \langle t, \beta\rangle :: mvrs_s \cdot mvrs_t & \text{if } \beta \neq \epsilon \end{cases}$$

**Adjoin** applies when the category of the adjunct is an adjunct of the category it is adjoining to, and if the adjunct is a $k$-level adjunct then the level of the phrase it is adjoining to is no higher than $k$. **Move** works as expected: the adjunct has negative licensing features left after it has had its category feature checked by **Adjoin**, it is added to the list of movers.

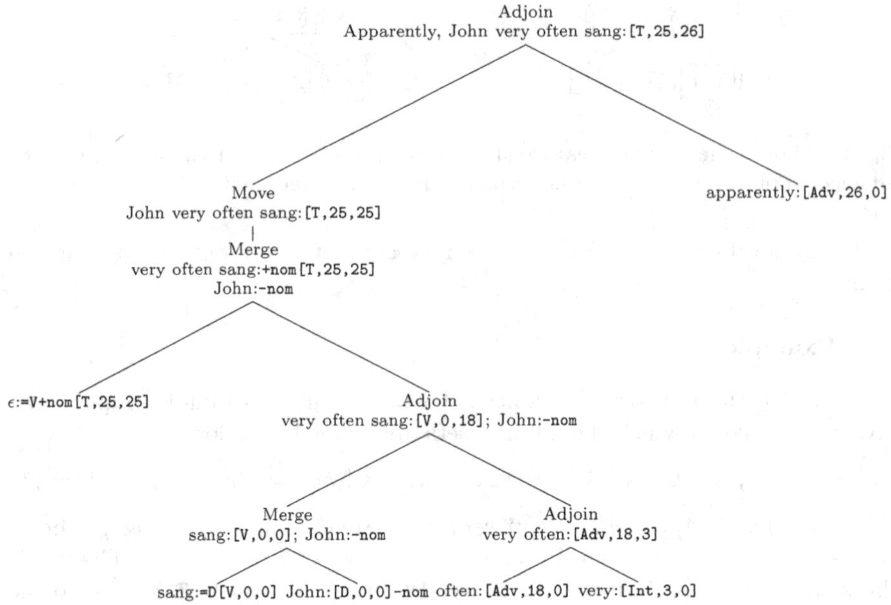


**Fig. 5.** Adjunct of adjunct; functional head merge; adjunction after functional head merge

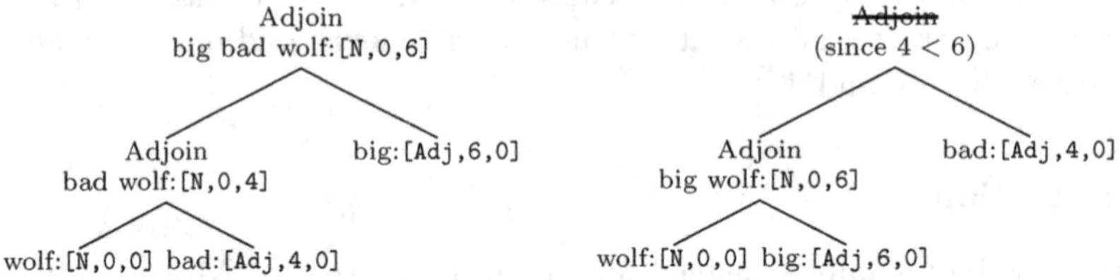


**Fig. 6.** Adjunct ordering: valid and invalid derivations

**Definition 5 (Adjoin).** *Let* $s, t \in \Sigma$ *be strings,* $\mathtt{Y}, \mathtt{X} \in \mathbf{sel}$ *be categories,* $i, j, n, m \in \mathbb{N}$, *mvrs* $\in (\Sigma^* \times F)^*$ *be a mover list, and* $\alpha, \beta \in F^*$.

$$\mathbf{Adjoin}(\langle s, [\mathtt{X}, \mathtt{i}, \mathtt{j}]\alpha :: mvrs\rangle, \langle t, [\mathtt{Y}, \mathtt{n}, \mathtt{m}]\beta\rangle)$$
$$= \begin{cases} \langle ts, [\mathtt{X}, \mathtt{i}, \mathtt{n}]\alpha\rangle :: mvrs & \textit{if } \mathtt{n} \geq \mathtt{j} \;\&\; \mathtt{Y} \in \mathbf{Ad}(\mathtt{X}) \;\&\; \beta = \epsilon \\ \langle s, [\mathtt{X}, \mathtt{i}, \mathtt{n}]\rangle :: \langle t, \beta\rangle :: mvrs & \textit{if } \mathtt{n} \geq \mathtt{j} \;\&\; \mathtt{Y} \in \mathbf{Ad}(\mathtt{X}) \;\&\; \beta \neq \epsilon \end{cases}$$

Notice that for Merge, there may be a mover list with both arguments ($movers_s$ and $movers_t$). Island constraints for adjoin are implemented by simply leaving out the mover list that would come with the adjunct. Adjoin is not defined when the adjunct has a mover.[11] This is not necessarily as stipulative as it sounds: [8] puts forth that for adjuncts to be truly optional they cannot have movers, or else the derivation tree without the adjunct would have an unchecked positive licensing feature. My definition of Adjoin is simply a way of conforming to this constraint.

**Definition 6 (MGA).** *A* ***Minimalist Grammar with Adjunction*** *is a six-tuple*
$G = \langle \Sigma, \mathbf{sel}, \mathbf{lic}, \mathbf{Ad}, Lex, M\rangle$. *$\Sigma$ is the alphabet.* $\mathbf{sel} \cup \mathbf{lic}$ *are the base features. Let* $F = \{+\mathtt{f}, -\mathtt{f}, =\mathtt{X}, [\mathtt{X}, \mathtt{n}, \mathtt{m}] | \mathtt{f} \in \mathbf{lic}; \mathtt{X}, \mathtt{Y} \in \mathbf{sel}; \mathtt{m}, \mathtt{n}\mathbb{N}\}$. $\mathbf{Ad} : \mathbf{sel} \rightarrow \mathcal{P}(\mathbf{sel})$ *maps categories to their adjuncts.* $Lex \subseteq_{fin} \Sigma \times F^*$, *and $M$ is the set of operations* ***Merge, Move,*** *and* ***Adjoin****. The language $L_G$ is the closure of Lex under $M$. A set $C \subseteq$* **sel** *of designated features can be added;* $\{[\mathtt{c}, \mathtt{i}, \mathtt{j}] | \mathtt{c} \in C; i, j \in \mathbb{N}\}$ *are the types of complete sentences.*

### 6.3 Adverbs and Functional Heads

Contra Cinque [4], I model adverbs as separate from functional heads. Adverbs and adjectives differ from functional heads in two ways. First, they are not themselves adjoined to, while adverbs and adjectives are (*very blue*). Second, functional heads are sometimes required and sometimes optional. For example, English requires `T`, but not, perhaps, $\mathtt{Mod}_{\text{epistemic}}$ in every sentence. To model this, I give adjectives and adverbs category triples with their second number set to 0. This allows adjuncts to adjoin to them, starting at the bottom of that hierarchy. Functional heads, on the other hand, will start with their second number equal to their first number. This means that when they Merge, the resulting phrase is at the right level in the hierarchy, preventing low adjuncts from adjoining after the merger of a high functional head.[12]

For example, in Figure 5, *very* adjoins to *often*, which is possible since the second number of *often* is 0. Later, functional head T Merges to the VP. Its second number is 25. This is important because we want to say that *apparently* can only adjoin here because its first number is 26, which is higher than 25. A low adverb such as *again*:: `[Adv,3,0]` cannot adjoin to T.

### 6.4 Properties

Let us consider the desiderata laid out in section 4.

[11] This is possible only because Adjoin and Merge are separate operations, as they are in [7]. A close look at the definitions of Merge and Adjoin reveals that there is nothing formally stopping Adjoin from being a case of Merge, one defined when both phrases display a category feature. I have chosen to keep them as separate operations so that Adjoin may have different properties from Merge, such as island effects, and to maintain a certain type of locality for Merge, discussed in section 7.

[12] There is nothing in this formalism that prevents adjunction to a functional head. If the function **Ad** assigns adjuncts to a functional head, then it has adjuncts. They just behave a little oddly: e.g. `[F,3,3]` requires adjuncts above level 3. Note also a shortcoming in the present model: while Merge of a high functional head will prevent later adjunction of a low adverb, nothing prevents a low functional head that selects, say, `V`, from merging after the adjunction of a high adverb.

1. **Optionality:** ✓the original category is kept as the first element of the category triple
2. **Transparency to selection:** ✓the original category is kept
3. **Order:** ✓The third element of the category is the level of the last adjunct adjoined. The Adjoin rule requires that the adjunct be higher in the order than that third element of the category triple.
4. **Selectability** ✓Adjuncts have regular categories.

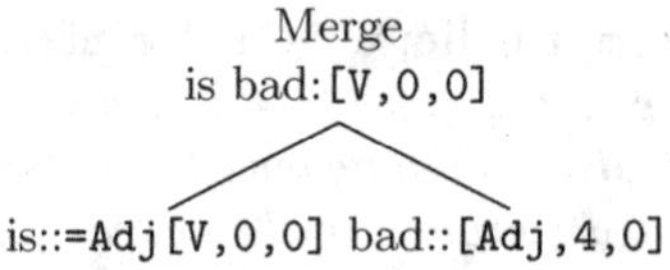


   (a) **Efficiency:** ✓Many adjuncts have the same category, so they have the same adjuncts. For example, **Ad**(`Adj`) = {`Adv`, `Int`}

5. **Adjuncts of adjuncts** ✓Adjunct categories are ordinary categories so they can have adjuncts too (Figure 5).
   (a) **Efficiency:**✓Many adjuncts have the same category, so they are selected by the same LI. For example, in the derivation of *is bad* above, *is* selects anything of category `Adj`.
6. **Unordered** ✓See section 6.5 below.
7. **Obligatory adjuncts: Maybe.** See Section 6.6
8. **Islands** ✓Since Adjoin is a separate operation, it can be defined so that there is no case for adjuncts with movers.

### 6.5 Unordered Adjuncts

As it stands, adjuncts such as PPs can be modelled as adjuncts, but they must all adjoin at the same level of the hierarchy, or else be cross-classified for each level of the hierarchy you want them to adjoin at. The former allows them to be freely ordered with respect to each other; the latter gives them freedom with respect to all adjuncts.

An expansion of this model[13] could add a non-number to the set of possible indicies, call it $\emptyset$, and Adjoin could be defined to disregard the hierarchy and asymmetrically check the features for $\emptyset$-indexed adjuncts. Any distinct index also opens the door to adjoining on a different side of the head than other adjuncts; the definition I will give here models Engish PPs, which are post-head, unlike adjectives and many adverbs.

In definition 7, the first and third cases are for adjuncts with number indicies, and the second and fourth are for adjuncts with $\emptyset$ indicies.

**Definition 7 (Adjoin 2).** *Let* $s, t \in \Sigma$ *be strings,* $Y, X \in \mathbf{sel}$ *be categories,* $i, j, n, m \in \mathbb{N}$*,* $mvrs \in (\Sigma^* \times F)^*$ *be a mover list, and* $\alpha, \beta \in F^*$*.*

[13] I thank an anonymous reviewer for this suggestion.

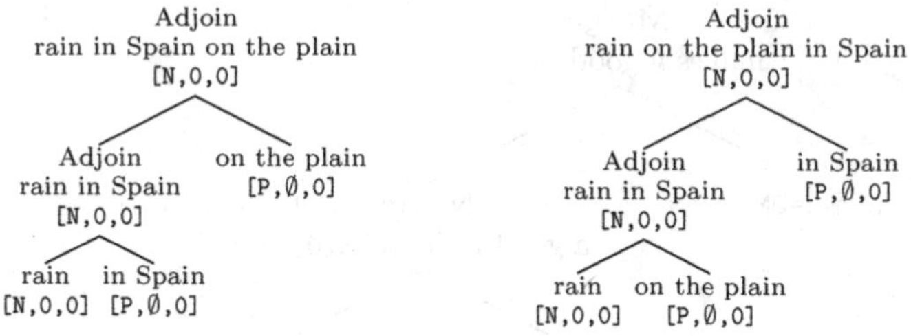


**Fig. 7.** Unordered English PPs

$$\mathbf{Adjoin}(\langle s, [\mathtt{X},\mathtt{i},\mathtt{j}]\alpha\rangle, \langle t, [\mathtt{Y},\mathtt{m},\mathtt{n}]\beta\rangle :: mvrs)$$
$$= \begin{cases} \langle ts, [\mathtt{X},\mathtt{i},\mathtt{m}]\alpha\rangle :: mvrs & \textit{if } \mathtt{m} \geq \mathtt{j} \;\&\; \mathtt{Y} \in \mathbf{Ad}(\mathtt{X}) \;\&\; \beta = \epsilon \\ \langle st, [\mathtt{X},\mathtt{i},\mathtt{j}]\alpha\rangle :: mvrs & \textit{if } \mathtt{m} = \emptyset \;\&\; \mathtt{Y} \in \mathbf{Ad}(\mathtt{X}) \;\&\; \beta = \epsilon \\ \langle s, [\mathtt{X},\mathtt{i},\mathtt{m}]\alpha\rangle :: \langle t, \beta\rangle :: mvrs & \textit{if } \mathtt{m} \geq \mathtt{j} \;\&\; \mathtt{Y} \in \mathbf{Ad}(\mathtt{X}) \;\&\; \beta \neq \epsilon \\ \langle s, [\mathtt{X},\mathtt{i},\mathtt{j}]\alpha\rangle :: \langle t, \beta\rangle :: mvrs & \textit{if } \mathtt{m} = \emptyset \;\&\; \mathtt{Y} \in \mathbf{Ad}(\mathtt{X}) \;\&\; \beta \neq \epsilon \end{cases}$$

### 6.6 Obligatory Adjuncts

Recall that some elements which really seem to be adjuncts are not optional, for example *He makes a *(good) father.* In MGAs there is a featural difference between nouns that have been modified and nouns that have not. For example, *father* is of category [N,0,0] and *good father* has category [N,0,4]. **Merge** is defined to ignore everything but the first element, N. However, the architecture is available to let **Merge** look at the whole category triple, by way of a positive selectional feature of the form =[N,_,1], which selects anything of category [N,i,j] with $\mathtt{j} \geq 1$.

**Definition 8 (Merge 2).** *For $\alpha, \beta$ sequences of negative* **lic** *feature;, $s, t$ strings; $X \in$* **sel**; $\mathtt{i},\mathtt{j},\mathtt{m} \in \mathbb{N}$; $\mathbf{C} = \mathtt{X}$ *or* $\mathbf{C} = [\mathtt{X},\_,\mathtt{m}]$ *&* $\mathtt{j} \geq \mathtt{m}$*:*

$$\textit{Merge}(\langle s, \texttt{=}\mathbf{C}\alpha\rangle :: mvrs_s,\ \langle t, [\mathtt{X},\mathtt{i},\mathtt{j}]\beta\rangle :: mvrs_t) = \begin{cases} \langle ts, \alpha\rangle :: mvrs_s \cdot mvrs_t & \textit{if } \beta = \epsilon \\ \langle s, \alpha\rangle :: \langle t, \beta\rangle :: mvrs_s \cdot mvrs_t & \textit{if } \beta \neq \epsilon \end{cases}$$

However, such an expansion of the definition of Merge is not of immediate help in all cases. In the case of *He makes a good father*, the NP *good father* is selected by D before the resulting DP is selected by *makes*, which is the verb that cares about whether the noun is modified. One solution is to cross-list *a* with a new determiner category only for modified NPs, and let *makes* select that category, as in Figure 8.

Obligatory adjuncts are not the only reason to suspect that the tighter relationship is between the verb and the noun, not the verb and the determiner; i.e. that V should perhaps select N, not D. For one, it is well known that in terms of semantics, verbs select nouns. For example, *The man slept* makes sense, but *The table slept* does not, because men are the kinds of things that sleep and tables are not. Both DPs are headed by *the*, which does not carry the animacy information that the noun does. Another piece of evidence comes from noun

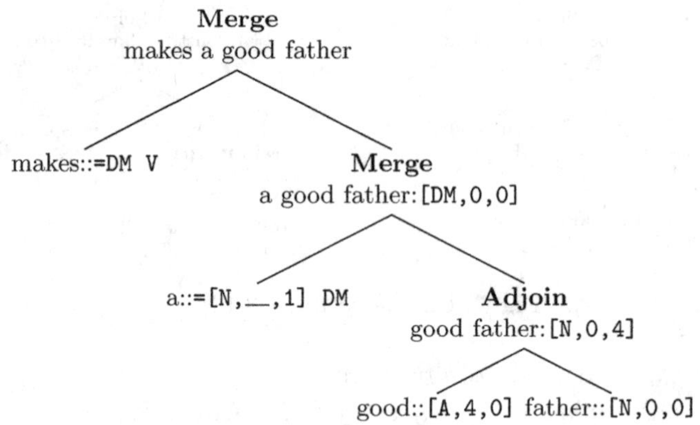


**Fig. 8.** Determiners of modified NPs could have their own category DM

incorporation. When a head is incorporated into a verb, normally it is the head that the noun selects that is incorporated, as in (13).

(13) a. He [**stab**bed me [$_{PP}$ in [$_{DP}$ the [$_{N}$ **back**]]]]
b. back-stabbing
c. *back-in, *back-the, *back-the-stabbing, *back-in-the-stabbing

[13] proposes that verbs select NPs, and the NPs move to their Ds, which are functional heads on the spine.

For example we might have something like the partial derivation in Fig 9.

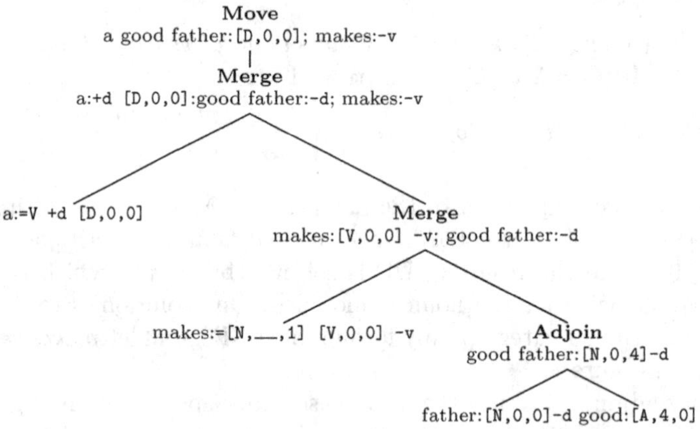


**Fig. 9.** Directly selecting N; moving NP up to functional projection D

## 7 Formal Properties

MGAs are clearly not strongly equivalent to traditional MGs, if we take *strong equivalence* to mean that the set of derivations trees are the same. This is of course is impossible since MGAs have an extra function, Adjoin. MGAs are, on the other hand, weakly equivalent to MGs, meaning that for every MGA, an MG can be defined that generates the same strings, and vice versa.

**Lemma 1.** $L(MG) \subseteq L(MGA)$

*Proof.* MGAs also include Merge and Move, and place no additional restrictions on their action. Any MGA language could have Adjoin stripped away and what remained would be an MG.

**Lemma 2.** $L(MGA) \subseteq L(MG)$

*Proof.* MGs are weakly (and indeed strongly) equivalent to Multiple Context Free Grammars (MCFGs) so it suffices to show that $L(MGA) \subseteq L(MCFG)$.

We translate an MGA into an MCFG is the normal way, following [9]: the nonterminals of the MCFG are sequences of feature sequences from the MGA. This translation is based on the basic grammar given in Definition 6, but it is easy to see how it could be expanded to include the extentions suggested in later sections.

Given MGA $G = \langle \Sigma, F = \mathbf{sel} \cup \mathbf{lic}, Lex, M, S, Ad$, define an MCFG $MCFG(G) = \langle \Sigma, N, P, S\rangle$ defining the language

$N = \{\langle \delta_0, \delta_1, ..., \delta_j\rangle | 0 \leq j \leq |\mathbf{lic}|, \text{ all } \delta_i \in \text{suffix}(Lex)\}$

Let $h = \text{Max}(\{i | \exists \mathtt{X} \in \mathbf{sel} : i = |\mathbf{Ad}(\mathtt{X})\}$

The rules $P$ are defined as follows, $\forall \alpha, \beta, \delta_0, ..., \delta_i, \gamma_0, ..., \gamma_j \in$ suffix(Lex). $s_0, ..., s_i, t_0, ..., t_j$ are variables over strings.

**Lexical rules:** $\alpha(s)$ $\qquad \forall \langle s, \alpha\rangle \in Lex$

**Merge-and-stay rules:** Here is the first case of the Merge rule for MGAs.

$$\mathbf{Merge}(\langle s, \mathtt{=X}\alpha\rangle :: \text{mvrs}_s, \langle t, [\mathtt{X}, \mathtt{m}, \mathtt{n}]\rangle :: \text{mvrs}_t) = \langle st, \alpha\rangle :: \text{mvrs}_s \cdot \text{mvrs}_t$$

It becomes a set of MCFG rules as follows. In the rule set below, $s = s_0, t = t_0$,the tree parts of mvrs$_s$ and mvrs$_t$ are $s_1, ..., s_i$ and $t_1, ..., t_j$ respectively, and their features become $\delta_1, ..., \delta_i$ and $\gamma_1, ... \gamma_j$. One rule is made for each index less than the maximum possible index $h$ for the grammar. (Any rule indicies that fall outside the set of indicies for that particular category simply go unused in practice.)

Here is the description of the MCFG rules corresponding to this Merge rule:

$$\langle \alpha, \delta_1, ..., \delta_i, \gamma_1, ... \gamma_j\rangle(s_0 t_0, s_1, ..., s_i, t_1, ..., t_j)$$
$$:\!-\ \langle = \mathtt{X}\alpha, \delta_1, ..., \delta_i\rangle(s_0, ..., s_i)\ \langle [\mathtt{X}, \mathtt{m}, \mathtt{n}], \gamma_1, ... \gamma_j\rangle(t_0, ..., t_j)$$
$$\forall \mathtt{X} \in \mathbf{sel}, \forall \mathtt{n}, \mathtt{m} \leq h$$

The rest of the MCFG rules are formed similarly.

**Merge-and-move rules:** $\forall \mathtt{X} \in \mathbf{sel}, \forall \mathtt{n}, \mathtt{m} \leq h$

$\langle \alpha, \beta, \delta_1, ..., \delta_i, \gamma_1, ...\gamma_j \rangle (s_0, t_0, s_1, ..., s_i, t_1, ..., t_j)$
:- $\langle = \mathtt{X}\alpha, \delta_1, ..., \delta_i \rangle (s_0, ..., s_i)$ $\langle [\mathtt{X}, \mathtt{m}, \mathtt{n}]\beta, \gamma_1, ...\gamma_j \rangle (t_0, ..., t_j)$

**Adjoin-and-stay rules:** $\forall \mathtt{X}, \mathtt{Y} \in \mathbf{sel}$ *s.t.* $\mathtt{Y} \in \mathbf{Ad}(\mathtt{X}), \forall \mathtt{k}, \mathtt{l}, \mathtt{n}, \mathtt{m} \leq h$ *s.t.* $\mathtt{n} \geq \mathtt{k}$

$\langle [\mathtt{X}, \mathtt{m}, \mathtt{n}], \delta_1, ..., \delta_i, \gamma_1, ...\gamma_j \rangle (s_0 t_0, s_1, ..., s_i, t_1, ..., t_j)$
:- $\langle [\mathtt{X}, \mathtt{m}, \mathtt{k}], \delta_1, ..., \delta_i \rangle (s_0, ..., s_i)$ $\langle [\mathtt{Y}, \mathtt{n}, \mathtt{l}], \gamma_1, ...\gamma_j \rangle (t_0, ..., t_j)$

**Adjoin-and-move rules:** $\forall \mathtt{X}, \mathtt{Y} \in \mathbf{sel}$ *s.t.* $\mathtt{Y} \in \mathbf{Ad}(\mathtt{X}), \forall \mathtt{k}, \mathtt{l}, \mathtt{n}, \mathtt{m} \leq h$ *s.t.* $\mathtt{n} \geq \mathtt{k}$

$\langle [\mathtt{X}, \mathtt{m}, \mathtt{n}], \beta, \delta_1, ..., \delta_i, \gamma_1, ...\gamma_j \rangle (s_0, t_0, s_1, ..., s_i, t_1, ..., t_j)$
:- $\langle [\mathtt{X}, \mathtt{m}, \mathtt{k}], \delta_1, ..., \delta_i \rangle (s_0, ..., s_i)$ $\langle [\mathtt{Y}, \mathtt{n}, \mathtt{l}]\beta, \gamma_1, ...\gamma_j \rangle (t_0, ..., t_j)$

**Move-and-stop rules:** $\forall \mathtt{f} \in \mathbf{lic}$

$\langle \alpha, \delta_1, ..., \delta_{i-1}, \delta_{i+1}, ..., \delta_j \rangle (s_i s_0, s_1, ..., s_{i-1}, s_{i+1}, ..., s_j)$
:- $\langle +\mathtt{f}\alpha, \delta_1, ..., \delta_{i-1}, -\mathtt{f}, \delta_{i+1}, ..., \delta_j \rangle (s_0, ..., s_j)$

**Move-and-keep-moving rules:** $\forall \mathtt{f} \in \mathbf{lic}$

$\langle \alpha, \delta_1, ..., \delta_{i-1}, \beta, \delta_{i+1}, ..., \delta_j \rangle (s_0, ..., s_j)$
:- $\langle +\mathtt{f}\alpha, \delta_1, ..., \delta_{i-1}, -\mathtt{f}\beta, \delta_{i+1}, ..., \delta_j \rangle (s_0, ..., s_j)$

These rule sets are finite since MGAs never add anything to feature sequences, but only either remove features or change just the indicies of `[X,i,j]` features. As such, the suffixes $\alpha, \beta, \delta, \gamma$ are limited in number. Since any given grammar has a maximal hierarchy depth $h$, the indicies `k,l,m,n` in the rules are defined to be limited by $h$.

**Theorem 1 (Weak equivalence of MGAs and MGs).** *For any MGA* $G = \langle \Sigma, \mathbf{sel}, \mathbf{lic}, \mathbf{Ad}, \mathit{Lex}, \{\mathbf{Merge}, \mathbf{Move}, \mathbf{Adjoin}\} \rangle$*, there is a weakly equivalent MG* $G' = \langle \Sigma, \mathbf{sel}_{MG}, \mathbf{lic}, \mathit{Lex}_{MG}, \{\mathbf{Merge}, \mathbf{Move}\} \rangle$.

*Proof.* By lemmas 1 and 2

## References


1. Adger, D.: A minimalist theory of feature structure 30, 06–08 (2007), http://ling.auf.net/lingBuzz/000583
2. Bernardi, R., Szabolcsi, A.: Optionality, scope, and licensing: An application of partially ordered categories. Journal of Logic, Language and Information 17(3), 237–283 (2008)
3. Chomsky, N.: The Minimalist Program. MIT Press, Cambridge (1995)
4. Cinque, G.: Adverbs and functional heads: a cross-linguistic perspective. Oxford studies in comparative syntax. Oxford University Press, Oxford (1999)
5. Cinque, G.: The syntax of adjectives: a comparative study. Linguistic Inquiry monographs. MIT Press, Cambridge (2010)
6. Fowlie, M.: Order and optionality: Minimalist grammars with adjunction. In: The 13th Meeting on the Mathematics of Language, p. 12 (2013)
7. Frey, W., Gärtner, H.M.: On the treatment of scrambling and adjunction in minimalist grammars. In: Proceedings of the Conference on Formal Grammar (FGTrento), Trento, pp. 41–52 (2002)
8. Graf, T.: The price of freedom: Why adjuncts are Islands. Slides of a talk given at the Deutsche Gesellschaft für Sprachwissenschaft, pp. 12–15 (2013)

9. Harkema, H.: A characterization of minimalist languages. In: de Groote, P., Morrill, G., Retoré, C. (eds.) LACL 2001. LNCS (LNAI), vol. 2099, pp. 193–211. Springer, Heidelberg (2001)
10. Keenan, E.L., Stabler, E.P.: Bare Grammar. CSLI Publications, Stanford (2003)
11. Kobele, G.M., Retoré, C., Salvati, S.: An automata-theoretic approach to minimalism. In: Rogers, J., Kepser, S. (eds.) Model Theoretic Syntax at ESSLLI 2007. ESSLLI (2007)
12. Rizzi, L.: Locality and left periphery. In: Belletti, A. (ed.) Structures and Beyond: The Cartography of Syntactic Structures, vol. 3, pp. 223–251. Oxford University Press, Oxford (2004)
13. Sportiche, D.: Division of labor between merge and move: Strict locality of selection and apparent reconstruction paradoxes. LingBuzz (2005)
14. Stabler, E.: Derivational minimalism. In: Retoré, C. (ed.) LACL 1996. LNCS (LNAI), vol. 1328, pp. 68–95. Springer, Heidelberg (1997)

# Models of Adjunction in Minimalist Grammars

Thomas Graf

Department of Linguistics
Stony Brook University, NY, USA
mail@thomasgraf.net
http://thomasgraf.net

**Abstract.** Three closely related proposals for adding (cyclic) adjunction to Minimalist grammars are given model-theoretic definitions and investigated with respect to their linguistic and formal properties. While they differ with respect to their linguistic adequacy, they behave largely the same on a computational level. Weak generative capacity is not affected, and while subregular complexity varies betweeen the three proposals, it does not exceed the complexity imposed by Move. The closure of Minimalist derivation tree languages under intersection with regular tree languages, however, is lost.

**Keywords:** Minimalist grammars, adjunction, derivation trees, subregular tree languages, closure properties.

## Introduction

The distinction between arguments and adjuncts is recognized by a variety of linguistic formalisms. Although a number of empirical properties have been identified — for instance, adjuncts can be freely iterated and dropped from sentences — there is little consensus as to how adjuncts should be implemented. In the case of Minimalist grammars (MGs; [12]), a formalization of contemporary Chomskyan syntax, at least three different mechanisms have been proposed: adjunction as category-preserving selection, adjunction as asymmetric feature checking [3], and adjunction without feature checking [2].

This paper evaluates these three proposals with respect to their formal properties and linguistic adequacy. Building on [5, 6], I give a model-theoretic definition of each system in terms of constraints on Minimalist derivation trees and a mapping from these derivations to derived trees. The linguistic adequacy of these proposals is then evaluated with respect to a number of fundamental properties of adjuncts such as optionality and iterability, extending previous observations by Fowlie [2]. On the formal side, I compare Graf's results on subregular complexity of the derivation tree languages of standard MGs [5] to that of MGs with adjunction.

As summarized in Tab. 1 and 2 at the end of the paper, only the implementation without feature checking satisfies all linguistic criteria, but it is also the most complex, which is reflected by its higher subregular complexity. Furthermore, the closure under intersection with regular tree languages enjoyed by

G. Morrill et al. (Eds.): Formal Grammar 2014, LNCS 8612, pp. 52–68, 2014.

standard MGs [4, 9] is lost with all three variants of adjunction. Intuitively, this is due to the optionality and iterability of adjuncts, which each implementation needs to capture.

I proceed as follows: Section 1.1 recapitulates the constraint-based definition of MGs in [5], while Sec. 1.2 lists some formal properties of standard MGs. Sections 2.1–2.3 then look at each implementation of adjunction in greater detail. Even though the discussion is not particularly technical, the reader is expected to already be familiar with MGs and their constraint-based definition. All relevant details can be found in [4–6].

# 1 Minimalist Grammars

## 1.1 Definition in Terms of Derivation Tree Languages

I follow [5, 6] in defining MGs in terms of their Minimalist derivation tree languages (MDTLs) and a mapping from derivation trees to derived trees. This perspective will make it a lot easier later on to add adjunction operations to MGs and reason about their generative capacity, "derivational" complexity, and linguistic adequacy.

The definition of MDTLs is rather intuitive. Each lexical item (LI) is translated into a tree that corresponds to the contiguous subpart of the derivation that the LI controls via its positive polarity features. These trees are called *slices*. An MDTL is the largest set of trees that can be assembled from a finite number of slices without violating any constraints imposed by the MG feature calculus. As we will see in the next section, adding adjunction amounts to the introduction of new slices and modifying the constraints that regulate their distribution.

We start by defining features in a modular way.

**Definition 1.** *Let* BASE *be a non-empty, finite set of* feature names. *Furthermore,* OP $:= \{merge, move\}$ *and* POLARITY $:= \{+, -\}$ *are the sets of* operations *and* polarities, *respectively. A* feature system *is a non-empty set* $Feat \subseteq \text{BASE} \times \text{OP} \times \text{POLARITY}$.

Negative Merge features are called *category features* (denoted $f$), positive Merge feature *selector features* (=f), negative Move features *licensee features* ($-f$), and positive Move features *licensor features* ($+f$). In the following, $\nu(f)$, $\omega(f)$ and $\pi(f)$ denote the name, operation, and polarity of $f$, respectively.

**Definition 2.** *Given a string alphabet* $\Sigma$ *and feature system Feat, a* $(\Sigma, Feat)$-lexicon *is a finite subset of* $\Sigma \times \{::\} \times \{f \mid \pi(f) = +\}^* \times \{f \mid \omega(f) = merge, \pi(f) = -\} \times \{f \mid \omega(f) = move, \pi(f) = -\}^*$.

The ordering restriction on features is actually a corollary of the Minimalist feature calculus (see [4, 9]) and thus usually omitted. In anticipation of the modifications brought about by adjunction in the next section, though, I opt for explicitness over succinctness.

Next LIs are converted into slices in a top-down fashion:

**Definition 3.** *Let Lex be a $(\Sigma, Feat)$-lexicon and $Lex_\star := \{\sigma :: f_1 \cdots f_n \star \mid \sigma :: f_1 \cdots f_n \in Lex\}$. Then the* slice lexicon *of Lex is* slice$(Lex) := \{\zeta(l) \mid l \in Lex_\star\}$, *where $\zeta$ is given by*

$$\zeta(\sigma :: f_1 \cdots f_i \star f_{i+1} \cdots f_n) := \begin{cases} \sigma :: f_1 \cdots f_n \\ \qquad \text{if } f_1 \cdots f_i = \varepsilon \\ \zeta(\sigma :: f_1 \cdots f_{i-1} \star f_i \cdots f_n) \\ \qquad \text{if } \pi(f_i) = - \\ move(\zeta(\sigma :: f_1 \cdots f_{i-1} \star f_i \cdots f_n)) \\ \qquad \text{if } \tau(f_i) = move \text{ and } \pi(f_i) = + \\ merge(\Box, \zeta(\sigma :: f_1 \cdots f_{i-1} \star f_i \cdots f_n)) \\ \qquad \text{if } \tau(f_i) = merge \text{ and } \pi(f_i) = + \end{cases}$$

This definition uses the functional representation of trees, i.e. $f(t_1, \ldots, t_n)$ denotes the tree whose root is labeled $f$ and whose $i$-th daughter is the root of tree $t_i$, $1 \leq i \leq n$. The symbol $\Box$ indicates a possible tree substitution site: For node $u$ of tree $s$, $s[u \leftarrow t]$ is the result of substituting $t$ for the subtree in $s$ that is rooted in $u$. This is also called a *concatentation of s and t*. For slices $s$ and $t$, $s[u \leftarrow t]$ is defined iff $u$ is labeled $\Box$.

Given a slice lexicon slice$(Lex)$, the *free slice language* FSL(slice$(Lex)$) is the smallest set that contains every tree $t$ that is the result of concatenating finitely many $s \in$ slice$(Lex)$. The set of well-formed derivations is the largest subset of the free slice language whose trees obey certain tree-geometric conditions. These conditions correspond to constraints imposed by the Minimalist feature calculus.

An interior node $m$ of $\zeta(l)$ is *associated to feature $f_i$ on $l$* iff $m$ is the root of $\zeta(\sigma :: f_1 \cdots f_i \star f_{i+1} \cdots f_n)$. Two features $f$ and $g$ *match* iff they have identical names and operations but opposite feature polarities. An interior node $m$ matches a feature $g$ iff the feature $m$ is associated to matches $g$. Finally, the *slice root* of LI $l := \sigma :: f_1 \cdots f_n$ is the unique node of $\zeta(l)$ reflexively dominating every node in $\zeta(l)$.

**Merge** For every $t \in$ FSL(slice$(Lex)$) and node $m$ of $t$, if $m$ is associated to selector feature $= f$, then its left daughter is the slice root of an LI with category feature $f$.

More succinctly, the selector features of an LI must be checked by an LI with a matching category feature.

**Final** Let $F \subseteq$ Base be a distinguished set of *final categories*. For every $t \in$ FSL(slice$(Lex)$) and LI $l$, if the slice root of $l$ is also the root of $t$, then the category feature $c$ of $l$ is a final category, i.e. $\nu(c) \in F$.

The conditions on *move* are only of ancillary importance to this paper. Consequently, I content myself with the bare definitions and do not further explore the reasoning behind them; the interested reader is referred to [5]. For every $t \in$ FSL(slice$(Lex)$) and LI $l$ in $t$ with string $-f_1 \cdots - f_n$ of licensee features, $n \geq 0$, the *occurrences* of $l$ in $t$ are defined as follows:

- $occ_0(l)$ is the mother of the slice root of $l$ in $t$ (if it exists).
- $occ_i(l)$ is the unique node $m$ of $t$ labeled *move* such that $m$ matches $-f_i$, properly dominates $occ_{i-1}$, and there is no node $n$ in $t$ that matches $-f_i$, properly dominates $occ_{i-1}$, and is properly dominated by $m$.

Intuitively, node $m$ is an occurrence of LI $l$ iff it denotes an operation that checks one of $l$'s negative polarity features.

For $t$ and $l$ as before, and every node $m$ of $t$:

**Move** There exist distinct nodes $m_1, \ldots, m_n$ such that $m_i$ (and no other node of $t$) is the $i^{\text{th}}$ occurrence of $l$, $1 \leq i \leq n$.
**SMC** If $m$ is labeled *move*, there is exactly one LI for which $m$ is an occurrence.

With these four constraints we can finally define MDTLs: Given an MG $G$ with lexicon *Lex*, the MDTL of $G$ is the largest $L \subseteq \text{FSL}(Lex)$ such that every tree in $L$ satisfies the four constraints above. Equivalently, each constraint can be taken to be the largest set of trees that satisfy the respective constraint, so that $G$'s MDTL is the intersection of $G$'s free slice language and the tree languages defined by **Final**, **Merge**, **Move** and **SMC**. Since all these tree languages are regular, it follows that MDTLs are too.

As mentioned at the beginning of this section, MDTLs must be mapped to derived tree languages. With the exception of *move*, this mapping is rather simple. The following is an informal description of the standard translation from derivation trees to multi-dominance trees (directed acyclic graphs), where the *phrasal root* of an LI is the same as its slice root:

1. **Linearize.** Switch the order of siblings $l$ and $n$ if $l$ is an LI whose mother belongs to the slice of $l$.
2. **Project.** If $n$ is a Merge node whose left daughter is an LI with at least one selector feature, relabel it $<$. All other interior nodes are labeled $>$.
3. **Add Branches.** For every LI $l$, add branches from the phrasal root of $l$ to each $occ_i(l)$, $i \geq 1$.
4. **Delete Features.** Relabel every LI $\sigma :: f_1 f_2 \cdots f_n$ by $\sigma$.

These steps can be carried out by a tree-to-graph transduction $\Phi_{\text{gr}}$ that is definable in monadic second-order logic (MSO). The mapping $\Phi_{\text{tr}}$ from derivations to derived trees with traces is also MSO-definable. See [6, 11] for details.

**Definition 4.** *A* Minimalist Grammar *is a 5-tuple* $G := \langle \Sigma, Feat, Lex, F, \mathcal{R} \rangle$ *such that*

- *Lex is a* $(\Sigma, Feat)$*-lexicon, and*
- $F \subseteq \text{BASE}$ *is the set of final features, and*
- $\mathcal{R}$ *is the set of regular tree languages* ***Final, Merge, Move, SMC****.*

*The MDTL of G is* $\text{FSL}(\text{slice}(Lex)) \cap \bigcap_{R \in \mathcal{R}} R$*. The tree language* $L(G)$ *generated by G is the image of its MDTL under the MSO transduction* $\Phi_{\text{tr}}$ *of [6]. Its string language is the string yield of* $L(G)$*.*

### 1.2 Formal Properties of MGs

MGs are weakly equivalent to MCFGs [10]. That MGs are at most as powerful as MCFGs actually follows from the fact that their string languages are the string yield of the image of regular tree languages under an MSO-definable transduction [11]. For the purpose of adding adjunction, this means that weak generative capacity is preserved as long as MDTLs are still regular and the mapping to derived trees is MSO-definable.

In [5], it is shown that MDTLs are actually subregular. They are definable in first-order logic with proper dominance (abbreviated FO[$<$]), and *homogeneous.*

**Definition 5 (Homogeneity).** *For $s_1, s_2, t_1, t_2$ arbitrary trees and $L$ a regular tree language, $L$ is* homogeneous *iff $t[u \leftarrow a(t_1, t_2)] \in L$, $t[u \leftarrow a(s_1, t_2)] \in L$, and $t[u \leftarrow a(t_1, s_2)] \in L$ jointly imply $t[u \leftarrow a(s_1, s_2)] \in L$.*

For MGs without movement (i.e. no LI with licensee features occurs in a well-formed derivation), the MDTLs are also definable in first-order logic with immediate dominance (FO[S]), closed under $k$-guarded vertical swap, and strictly local.

**Definition 6 (Vertical swap).** *Let $t$ be the concatenation of trees $t_1$, $t_2$, $t_3$, $t_4$, $t_5$ such that for $1 < i \leq 5$, the root of $t_i$ is immediately dominated by some node $n_{i-1}$ of $t_{i-1}$. The vertical swap of $t$ between $t_2$ and $t_4$ is the result of switching $t_2$ and $t_4$ in $t$ such that the roots of $t_2, t_3, t_4, t_5$ are immediately dominated by $n_3$, $n_4$, $n_1$, and $n_2$, respectively. The vertical swap is $k$-*guarded *iff it holds that $t_2$ and $t_4$ are identical up to depth $k$, and so are $t_3$ and $t_5$.*

**Definition 7 (Strictly local).** *Given a tree $t$ over alphabet $\Sigma$, its $k$-*augment *is the result of adding nodes $n_1, \ldots, n_k$ above the root and below each leaf such that $n_i$ immediately dominates $n_{i+1}$, $1 \leq i < k$ and each $n_i$ has the distinguished label $\diamond \notin \Sigma$. A $k$-*factor *of $t$ is a subtree of $t$ that has been truncated at depth $k$. That is to say, if $s$ is a subtree of $t$ with $m$ nodes $n_1, \ldots n_m$ that are labeled $l_1, \ldots, l_m$, respectively, and properly dominated by $k-1$ nodes, then $s[n_1 \leftarrow l_1] \cdots [n_m \leftarrow l_m]$ is a $k$-factor of $t$. The set of $k$-factors of $t$ is denoted $F_k(t)$. A regular tree language $L$ over alphabet $\Sigma$ is* strictly local *iff there is some $k \in \mathbb{N}$ and finite set $S$ of trees over $\Sigma \cup \{\diamond\}$ with depth at most $k$ such that $t \in L$ iff $F_k(t') \subseteq S$, where $t'$ is the $k$-augment of $t$.*

These properties are interesting because they tell us something about the complexity of the constraints imposed by the feature calculus. Closure under $k$-guarded vertical swaps implies local threshold testability, while homogeneity is equivalent to recognizability by a particular kind of deterministic top-down tree automaton. Strict locality tells us that no dependency is unbounded.

## 2 Three Models of Adjunction

### 2.1 Category Preserving Selection

The simplest model of adjunction is naturally one that does not require any modifications to the formalism. Taking inspiration from Categorial Grammar, adjuncts can be implemented as LIs whose category feature has the same feature name as their selector feature. For example, the VP-adjunct *quickly* would correspond to quickly :: $=$V V. Such cases of selection are *category preserving.* This model was first discussed in [2], but it has been part of the general "MG folklore" for a long time.

**Linguistic Properties.** Despite its simplicity, this account captures several properties of adjuncts. Since an LI can only be selected once all its positive polarity features have been checked — which corresponds to projection in the derived tree — category preserving LIs may only select LIs that have already projected their full structure. Consequently, adjuncts are correctly predicted to adjoin only to maximal projections.

The category preserving nature of adjuncts in this model also entails that adjuncts are optional with respect to Merge. If an LI can select an adjunct $a$, then it can also directly select the LI $l$ that $a$ adjoins to as both have the same category feature. Category preservation also implies that multiple adjuncts can adjoin one after another, that is to say, adjuncts are iterable.[1]

Certain important aspects of adjuncts are missed, though. First of all, nothing in this system rules out the existence of adjuncts that adjoin to multiple phrases at the same time. Consider an entry with two identical selector features, e.g. quickly :: $=$V $=$V V. This LI could be used to generate a structure where two VPs are simultaneously modified by a single adverb occurring between them, yet no such structures are attested.

Another problem is posed by ordering effects. These can of course be handled via subcategorization in the familiar fashion: if adjunct $a$ cannot precede adjunct $b$, then $a$ does not select $b$, nor are there LIs $c_1, \ldots, c_n$ such that $a$ selects $c_1$, $c_i$ selects $c_{i+1}$ $(1 \leq i < n)$, and $c_n$ selects $b$ (cf. [2]). However, since these categories must be pair-wise distinct, the LIs are not category-preserving. Hence they are not necessarily optional or iterable. As was already pointed out by Fowlie [2], these properties can be enforced globally in the lexicon, but the cost is lexical redundancy that hides the important generalizations.

Movement is also challenging. Since the adjunct selects the phrase it supposedly adjoins to, it is not included in the phrase projected by the latter. So if the adjoinee undergoes movement, the adjunct is left behind, just like a complement DP that undergoes movement leaves behind the VP containing it. This can be

[1] Notice that these properties break down as soon as movement is involved — removal of an adjunct that contains a licensee feature might render the derivation ill-formed, and having multiple instances of such an adjunct may trigger SMC violations. However, if movement out of adjuncts is prohibited by something like the Adjunct Island Constraint, for which there is a lot of empirical support, then problems arise only where the adjunct itself undergoes movement. Displacement of adjuncts, though, might not involve movement at all and may simply be an instance of a phrase or LI being base-generated in a different position than where it is interpreted. For this reason — and because all conceivable accounts of adjunction have similar problems reconciling movement with optionality [7] — I will ignore movement dependencies in the remainder of this paper as far as optionality and iterability are concerned.

fixed by I) instantiating the licensee features on lexical heads (suggested by an anonymous reviewer), or II) adding a limited form of pied-piping to the mapping from derivation trees to derived trees such that if $l$ undergoes movement, the movement branches connect the occurrences of $l$ to the slice root of the highest adjunct of $l$, rather than the slice root of $l$ itself.

The first solution quickly leads to massive redundancy in the lexicon and once again misses generalizations. Suppose our grammar contains the LIs $l_1 :=$ s :: c $-$f and $l_2 :=$ t :: c. If licensee features are instantiated on empty heads instead, the lexicon would contain the LIs $l_1' :=$ s :: c, $l_2' :=$ t :: c, and $f := \varepsilon$ :: $=$c c $-$f. So if $l_1$ moves together with an adjunct $a$, this corresponds to $l_1$ being selected by $a$, which in turn is selected by the actual mover $f$. But keep in mind that $l_2$ should not be allowed to move, so the corresponding $l_2'$ must not be selected by $f$. This can only be guaranteed by changing the category of $l_2'$ to, say, $c'$. However, every LI except $f$ that selects an LI of category $c$ must still be able to select $l_2'$, so for each one of them we have to create a new variant whose selector feature is $=$c$'$ instead of $=$c. Not only does this unnecessarily increase the size of the lexicon, the fact that $l_1$ and $l_2$ have the same adjuncts is purely accidental in this revised grammar.

The second option avoids the lexical blow-up by enabling the transduction to pied-pipe adjuncts. But since adjuncts differ from non-adjuncts only in that they are category preserving, this may cause problems when non-adjuncts are also category preserving — for instance in the analysis of serial verb constructions in [1], where each verb except the most deeply embedded one has the feature component $=$V V. This actually highlights a more fundamental problem of this approach: adjuncts do not form a natural class since arguments may have the same feature make-up.

The inability to consistently distinguish arguments from adjuncts can also be seen in the case of recursive adjunction, i.e. adjunction to an adjunct. Without tricks, category-preserving selection cannot handle such configurations. Suppose that $b$ adjoins to $a$, which is an XP-adjunct and therefore has category feature X. In order to adjoin to $a$, LI $b$ must have selector feature $=$X, and by category preservation it also has category X. But then $b$ is an XP-adjunct just like $a$. So a phrase like *very red car* only has a structure where *very* modifies *car* rather than *red*.

As pointed out by an anonymous reviewer, this can be handled via empty heads, but the proposed solution serves only to highlight the fact that arguments cannot be separated from adjuncts in this system. Suppose that very :: adv and red :: a are selected by the empty head $\varepsilon$ :: $=$a $=$adv a, and this complex phrase is then combined with car :: n by the empty head $\varepsilon$ :: $=$n $=$a n. Then we obtain the structure [[very [$\varepsilon$ red]] [$\varepsilon$ car]], which is reasonably close to the intended [[very red] car]. But now consider cases where the adjunct follows the noun, e.g. *my cousin twice removed*. Here the empty head should be $\varepsilon$ :: $=$a $=$n n. This very feature template also arises in the standard DP-analysis of English possessor phrases such as *John's car*, where the possessive marker is given by 's :: $=$n $=$d d. Yet this structure is not assumed to involve adjuncts of any kind.

We see, then, that adjuncts cannot be reliable identified with a specific type of feature component.

**Formal Properties.** While modeling adjunction as category preserving selection has some important drawbacks from a linguistic perspective, it has the advantage of being compatible with the standard MG formalism. Consequently, all formal properties of MDTLs are unaffected. Only closure under intersection with regular tree languages can be lost under very specific assumptions. The proofs in [4, 9] rely on the ability to subscript category and selector features with specific states of a tree automaton. In many cases, an LI will have different state-suffixes on its selector and category features. As a result an LI may no longer be category preserving after its features have been suffixed with states. Hence, if suffixation must respect category preservation, then the class of MDTLs is no longer closed under intersection with regular tree languages. Or the other way round: the property of being an adjunct is not preserved under intersection with regular tree languages.

*Example 1.* Consider the regular tree language that includes a tree $t$ iff $t$ contains at most three LIs whose phonetic exponent is *quickly*. If the LI $l$ for *quickly* must be category preserving, then irrespective of how its features are altered, one instance of $l$ can always be selected by another one. Hence if the MDTL contains a tree with at least one instance of *quickly*, it also contains trees with more than three of them.

## 2.2 Asymmetric Feature Checking

A very different implementation of adjunction was presented by Frey and Gärtner [3]. Adjuncts are now formalized as LIs whose category feature $c$ has been replaced by an adjunction feature $\approx$a. An LI $l$ with feature $\approx$a can adjoin to any LI $l'$ of category $a$. Crucially, the adjunction operation is asymmetric in that it only checks the adjunction feature of $l$, whereas the category feature on $l$ persists, thereby allowing for multiple LIs to adjoin to it. This partial feature persistence of adjunction sets it apart from Merge, which always checks the relevant features of both the selector and the selectee. In addition, it is the phrase being adjoined to that projects, rather than the adjunct itself. An example derivation with the corresponding derived tree is given in Fig. 1.[2]

**Definition.** Only a few things have to be altered in our definition of MDTLs to incoporate Frey and Gärtner's version of adjunction. First, adjunction features must be added to the feature system and restricted to be in complementary distribution with category features. To this end, Def. 1 is revised such that $\text{OP} := \{\textit{merge}, \textit{move}, \textit{adjoin}\}$, and the defintion of LI is altered accordingly:

[2] A related system is developed by Hunter in [8], who uses the same feature checking mechanism but a different type of Minimalist derivations. As far as I can tell, my observations about Frey and Gärtner's system apply to Hunter's, too.

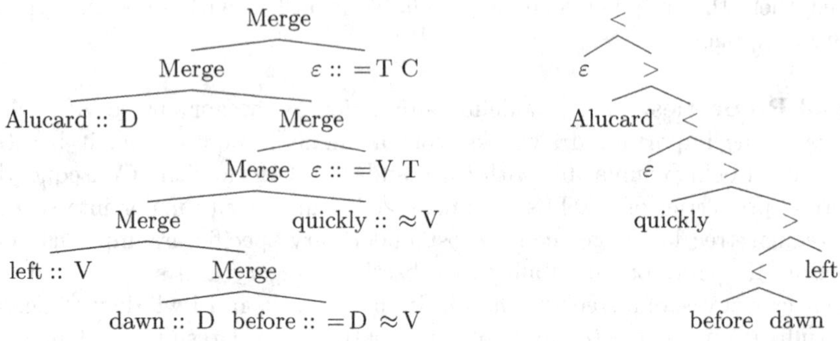


**Fig. 1.** Left: derivation tree with adjunction as asymmetric feature checking; Right: corresponding derived tree

**Definition 8.** *Given a string alphabet $\Sigma$ and feature system Feat, a $(\Sigma, Feat)$-*lexicon *is a finite subset of* $\Sigma \times \{::\} \times \{f \mid \pi(f) = +\}^* \times \{f \mid \omega(f) \in \{merge, adjoin\}, \pi(f) = -\} \times \{f \mid \omega(f) = move, \pi(f) = -\}^*$.

The only change from Def. 2 is that the unique category feature may be replaced by an adjunction feature of negative polarity.

The next step is to assign adjuncts an interpretation in terms of derivation trees. That is to say, both the translation from LIs to slices and the well-formedness conditions on MDTLs need to be amended. Def. 3 is extended to cover one more case: $merge(\Box, \zeta(\sigma :: f_1 \cdots f_{i-1} \star f_i \cdots f_n))$ if $\tau(f_i) = adjoin$. Note that even though adjuncts do not project in the derived tree, in the derivation tree the adjunction step belongs to the slice of the LI with the adjunction feature. So from a derivational perspective adjunction looks very similar to Merge (we do not even introduce a new label to distinguish the two).

A simple modification of the constraint **Merge** suffices to regulate the distribution of adjuncts.

**Merge** For every $t \in \text{FSL}(\text{slice}(Lex))$ and node $m$ of $t$, if $m$ is associated to selector feature $= f$ or adjunction feature $\approx f$, then its left daughter is the slice root of an LI with category feature $f$ or adjunction feature $\approx f$.

The only difference to the original definition is the presence of the two disjuncts "or adjunction feature $\approx f$". The first disjunct allows Merge to be triggered by adjunction features, too. The second one allows for configurations where a phrase has multiple adjuncts, as in Fig. 1. In this case, only the Merge node of the lowest adjunct is the mother of the slice root of an LI with category feature $f$. For a higher adjunct $a$, however, the slice root belongs to an LI $b$ with an adjunction feature $\approx f$. As long as the two adjunction features are the same, though, it follows by induction that there is some LI further down that both $a$ and $b$ adjoin to.

This completely characterizes the adjunction operation on a derivational level, so it only remains for us to modify the mapping from derivations to derived trees.

Two steps must be altered. **Linearize** now distinguishes two cases where siblings are switched around:

**Linearize** Switch the order of siblings $l$ and $n$ if
- $l$ is an LI whose mother belongs to the slice of $l$, or
- the mother of $l$ is a Merge node associated to an adjunction feature.

In the definition of **Add branches**, the phrasal root of LI $l$ is now defined as either the slice root of the highest adjunct of $l$ or the slice root of $l$ if the former does not exist. Without this change, our model-theoretic definition would differ from Frey and Gärtner's in that adjuncts would be stranded if the phrase they adjoin to undergoes movement (a problem already encountered with the category preservation model).

The change to how phrasal roots are determined also solves a minor problem in the definition of **Final**, which is no longer adequate because the slice root of the head of a tree may no longer be the root of the tree. However, its phrasal root still is, so it suffices to replace "slice root" by "phrasal root" in **Final**.

**Formal Properties.** Considering how little needs to be changed in the definitions, it is hardly surprising that most formal properties of MGs also hold after the introduction of a dedicated adjunction mechanism. The revised version of **Merge** only adds a few disjunctions, wherefore it is still MSO-definable and defines a regular tree language. The new clause in **Linearize** is also easily expressed in MSO, as is the new definition of phrasal root (cf. Sec. 2.2 of [6]). So both MDTLs and their mapping to derived trees are still MSO-definable, which entails that this variant of MGs generates at most MCFLs, and consequently adjunction has no effect on weak generative capacity.

Even the subregular complexity of MDTLs is unaffected. For MGs without Move, they are still strictly local since the domain for **Merge** comprises only two slices. This also implies that adding adjunction to MGs with movement does not negatively affect their definability in FO[$<$]. Finally, they are homogeneous and therefore can be recognized by lrDTDAs.

**Lemma 1.** *Let $G$ be an MG with adjunction as asymmetric feature checking. Then $G$'s MDTL is homogeneous.*

*Proof.* Recall that a tree language $L$ is homogeneous iff $t[u \leftarrow a(t_1, t_2)] \in L$ and $t[u \leftarrow a(s_1, t_2)] \in L$ and $t[u \leftarrow a(t_1, s_2)] \in L$ jointly imply $t[u \leftarrow a(s_1, s_2)] \in L$. We are only interested in subtrees whose root $a$ is a Merge node associated to an adjunction feature. All other cases are already covered by the homogeneity proof for standard MDTLs in [5].

Let $t[\approx f]$ and $t[= f]$ denote that the head of $t$ has $\approx f$ and $= f$ as its first feature, respectively, whereas $t[f]$ indicates that there is an LI $l$ in $t$ whose first unchecked feature is $f$ and every Merge node properly dominating the slice root of $l$ is the slice root of an LI with feature $\approx f$. Then $t[u \leftarrow a(t_1, t_2)] \in L$ only if $t_i[f]$ and $t_j[\approx f]$ for $i \neq j \in \{1, 2\}$. Assume w.l.o.g. that $i = 1$ and $j = 2$. Then it must also be the case that $s_1[f]$, and $s_2[\approx f]$ or $s_2[= f]$, so that $a(s_1, s_2)$ is well-formed with respect to **Merge**. Furthermore, we know that $s_1$ and $t_1$ on the one hand and $s_2$ and $t_2$ on the other have the same unchecked licensee features since substituting one for the other preserves well-formedness. It follows that $t[u \leftarrow a(s_1, s_2)] \in L$.

As in the case of category-preserving selection, the status of closure under regular intersection is not entirely straight-forward. Simply applying the suffixation strategy of [4, 9] is insufficient. Once again this is illustrated by the regular tree language in example 1, which contains only trees with at most three occurrences of *quickly*. If one instance of *quickly* can adjoin to a given LI $l$, then arbitrarily many instances of it may adjoin to $l$. Hence there can be no MG $G$ whose MDTL contains trees with up to three occurrences of $a$, but not more than that.

It is possible, however, to switch from asymmetric feature checking to standard symmetric Merge using category-preserving selection without changing the shape of the derivation tree — both are just instances of Merge. To the extent that this is a licit step, closure under regular intersection would once again hold if adjunct status need not be preserved. But even though the derivation trees would be identical, the derived structures are not: an adjunct in Frey and Gärtner's system does not project, whereas a category-preserving head does. And since we already saw in the discussion of the category-preservation account that not every instance of category-preserving selection constitutes adjunction, a derivation with adjunction cannot be uniquely recovered from an isomorphic one with selection. This illustrates once again that closure under regular intersection can be obtained only if LIs may lose their adjunct-status.

**Linguistic Properties.** Just like the category preservation approach, the implementation of adjunction as asymmetric feature checking captures several core properties of adjuncts. Since category features are unaffected by adjunction, adjuncts are correctly predicted to be optional and iterable (*modulo* movement dependencies). In addition, adjunction features behave similar to selector features in that they are checked by category features. An LI therefore can be adjoined to only after it has discharged all its positive polarity features, i.e. projected a full phrase in the derived tree, which rules out X′-adjuncts.

In contrast to the category preservation approach, Frey and Gärtner's implementation also behaves correctly with respect to Move — an adjunct moves together with the phrase it is adjoined to. Admittedly this has to be explicitly stipulated in the definition of phrasal root, but the clear division between adjunction and selection in the feature system means that this modification does not bring about any unexpected side-effects. Another welcome property is that thanks to the limit to one adjunction feature, adjoining to multiple phrases at once is impossible.

But Frey and Gärtner's approach is not without shortcomings, either. Since adjunction features replace category features, adjuncts lack category features and thus cannot be adjoined to. So just like the category preservation implementation, asymmetric feature checking cannot assign the correct structure to *very red car*. Even empty heads only solve the problem if one treats the adjuncts as arguments of an unpronounced adjunct, which defeats the purpose of having an explicit argument-adjunct distinction in the system.

Ordering effects are also unaccounted for. Since adjunction to an LI $l$ with category feature $f$ can be triggered only by the feature $\approx f$, all adjuncts of $l$ have said category feature. From this it follows immediately that these adjuncts may adjoin in any given order and **Merge** will still be satisfied. Fowlie [2] sketches a workaround based on a Cinque-style hierarchy of empty heads, each of which serves as an adjunction site for adjuncts of a particular type. Still, the order is enforced by selection rather than the adjunction mechanism itself, which reintroduces many problems of treating adjunction as selection, for instance regarding phrasal projection and the interaction with movement.

### 2.3 Adjunction Hierarchies

A third approach has been recently proposed by Fowlie in [2]. Fowlie's primary interest is to reconcile optionality with ordering effects. Technically this is an easy task if one uses standard selection: if LI $l$ has selector feature $=$c, and LIs of category $c$ may be adjoined to by adjuncts $a_1, \ldots a_n$ as long as each $a_i$ is structurally lower than $a_j$, $1 \leq i < j \leq n$, then we have $n$ additional versions of $l$ where $=$c has been replaced by $=$x, where x is the category feature of some $a_i$, $1 \leq i \leq n$. But this solution comes at the price of a significantly larger lexicon.

Fowlie proposes to put a partial order $R$ on the set of categories instead. Moreover, there are no adjunction features; adjunction of $a$ to $l$ has no effect on the feature make-up of either LI and may take place as long as

- for $c_a$ and $c_l$ the category features of $a$ and $l$, respectively, and $\alpha : \textsc{Base} \rightarrow \wp(\textsc{Base})$ a map from categories to sets of categories that may adjoin to them, it must hold that $c_a \in \alpha(c_l)$, and
- $c_a \not R\, c_l$, and
- for every LI $b$ of category $c_b$ that adjoined to $l$ before $a$, $c_a \not R\, c_b$.

Notice that $x \not R\, y$ is compatible with $y \not R\, x$ and $y\, R\, x$. This is used by Fowlie to make a distinction between order-insensitive and order-sensitive adjuncts, respectively, with the former being linearized to the right of the adjoinee and the latter to the left (the split is motivated by empirical observations). For the sake of simplicity I ignore this distinction in what is to follow.

**Definition.** While a literal implementation of the formalism in [2] is tedious because of the way non-local information is passed around via feature pairs, this mechanism can safely be ignored for a model-theoretic definition. As a matter of fact, this is preferable from a formal perspective since local book-keeping of non-local information could obscure the subregular complexity of adjunction in derivation trees. Without Fowlie's feature pairs, the feature system is exactly the one of standard MGs. Consequently, all modifications take place on the level of derivation trees and the mapping to derived trees.

For derivation trees, a viable strategy is to insert Adjoin nodes at arbitrary points and then filter out the illicit derivations created this way. Technically, this is achieved by a minimal change in the definition for slice lexicons such that slice(*Lex*) := $\{\zeta(l) \mid l \in Lex_*\} \cup \{adjoin(\square_1, \square_2)\}$. As a result, the free slice language not only contains combinations of lexical slices, but also trees where binary branching *adjoin* nodes occur in random positions (but not within lexical slices).

For movement, the presence of adjunction nodes causes no problems because the constraints **Move** and **SMC** are non-local. But **Merge** was stated under the assumption that the left daughter is the slice root of an LI with the matching category feature. Adjunction destroys this local relation. In the system of Frey and Gärtner, this could still be worked around due to the feature-driven nature of adjunction. An LI with feature $\approx f$ could only adjoin to LIs of category $f$. But this isn't necessarily the case in this system, where $\alpha^{-1}$ — which determines for every category the categories it may adjoin to — is not guaranteed to be a function. Prepositional phrases, for example, can adjoin to both nouns and verbs. So if a Merge node is associated to feature $=$N and its left daughter is a node indicating adjunction of a PP, it is still unclear whether a matching feature N can be found further down the tree. But suppose that we always interpret *adjoin* nodes in a fashion such that the adjoinee is found along the left branch and the adjunct along the right branch. Suppose $m$ *left-dominates* $n$ iff $m$ properly dominates $n$ and there is no $z$ such that $z$ is properly dominated by $m$, reflexively dominates $n$, and has a left sibling. Then the following definition will do the trick:

**Merge** For every $t \in$ FSL(slice(*Lex*)) and node $m$ of $t$, if $m$ is associated to selector feature $=f$, then the highest node that is left-dominated by $m$ and not labeled *adjoin* is the slice root of an LI with category feature $f$.

This takes care of Merge nodes, but it still remains for us to regulate the distribution of *adjoin* nodes. The first step is to determine the arguments of each adjunction step, i.e. the adjoining phrase and the phrase being adjoined to. If adjuncts cannot be adjoined to, this is very easy. For then the right daughter is the slice root of the adjunct, and the slice root of the phrase being adjoined to is the highest Merge node that is left-dominated by the adjoin node (once again left-dominance is used to account for the fact that there may be other adjoin nodes along the path). Somewhat surprisingly, though, allowing adjuncts to be adjoined to is an easy process once left-dominance has been defined.

First, an adjunction node $m$ is an *adjunction occurrence* of LI $l$ in derivation tree $t$ iff $m$ is the lowest node in $t$ that properly dominates the slice root of $l$ but does not left-dominate the slice root of $l$. Second, an adjunction node $m$ is associated to category feature $c$ iff $m$ is an adjunction occurrence of LI $l$ with category feature $c$.

**Adjoin** For every $t \in$ FSL(slice(*Lex*)) and node $m$ of $t$, if $m$ is associated to category feature $c_m$, then
- the highest Merge node in $t$ left-dominated by $m$ is the slice root of an LI $l$ with category feature $c_l$, and

- $c_m \in \alpha(c_l)$, and
- for every node $n$ that is properly dominated by $m$, reflexively dominates $l$, and is associated to category feature $c_n$, it holds that $c_m \not R\, c_n$.

As with Frey and Gärtner's system, we also have to make changes to the mapping from derivations to derived trees.

**Linearize** Switch the order of siblings $l$ and $n$ if
- $l$ is an LI whose mother belongs to the slice of $l$, or
- the mother of $l$ is an adjunction node.

The phrasal root of an LI $l$ (referenced in **Add Branches** is now the highest adjunction node $n$ such that $n$ properly dominates the slice root of $l$ and no Merge properly dominated by $n$ properly dominates the slice root of $l$. Once again **Final** is easily adapted to the new system by replacing "slice root" by "phrasal root".

Figure 2 gives an example of what derivations with multiple adjunctions look like in this system, and what kind of structures they yield.

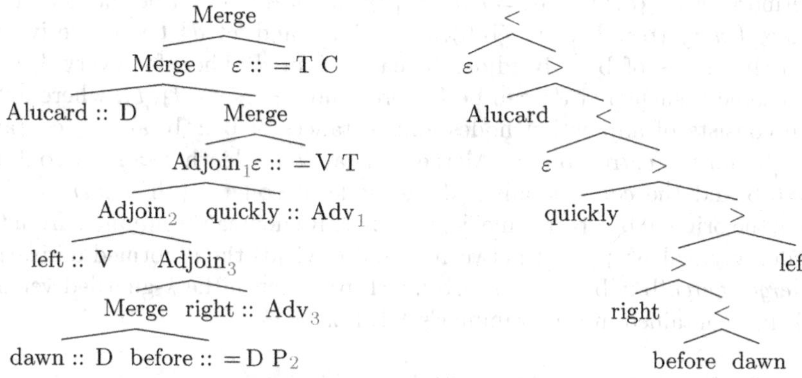


**Fig. 2.** Left:derivation tree with recursive adjunction, suffixes indicate adjunction occurrences; Right: corresponding derived tree

**Formal Properties.** As with the previous model the changes in the definitions are innocuous enough to see that the formal properties of MDTLs are mostly unaffected. The major change is the introduction of left-dominance, which can easily be defined in first-order logic with proper dominance. The function $\alpha$ and the relation $R$ are both finite by virtue of Minimalist lexicons being finite, so they, too, are first-order definable. From all this it follows that both the MDTLs and their mapping to derived trees are MSO-definable and weak generative capacity is not increased.

The definability of MDTLs in FO[<] is not endangered either, because the non-local dependencies established by adjunction are no more complex than those regulating Move. For MGs without movement, however, adjunction does increase subregular complexity significantly. First of all, the dependence on left-dominance, an unbounded relation, means that MDTLs are no longer strictly local. This can only be avoided by banning adjunction to adjuncts, which allows for an unbounded number of adjoin nodes to occur between an LI and its adjunction occurrence. But even then $\alpha$ and $R$ have to be chosen very carefully to ensure that MDTLs are strictly local: the category of a phrase must be predictable from the categories of its adjuncts, which is not the case for natural language (*very*, for example, is freely iterable and may be an adjunct of adjectives or adverbs).

Without strong restrictions on $\alpha$ and $R$, MDTLs are not even homogeneous or closed under $k$-guarded vertical swap.

**Lemma 2.** *The MDTLs of MGs with hierarchical adjunction are not closed under k-guarded vertical swap.*

*Proof.* Consider a grammar containing (at least) the following items:

$$\mathrm{b} :: \mathrm{b} \qquad \mathrm{a} :: =\mathrm{b}\ \mathrm{a} \qquad \mathrm{b} :: =\mathrm{b}\ \mathrm{b}$$

Furthermore, $F := \{a, b\}$, $\alpha(a) = \alpha(b) = \{b\}$, and $b \not R\, b$. Let $d$ be the derivation tree $merge(merge(\mathrm{b} :: \mathrm{b}, \mathrm{b} :: =\mathrm{b}\ \mathrm{b}), \mathrm{a} :: =\mathrm{b}\ \mathrm{a})$, and let $d^n$ be the derivation where $n$ instances of b :: b adjoin to each LI in $d$. Then for every $k \in \mathbb{N}$ there is some $n$ such that $d^n$ can be factored into $t_1$, $t_2$, $t_3$, $t_4$, $t_5$, where every subtree consists of adjunction nodes and instances of b :: b, and $t_2$ contains a :: =b b and the corresponding Merge node at some depth $f > k$, $t_3$ contains b :: =b b and the corresponding Merge node at some depth $g > k$, and $t_4$ contains the original b :: b at some depth $h > k$. Removing the adjuncts from the $k$-guarded vertical swap of $d^n$ between $t_2$ and $t_4$ yields the ill-formed derivation tree $merge(merge(\mathrm{b} :: \mathrm{b}, \mathrm{a} :: =\mathrm{b}\ \mathrm{a}), \mathrm{b} :: =\mathrm{b}\ \mathrm{b})$, whence the $k$-guarded vertical swap is not contained in the grammar's MDTL.

**Lemma 3.** *The MDTLs of MGs with hierarchical adjunction are not homogeneous.*

*Proof.* Consider the following grammar:

$$\mathrm{a} :: \mathrm{a} \qquad \mathrm{b} :: \mathrm{b} \qquad \mathrm{c} :: \mathrm{c} \qquad \mathrm{d} :: \mathrm{d}$$

Suppose $F := \{a, b\}$, $\alpha(a) := \{c, d\}$ and $\alpha(b) := \{c\}$. Now let $a$ = adjoin, $t_1 :=$ a :: a, $t_2 :=$ c :: c, $s_1 :=$ b :: b, and $s_2 :=$ d :: d. Then $a(t_1, t_2)$, $a(t_1, s_2)$, and $a(s_1, t_2)$ are all well-formed derivations, but $a(s_1, s_2)$ is not.

Closure under intersection with regular tree languages is also lost. Admittedly our standard example — the regular language of trees that contain at most three instances of *quickly* — can be accommodated in this system (use three different categories $c_i$ for *quickly*, such that $1 \leq i \leq 3$ and $c_j$ may adjoin to $c_i$ iff $j > i$). But we still run into problems with its dual, the regular language of trees that contain at least three instances of *quickly*. Since adjuncts are completely decoupled from the feature checking mechanism, they are not required in order for a derivation to be well-formed, wherefore the presence of even one instance of *quickly* cannot be enforced.

**Linguistic Properties.** The hierarchical approach improves significantly on both previous proposals, to the extent where it passes all criteria discussed in this paper: optionality, iterability, recursive adjunction, correct behavior with respect to Move, the ability to capture ordering effects, and the prohibition against X′-level adjunction. It should be noted, though, that some properties do not fall out naturally under the model-theoretic perspective and are simply a matter of how we phrase our definitions. The interaction with Move, for example, depends purely on the how *phrasal root* is defined, and hence could easily be altered. It would also be a rather easy technical exercise to allow for X′-level adjunction. In addition, some properties depend on the choice of $R$. If $R$ is reflexive, for instance, then adjuncts cannot be iterated because the condition $c_m \not R\, c_n$ in **Adjoin** would be trivially violated. Optionality, however, is a robust property of this system thanks to the decoupling of adjunction and feature checking.

## Conclusion

An overview of the formal and linguistic properties of the three models of adjunction are given in Tab. 1 and 2. The emerging picture is that all accounts capture the most basic properties of adjuncts — optionality, iterability, the lack of X′ adjuncts — but diverge once one considers other aspects such as adjunction to adjuncts, the interaction with Move, and ordering effects. Only Fowlie's hierarchical approach performs well across the board, but does so at the expense of increasing the subregular complexity of MDTLs, even with respect to standard MGs. Loss of homogeneity and closure under $k$-guarded vertical swap, however, are unavoidable in any implementation of adjunction where adjuncts are iterable and can adjoin to phrases with different categories, both of which seem to be empirical necessities. Similarly, closure under intersection with regular tree languages is incompatible with the optionality of adjuncts.

**Table 1.** Linguistic properties of adjunction implementations

| | Cat. Preserv. | Asymm. Checking | Hierarchical |
|---|---|---|---|
| no X′ adjuncts | ✓ | ✓ | ✓ |
| optional | ✓ | ✓ | ✓ |
| iterable | ✓ | ✓ | ✓ |
| recursive | ∼ | ∼ | ✓ |
| no double adjunction | | ✓ | ✓ |
| ordering effects | ✓ | ∼ | ✓ |
| correct projection | | ✓ | ✓ |

**Table 2.** Formal properties of adjunction implementations (without Move)

| | Cat. Preserv. | Asymm. Checking | Hierarchical | Move |
|---|---|---|---|---|
| strictly local | ✓ | ✓ | | |
| vertical swap | ✓ | ✓ | | |
| homogeneous | ✓ | ✓ | | ✓ |
| FO[S] | ✓ | ✓ | | |
| FO[<] | ✓ | ✓ | ✓ | ✓ |
| reg ∩ | | | | ✓ |
| preserves gen. capacity | ✓ | ✓ | ✓ | NA |

# References


[1] Collins, C.: Argument sharing in serial verb constructions. Linguistic Inquiry 28, 461–497 (1997)

[2] Fowlie, M.: Order and optionality: Minimalist grammars with adjunction. In: Kornai, A., Kuhlmann, M. (eds.) Proceedings of the 13th Meeting on the Mathematics of Language (MoL 2013), pp. 12–20 (2013)

[3] Frey, W., Gärtner, H.M.: On the treatment of scrambling and adjunction in minimalist grammars. In: Proceedings of the Conference on Formal Grammar (FGTrento), Trento, pp. 41–52 (2002)

[4] Graf, T.: Closure properties of minimalist derivation tree languages. In: Pogodalla, S., Prost, J.-P. (eds.) LACL 2011. LNCS (LNAI), vol. 6736, pp. 96–111. Springer, Heidelberg (2011)

[5] Graf, T.: Locality and the complexity of minimalist derivation tree languages. In: de Groote, P., Nederhof, M.-J. (eds.) Formal Grammar 2010/2011. LNCS, vol. 7395, pp. 208–227. Springer, Heidelberg (2012)

[6] Graf, T.: Local and Transderivational Constraints in Syntax and Semantics. Ph.D. thesis, UCLA (2013)

[7] Graf, T.: The syntactic algebra of adjuncts. In: Proceedings of CLS49 (to appear)

[8] Hunter, T.: Deriving syntactic properties of arguments and adjuncts from neo-davidsonian semantics. In: Ebert, C., Jäger, G., Michaelis, J. (eds.) MOL 10/11. LNCS (LNAI), vol. 6149, pp. 103–116. Springer, Heidelberg (2010)

[9] Kobele, G.M.: Minimalist tree languages are closed under intersection with recognizable tree languages. In: Pogodalla, S., Prost, J.-P. (eds.) LACL 2011. LNCS (LNAI), vol. 6736, pp. 129–144. Springer, Heidelberg (2011)

[10] Michaelis, J.: Transforming linear context-free rewriting systems into minimalist grammars. In: de Groote, P., Morrill, G., Retoré, C. (eds.) LACL 2001. LNCS (LNAI), vol. 2099, pp. 228–244. Springer, Heidelberg (2001)

[11] Mönnich, U.: Grammar morphisms, ms. University of Tübingen (2006)

[12] Stabler, E.P.: Derivational minimalism. In: Retoré, C. (ed.) LACL 1996. LNCS (LNAI), vol. 1328, pp. 68–95. Springer, Heidelberg (1997)

# Quantifiers in Frame Semantics

Laura Kallmeyer[1] and Frank Richter[2]

[1] Abteilung für Computerlinguistik, Institut für Sprache und Information, Heinrich-Heine-Universität Düsseldorf, Universitätsstr. 1, 40225 Düsseldorf, Germany
[2] Goethe Universität Frankfurt a.M., Institut für England- und Amerikastudien, Abteilung Linguistik, Grüneburgplatz 1, 60323 Frankfurt am Main, Germany

**Abstract.** We present a Lexicalized Tree Adjoining Grammar with lexical meaning specifications in Frame Semantics for the analysis of complex situations. Frame Semantics is extended by a notion of quantifier frames, which provide the basis for a translation function from frames to underspecified type-logical representations. An analysis of repetitive and restitutive readings of achievements with the adverb *again* demonstrates the interaction of all components of the new semantic architecture.

**Keywords:** frame semantics, quantifier scope, event semantics, Tree Adjoining Grammar, underspecification, syntax-semantics interface.

## 1 Introduction

The goal of this paper is to develop an architecturally and conceptually clear grammar architecture that unifies lexical meaning specifications in Frame Semantics and a truth-conditional sentential semantics with generalized quantifiers and other operators favored by many formal semanticists. To obtain a framework with computational properties that are amenable to implementation, Frame Semantics and formal semantics are combined in a Tree Adjoining Grammar, and the type-logical logical representations of semantics are phrased as underspecified representations with dominance constraints.

The framework of Frame Semantics, nowadays most prominently known from the Berkeley FrameNet project [2], takes a lexicographically oriented approach to the investigation of meaning. It perceives word meanings as expressible by schematic representations of conceptual structures that stand in a web of mutual relationships, not unlike the (multiple) inheritance hierarchies of feature logical grammars. The lexical cognitive structures encode the speakers' knowledge of situations or states; moreover, they record the relationships between word senses and morphosyntactic realization patterns. With the increasing inclusion of grammatical constructions, frame semantic descriptions have decidedly moved beyond their lexical roots. Inspired by work from cognitive psychology (Barsalou, [3]), Löbner [4] takes the empirical scope of frames even further and hypothesizes that in fact the entire human cognitive system employs frames as an all-comprising single data format.

To test the viability of this research program for linguistics, a number of recent investigations have used frame semantic insights in the formulation

G. Morrill et al. (Eds.): Formal Grammar 2014, LNCS 8612, pp. 69–85, 2014.

of a syntax-semantics interface that combines the rich conceptual structures of frames with techniques of formal semantics: [5] conclude from their analysis of FrameNet that the systematicity and consistency of FrameNet's relations and its predictions concerning linking generalizations could benefit from frame representations whose structure reflects the internal stucture of events. They aim at a decompositional Frame Semantics that analyzes the structure of situations and events along the lines of a denotationally interpretable decompositional lexical semantics [6]. The idea of a denotational semantics for frames in terms of an intensional logic to analyze causation, situations, aktionsarten and result states is then taken up by [7] in a fragment of Lexicalized Tree Adjoining Grammar (LTAG) that captures the behavior of lexically and syntactically complex direct motion expressions, and of the dative alternation in English. The proposal combines LTAG with representations of lexical and constructional meaning, given in the form of frames that are taken to be representations of mental concepts. [7] define frames as feature structures, and the structure of their frames is suggestive about possible ways to obtain an event-logical characterization of the truth conditions of a sentence from its frame. However, the relation to truth conditions needs further clarification. Moreover, a considerable gap remains between this analysis and full-fledged Montagovian semantics: So far, frames do not foresee quantificational operators, which precludes the treatment of such standard constructs as negation or nominal phrases as generalized quantifiers.

Thus at the current stage, Frame Semantics is of limited interest to semanticists working on sentential semantics. Our main goal in this paper is to bridge this gap and connect the vision of a rich Frame Semantics embedded in a general theory of human cognition to the achievements of model-theoretic semantics. After outlining current assumptions about embedding Frame Semantics into LTAG (Section 2), we extend Barsalou frames by *quantifier frames* in Section 3. Intuitively, a quantifier frame embodies the idea of a concept of a quantifier, or the cognitive correlate of a quantifier. With a quantifier frame, we are able to specify frames for sentences with quantificational NPs. In order to determine the truth value of such sentences in a given model, a translation of frames into underspecified semantic representations is given, which receives its usual interpretation. The following sections demonstrate how the new architecture works: Section 4 discusses a case of a quantifier ambiguity with nested quantifiers and confirms that the syntax-semantics interface of simpler architectures without Frame Semantics can be preserved. Section 5 is dedicated to an analysis of scope ambiguities observed with repetitive and restitutive readings with the adverb *again* in which the lexical semantic analysis of causes and states in frames interacts in interesting ways with the quantificational properties of *again*. We obtain an architecture that is suitable for syntactic parsing with underspecified semantic representations with quantificational operators that benefits from lexical and constructional meaning analyses and linking in Frame Semantics.

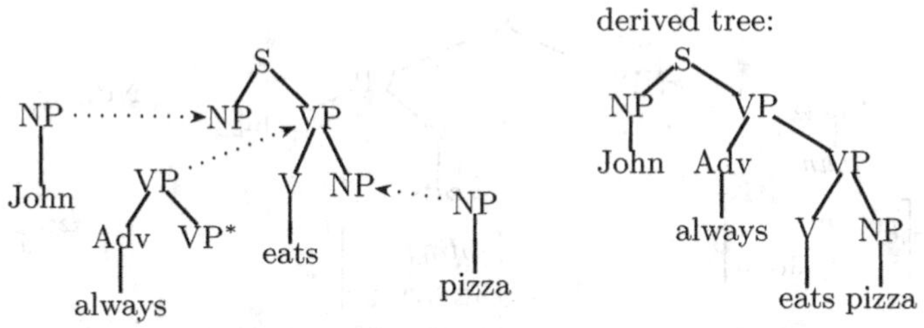


**Fig. 1.** A sample LTAG derivation

## 2 LTAG and Frames Semantics

A *Lexicalized Tree Adjoining Grammar* (LTAG [8, 9]) consists of a finite set of *elementary trees*. Starting from these trees, larger trees can be derived via the tree composition operations *substitution* (replacing a leaf with a new tree) and *adjunction* (replacing an internal node with a new tree). In Fig. 1 the trees for *John* and *pizza* substitute into the subject and the object slot of the tree for *eats*, and the tree of *always* adjoins to the VP node. An adjoining tree has a unique non-terminal leaf that is its *foot node* (marked with an asterisk). When adjoining such a tree to some node $v$, in the resulting tree, the subtree with root $v$ from the old tree ends up below the foot node.

In order to capture syntactic generalizations, the non-terminal node labels in elementary trees are usually enriched with feature structures [10]. Each node has a top and a bottom feature structure (except substitution nodes, which have only a top). Nodes in the same elementary tree can share features. Substitutions and adjunctions trigger the following unifications: In a substitution operation, the top of the root of the new tree unifies with the top of the substitution node. In an adjunction operation, the top of the root of the adjoining tree unifies with the top of the adjunction site and the bottom of the foot of the adjoining tree unifies with the bottom of the adjunction site. Furthermore, in the final derived tree, top and bottom must unify in all nodes.

Our architecture for the interface between TAG syntax and frame semantics follows ideas by [7], which in turn builds on previous approaches which link a semantic representation to an entire elementary tree and model composition by unifications triggered by substitution and adjunction [11–13]. One of the innovations of [7] is that their semantic representations are frames, expressed as typed feature structures as shown in Fig. 2. The feature I on the nodes is a syntax-semantics interface feature which stands for "individual". The assignment of semantic roles to syntactic arguments is handled by these interface features. In Fig. 2, the syntactic unifications $\boxed{1}\sqcup\boxed{3}$ and $\boxed{2}\sqcup\boxed{4}$ identify the semantic frames of the argument NPs with the semantic roles of the verbal frame.

Following [14], the frames in [7] are formalized as multi-rooted typed feature structures with multiple base labels. In other words, some of the nodes are labelled with base labels $\boxed{0}, \boxed{1}, \ldots$, which give access to these nodes.[1] Furthermore, there is

[1] Note that when using an elementary tree with its frame in a derivation, we always use a copy with fresh base labels.

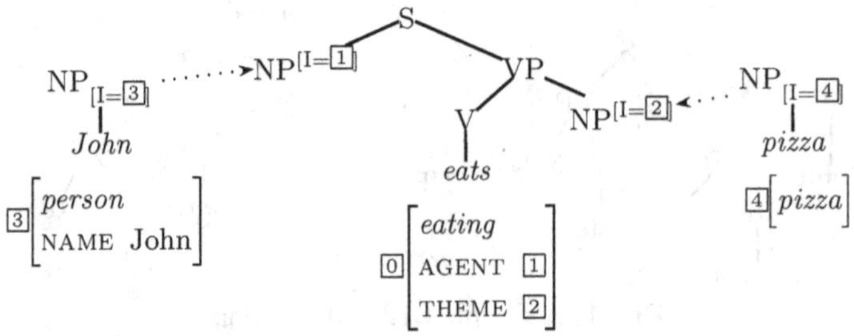


**Fig. 2.** Syntactic and semantic composition for *John eats pizza*

no explicit type hierarchy. Instead, nodes in the frames can have several types; dependencies between types such as subtype relations and type incompatibilities are formulated in constraints in the feature logic.

Concerning complexity, some restrictions on possible frame constraints are required in order to keep the system tractable. The frames that are linked to elementary trees are described within the metagrammar. Here, the possible constraints must be restricted in such a way that the existence of finite minimal models is guaranteed. To this end, we need for instance to avoid constraint loops (see also [15]). During parsing, we have to build larger frames via unifications triggered by substitution and adjunction. In order to keep this tractable, [7] assume that the constraints under consideration do not introduce new nodes to the structure. Then the complexity of unification is close to linear [14].

The extended domain of locality of LTAG, in combination with the rich factorization possibilities provided in the metagrammar through descriptions of elementary trees, permits a clean separation between lexical and constructional meaning contribution (cf. [7]).

## 3 Frames for Quantificational NPs

### 3.1 Quantifier Frames

Within a frame, some of the properties of a mental concept are captured by the type and others by attributes. We assume that the relation between the two arguments of binary quantifiers is captured in their frame type. This leads to the types *every*, *most*, *two*, etc. But what are the attributes that characterize a quantifier concept? We propose that the mental concept of a quantifier (in context) must minimally delimit the candidate concepts of its arguments, i.e. the concepts that occupy the restrictor and nuclear scope of the logical counterpart of a given quantifier concept. For this purpose, a quantifier frame contains the attribute RESTR for the maximal type of objects that the natural language quantifier in question lives on (in terms of logic: the restricting predicate), and the attributes MAXS and MINS that, in logical terms, characterize the scope window of the quantifier: The logical counterpart of the quantifier frame will scope at least over everything below the MINS value and at most over everything below the MAXS value. The embedding of the quantifier frame in a predicate frame expresses the semantic role of the syntactic constituent.

(1) Most dogs sleep.

$$\boxed{0}\begin{bmatrix} \textit{sleeping} \\ \text{SLEEPER} \ \boxed{1} \end{bmatrix} \sqcup \boxed{1}\begin{bmatrix} \textit{most} \\ \text{RESTR} \ \boxed{2} \\ \text{MAXS} \ \boxed{3} \\ \text{MINS} \ \boxed{0} \end{bmatrix} \sqcup \boxed{2}\begin{bmatrix} \textit{dog} \end{bmatrix} = \boxed{0}\begin{bmatrix} \textit{sleeping} \\ \text{SLEEPER} \ \boxed{1}\begin{bmatrix} \textit{most} \\ \text{RESTR} \ \boxed{2}\begin{bmatrix} \textit{dog} \end{bmatrix} \\ \text{MAXS} \ \boxed{3} \\ \text{MINS} \ \boxed{0} \end{bmatrix} \end{bmatrix}$$

Graph corresponding to the AVM:

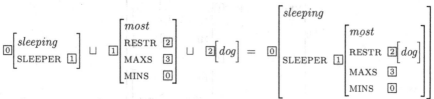

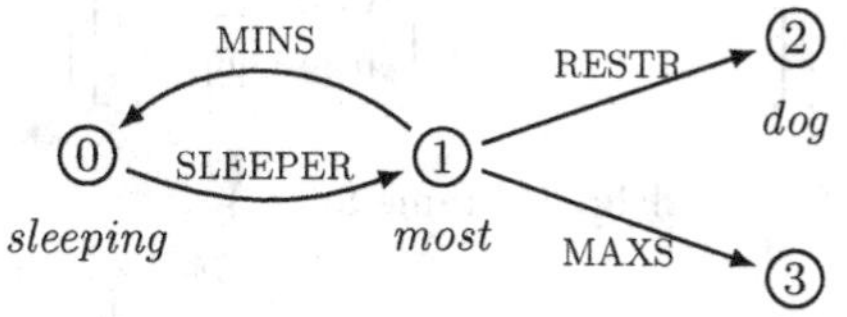


**Fig. 3.** Frame analysis of (1)

The cyclic structure of the frame in Fig. 3, resulting from the analysis of (1), reflects the two ways a quantifier contributes to meaning in Frame Semantics: On the one hand it (minimally) embeds some event, on the other hand it functions as an argument participant in this event and is therefore embedded in the event frame. Note that the quantifier frame does not fix the scope of the quantifier, it only records its minimal scope. In the case of (1), the minimal scope is also the actual scope. But in examples with several quantificational operators, the overall frame in effect resembles an underspecified representation of several scope orderings known from underspecified semantics.

(2) Every boy loves two girls.

In the frame in Fig. 4 of the logically ambiguous sentence (2) we also see that the semantic roles of the participants of the event are not necessarily unique: the frame labeled $\boxed{1}$ is not only the LOVER but also the EMOTER and the EXPERIENCER, depending on how abstract a characterization of the event is adopted. This is in line with the semantic role hierarchy proposed in [16].

### 3.2 Truth Conditions and Underspecification

The frames in Fig. 3 and 4 do not immediately encode truth conditions that come with a model-theoretic interpretation. They are mental representations of the concepts expressed by (1) and (2). However, we can extract a predicate-logical formula with holes, labels and dominance constraints from these frames that tells us what properties the world must have in a situation where the concepts represented by the frames get instantiated. The predicate logical formulas with

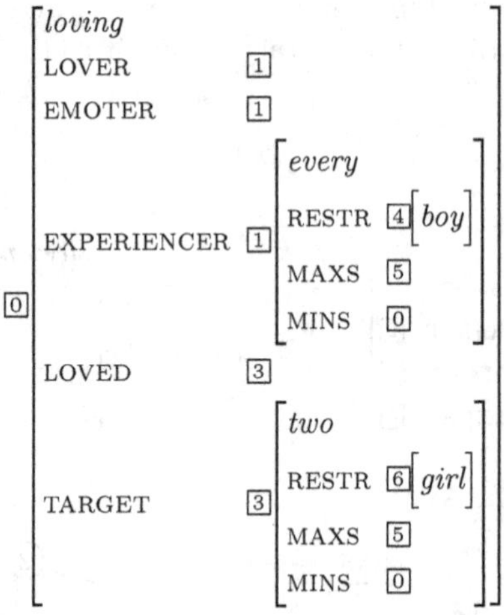


**Fig. 4.** Frame for (2)

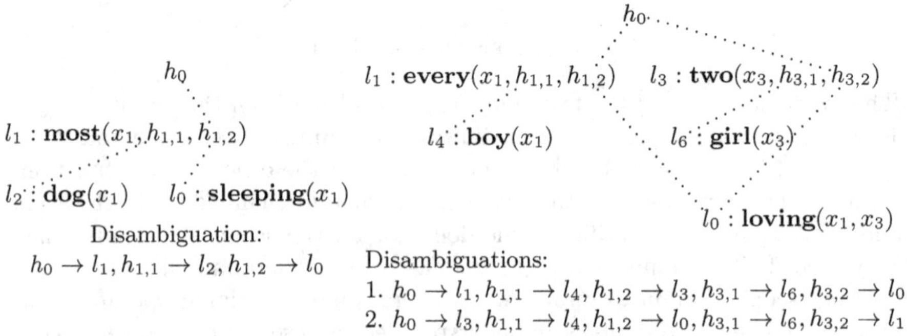


**Fig. 5.** Dominance constraints for (1) and (2)

holes and dominance constraints (in the sense of Hole Semantics [17]) for (1) and (2) are shown in Fig. 5.[2]

Only a part of the frame is relevant for the truth conditions. For instance, the fact that the first argument of *loving* in Fig. 4 is characterized not only as EXPERIENCER but also via the more specific roles of EMOTER and LOVER is without counterpart in the predicate-logical formula. Furthermore, frames contain not only knowledge originating from the frames paired with elementary trees (i.e., lexical and constructional meaning) but also world knowledge. The concept of a dog in the frame in Fig. 3 comprises much more than the type *dog* because it comes with an entire bundle of attributes characterizing entities

[2] Related ideas about semantic underspecification have been framed in Underspecified Discourse Representation Theory (UDRT, [18]), the Constraint Language for Lambda Structures (CLLS, [19]) and Minimal Recursion Semantics (MRS, [20]).

of type *dog*. Since these are not relevant for our paper, they are left aside. For extracting truth conditions, only meaning contributed by elementary trees is relevant and, furthermore, only nodes with base labels or attributes relating base labeled nodes play a role.

This leaves the task to specify how to read off underspecified predicate-logical formulas for frames of type *generalized-quantifier* (with subtypes *every*, *two*, etc.), *eventuality* (with subtypes *event* and *state*) and *entity* (subtypes *dog*, *boy*, etc.):

- Frames of type *eventuality* with argument roles ⟨arg1⟩, ⟨arg2⟩, ... (e.g., EXPERIENCER and THEME):

$$\boxed{i}\begin{bmatrix} \textit{pred} & \\ \langle\text{ARG1}\rangle & \boxed{j} \\ \langle\text{ARG2}\rangle & \boxed{k} \\ \ldots & \end{bmatrix} \rightsquigarrow l_i : \mathbf{pred}(x_j, x_k, \ldots)$$

  where *pred* is a subtype of *eventuality*.
- Frames of type *generalized-quantifier* and of type *entity*:[3]

$$\boxed{i}\begin{bmatrix} \textit{quant} & \\ \text{RESTR} & \boxed{j}\begin{bmatrix}\textit{pred}\end{bmatrix} \\ \text{MAXS} & \boxed{k} \\ \text{MINS} & \boxed{l} \end{bmatrix} \rightsquigarrow \begin{array}{l} l_i : \mathbf{quant}(x_i, h_{i,1}, h_{i,2}), \\ l_j : \mathbf{pred}(x_i), \\ h_k \lhd^* h_2, h_{i,1} \lhd^* l_j, h_{i,2} \lhd^* l_l \end{array}$$

  where *quant* is a subtype of *generalized-quantifier* and *pred* is a subtype of *entity*. **quant** and **pred** are the predicate logical constants corresponding to *quant* and *pred* respectively.

We assume the predicate-logical formulas to be typed. However, the types are of course not the types from our frames and there is no one-to-one correspondence between the types of frames and the type system of the predicate logic. Assuming that our variables $x_1, x_2, \ldots$ are of type $e$ in our predicate logic and holes and labels are of type $t$ (or $\langle s, t\rangle$ if we use propositions), all other types in the formulas we extract from the frames can be inferred.

Why do we pair syntactic trees with frames, generate a frame for a sentence during parsing and extract a predicate logical formula only later in a subsequent step? First of all, this architecture has advantages in terms of complexity and tractability. If we make sure that we use only feature constraints of a certain restricted type (see [7] for details), then we know that the frame unification that constitutes our semantic composition is almost linear in the size of the frames. At the same time, frame unification already acts as a filter since certain analyses will be excluded because of incompatibilities in the frames. Another important aspect of frames is that in addition to lexical and constructional meaning contributed by the building blocks of the sentence they also include world knowledge. We think it useful to be able to access this during syntactic-semantic composition and not only at some later point of interpreting a sentence in a given discourse

[3] The symbol $\lhd^*$ for the dominance relation is borrowed from the notation of dominance constraints in [21, 22].

model. Finally, following [4, 5] we believe that the mental concept expressed by a sentence exists independently of the actual situation in which the sentence is supposed to hold. This mental concept is represented by the frame.

### 3.3 The Syntax-Semantics Interface

In the previous sections, we presented the frames that we obtain from parsing and the way they (indirectly) characterize truth conditions. Concerning the syntax-semantics interface in LTAG, the analyses from [13] can be transferred to the LTAG frame semantics architecture. Fig. 6 shows the LTAG analysis of (1). The interface feature PRED serves to pass the embedded predication frame (here of type *dog*) to the restriction (feature RESTR) of the quantifier. The interface features MINS and MAXS enable passing the scope window from the verb tree to the quantifier tree. The syntax-triggered unifications are exactly the ones from Fig. 3.

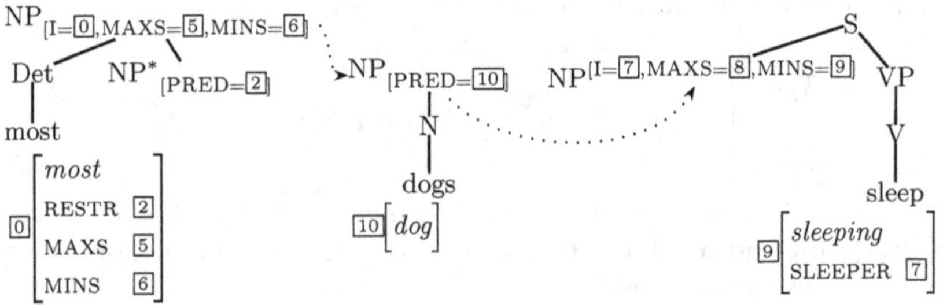


**Fig. 6.** Analysis for (1)

If there is more than one quantificational NP, all NPs find the same MAXS and MINS values at the substitution nodes, which entails the same scoping possibilities for all of them.

## 4 Quantifiers in Complex NPs

The feasibility of the quantifier frames in our grammar architecture depends on whether they provide enough structure to derive the correct truth conditions for all and only the intended interactions between the corresponding logical operators. To ascertain that this is the case we analyze a construction (see (3)) with complex NPs with two quantifiers where the embedded quantifier can scope over the embedding one (an order sometimes called inverse linking) but a third quantifier cannot scope in between the first two (cf. [23–26]). We will show that our Frame Semantics of quantifiers is capable of reproducing the exact same readings as the ones predicted by the analysis of [13].

(3) Two policemen spy on someone from every city.
 **every* > *two* > *some*
 **some* > *two* > *every*

The *from*-PP in (3) expresses the origin of a person or an object. We assume that *from every city* is an adjunct that adds a frame attribute ORIGIN whose value is contributed by the NP of the PP. Furthermore, it passes the label of the quantifier to which it adjoins (here [5]) as the new MAXS value to any embedded NP. The MINS of the embedded NP (here [9]) is the frame label to which the ORIGIN feature is added. The derivation is given in Fig. 7 and the resulting frame is shown in Fig. 8.

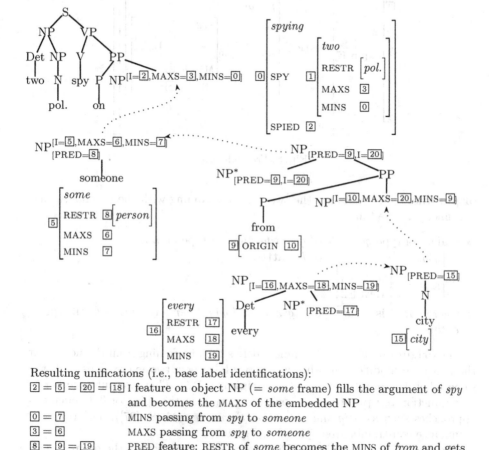

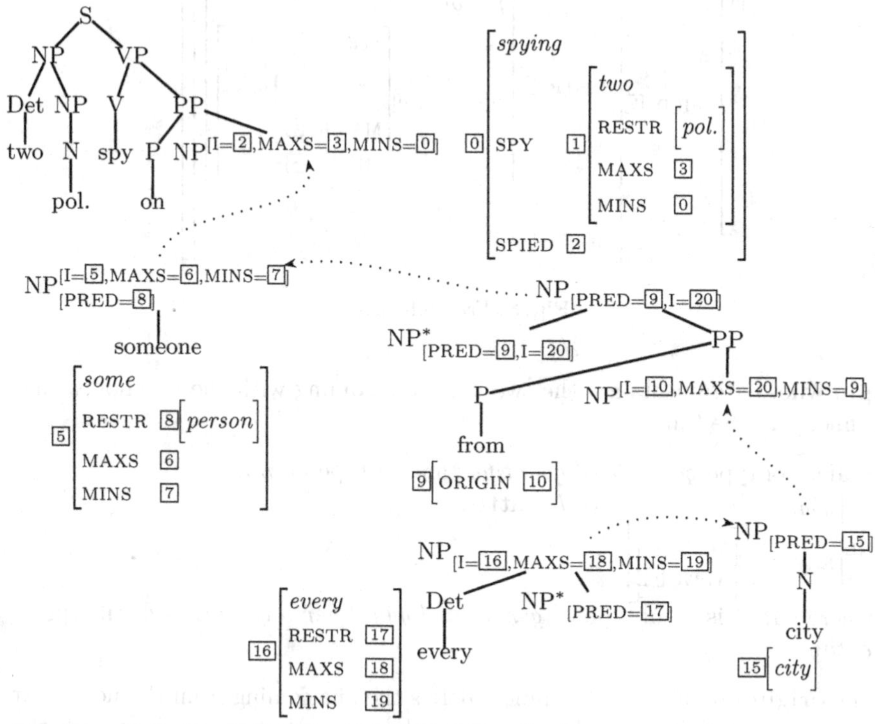


Resulting unifications (i.e., base label identifications):

| | |
|---|---|
| [2] = [5] = [20] = [18] | I feature on object NP (= *some* frame) fills the argument of *spy* and becomes the MAXS of the embedded NP |
| [0] = [7] | MINS passing from *spy* to *someone* |
| [3] = [6] | MAXS passing from *spy* to *someone* |
| [8] = [9] = [19] | PRED feature: RESTR of *some* becomes the MINS of *from* and gets passed to *every* |
| [15] = [17] | PRED feature: RESTR of *every* becomes the *city* frame |
| [10] = [16] | I feature: *every* frame fills the ORIGIN attribute |

**Fig. 7.** Analysis for (3)

For interpretation, the frame in Fig. 8 must again be transformed into an underspecified predicate logical representation. The only addition compared to the preceding section is that the attribute ORIGIN is translated to a binary predicate

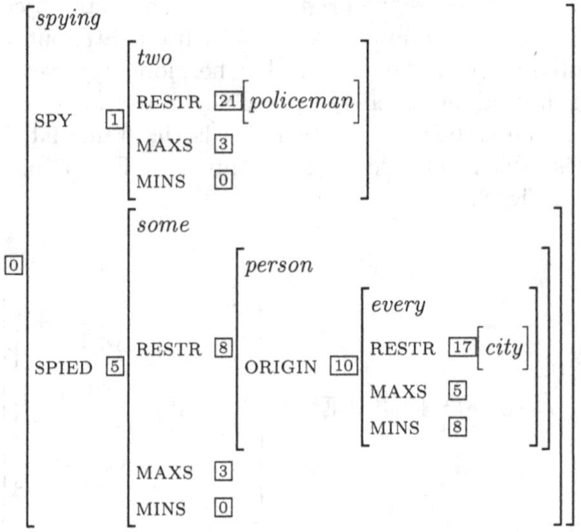

**Fig. 8.** Frame for (3)

**origin** whose arguments are the two variables coming with the two nodes that it connects in the frame.

- Frames of type *generalized-quantifier* and of type *entity*:

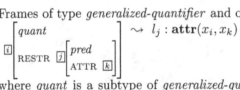

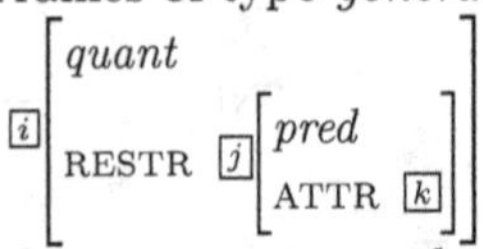

$\rightsquigarrow$ $l_j$ : **attr**$(x_i, x_k)$

where *quant* is a subtype of *generalized-quantifier* and *pred* is a subtype of *entity*.

The **origin**-formula has the same label as the one coming from the node that the attribute modifies. In other words, we have now two expressions with the same label, $l_8$ : **person**$(x_5)$ and $l_8$ : **origin**$(x_5, x_{10})$. They are put together in a conjunction $l_8$ : **person**$(x_5) \wedge$ **origin**$(x_5, x_{10})$, in the spirit of flat semantics approaches such as *Minimal Recursion Semantics* (MRS, [27]). The resulting dominance constraints are depicted in Fig. 9.

The dominance constraints in Fig. 9 are resolved to exactly the desired readings. Note that it is crucial for disambiguation that the resulting structures are trees. In particular, since they are trees, nothing can be dominated by the restriction and the scope of a quantifier at the same time. In our example, if **some** outscopes **every**, then we necessarily obtain that **every** is dominated by the restriction $h_{5,1}$. On the other hand, if **some** outscopes **two**, then the latter necessarily is part of the scope $h_{5,2}$. Consequently, if **some** has wide scope, there is no scope relation between the other two quantifiers and the order **some** > **two** > **every** is correctly excluded. The other impossible reading, **every** > **two** > **some**, is excluded because the constraints state that if

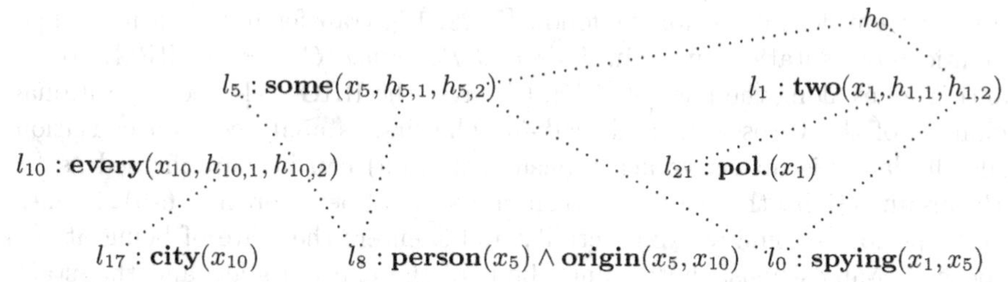


Resulting readings after disambiguation:
1. **some**$(x_5,$ **every**$(x_{10},$ **city**$(x_{10}),$ **person**$(x_5) \wedge$ **origin**$(x_5, x_{10})),$
   **two**$(x_1,$ **pol.**$(x_1),$ **spying**$(x_1, x_5)))$
2. **two**$(x_1,$ **pol.**$(x_1),$ **some**$(x_5,$ **every**$(x_{10},$ **city**$(x_{10}),$ **person**$(x_5) \wedge$ **origin**$(x_5, x_{10})),$
   **spying**$(x_1, x_5)))$
3. **every**$(x_{10},$ **city**$(x_{10}),$ **some**$(x_5,$ **person**$(x_5) \wedge$ **origin**$(x_5, x_{10}),$
   **two**$(x_1,$ **pol.**$(x_1),$ **spying**$(x_1, x_5))))$
4. **two**$(x_1,$ **pol.**$(x_1),$ **every**$(x_{10},$ **city**$(x_{10}),$ **some**$(x_5,$ **person**$(x_5) \wedge$ **origin**$(x_5, x_{10}),$
   **spying**$(x_1, x_5))))$

**Fig. 9.** Dominance constraints obtained from Fig. 8 and resulting readings

**every** scopes over **some**, then its scope argument must be the **some** formula ($h_{10,2} \rightarrow l_5$) itself, and no other quantifier may intervene.

## 5 Adverbs and Scope Ambiguities: The Case of "Again"

### 5.1 Repetitive and Restitutive Readings

After showing how to reconstruct operator scope ambiguities from quantifier frames, we will now turn to a phenomenon where the interaction of operator scope with the rich structure of semantic frames offers a natural basis for an interesting new analysis. When the adverb *again* adjoins to accomplishments like *open the door* or directed motion constructions like *walk to the hall* we observe an ambiguity between a repetitive reading (4-a) and a restitutive reading (4-b) [28].

(4) Bilbo opened the door again. (ex. from [28])
  a. Bilbo opened the door, and that had happened before.
  b. Bilbo opened the door, and the door had been open before.

(5) Bilbo walked to the hall again. (ex. from [28])
  a. Bilbo walked to the hall and that had happened before.
  b. Bilbo walked to the hall and he had been there before.

(4) actually has a third reading that is weaker than (4-a) but stronger than (4-b), namely that Bilbo opened the door and the door had been opened before. This means it was not necessarily Bilbo who opened the door before but it is at least the second time that the state of the door changes from closed to open.

In our event decomposition we follow [6, 29, 16], transferring the semi-formal semantic representations used in *Role and Reference Grammar* (RRG) to semantic frames along the lines of [7, 5]. (6) gives the RRG-style decompositional semantics of the verbs *open* and *walk-to*. The first is analyzed as a causation while the latter is an active accomplishment. (6-a) can be paraphrased as "$x$ performs an activity that causes $y$ to enter a state of being open". (6-b) signifies that "$x$ performs some walking activity and $x$ enters the state of being at $y$". In both decompositions, the coming about of the effected state and this state itself are present. This structure can be exploited to account for the restitutive readings of (4) and (5). Fig. 10 shows the translation of (6) to frames.

(6) a. $[\mathbf{do}(x, \emptyset)]$ CAUSE $[$INGR $\mathbf{open}(y)]$ (causation)
b. $\mathbf{do}(x, \mathbf{walk}(x))$ & INGR $\mathbf{be} - \mathbf{at}(y, x)$ (active accomplishment)

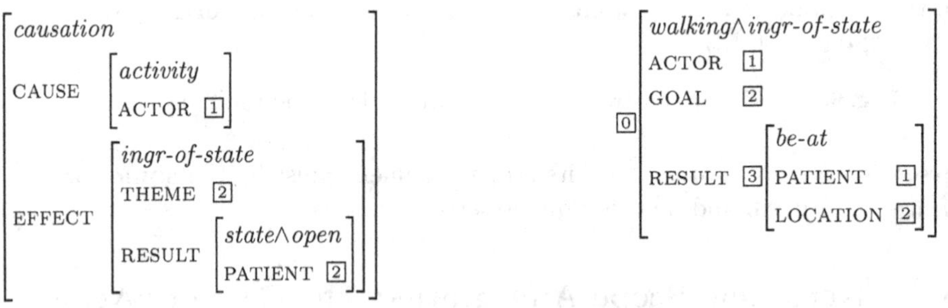


**Fig. 10.** Frames for *open* and *walk-to*

According to the event decomposition in Fig. 10 the frame of *open* comprises three (sub-)frames for eventualities (states or events) that *again* can apply to, a *causation* frame, an *ingr-of-state* frame and the frame of the *open* state. Thus, the structure of the frame provides the basis for the observed ambiguity. In the *walk-to* frame, we only have two such nodes, the *motion* frame and the *be-at* frame, and there are only two readings. One of the advantages of frames is that the difference in decomposition is made explicit and the different eventualities become accessible for adverbial modification. The embedding structure explains why the restitutive reading is weaker than the intermediate reading which, in turn, is weaker than the repetitive reading.

For the semantics of *again*, we follow [28, 30] in assuming that the ambiguity between the repetitive and the restitutive reading is not due to a lexial ambiguity of *again*; *again* always indicates repetition. However, unlike these approaches we assume that the ambiguity is not a syntactic ambiguity. Instead, we derive a single syntactic tree paired with a single frame for examples such as (4) and (5). As in the quantifier scope analyses proposed in the preceding sections, the frame does not fix the ultimate scope of *again*. It states that the adverb minimally modifies the embedded *state∧open* or *be-at* frame, respectively. Our analysis of

(4) is shown in Fig. 11. The adverb contributes a frame of type *repetition*. Its minimal frame argument (the MINS value) is determined by the interface feature MINS on the VP node. This MINS value is different from the one provided for the quantifier frames at NP nodes, since the latter concern the entire event. The minimal argument of *again* is the resulting *open* state that is embedded under the effected change of state. Similarly, in the case of (5), where we do not have a causation but a walking event that is an *ingr-of-state*, the minimal argument provided in the construction tree is the resulting state of being at the goal of the walking.[4]

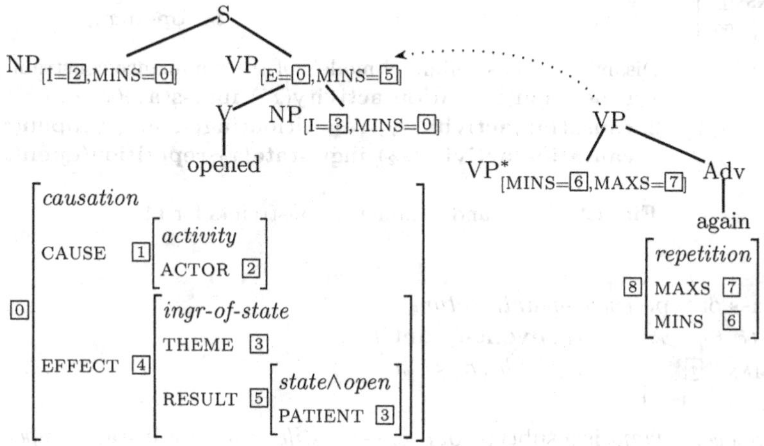


**Fig. 11.** Analysis of (4)

As a result of the derivation in Fig. 11, we obtain the frame and the dominance constraints in Fig. 12. Note that the frame does not have a unique root. Both the *causation* node and the *repetition* node lack incoming edges. This is different from the earlier examples where the quantifier was embedded under the event since it contributed a participant to the event.

To obtain the dominance constraints from the frames, we assume the following additional rules:

[4] Other authors (and an anonymous reviewer) note that only few adverbs seem to have access to sublexical modification of result states (for an overview and discussion, see [28]). For example, (i) cannot mean that Bilbo opened the door once and then it always stayed open.

(i) Bilbo always opened the door.

Appropriate restrictions can be stated in our framework, but we do not want to commit to a particular formulation here. For other relevant examples of sublexical modification such as the ambiguous noun phrase *a beautiful dancer* (a dancer who dances beautifully vs. a beautiful-looking dancer), see [31].

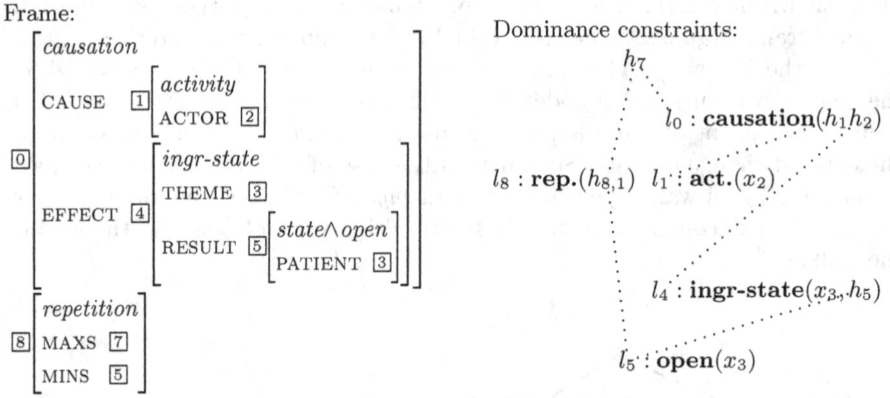


Disambiguations (minimal models of the dominance constraints):
1. **repetition**(**causation**(**activity**($x_2$), **ingr-state**($x_3$, **open**($x_3$))))
2. **causation**(**activity**($x_2$), **repetition**(**ingr-state**($x_3$, **open**($x_3$))))
3. **causation**(**activity**($x_2$), **ingr-state**($x_3$, **repetition**(**open**($x_3$))))

**Fig. 12.** Frame and dominance constraints for (4)

- Frames of type *event-quantification*:

$$\boxed{i}\begin{bmatrix} \textit{event-quant} & \\ \text{MAXS} & \boxed{j} \\ \text{MINS} & \boxed{k} \end{bmatrix} \rightsquigarrow \begin{array}{l} l_i : \textbf{event-quant}(h_i), \\ h_j \lhd^* h_i, h_i \lhd^* l_k \end{array}$$

  where *event-quant* is a subtype of *event-quantification* (for instance *repetition.*
- Frames of type *eventuality* with entity-valued argument roles ⟨arg1⟩, ⟨arg2⟩, ... (e.g., EXPERIENCER and THEME) and eventuality-valued argument roles ⟨event1⟩, ⟨event2⟩, ... (e.g., CAUSE and EFFECT):

$$\boxed{i}\begin{bmatrix} \textit{pred} & \\ \langle\text{ARG1}\rangle & \boxed{j} \\ \langle\text{ARG2}\rangle & \boxed{k} \\ \dots & \\ \langle\text{EVENT1}\rangle & \boxed{l} \\ \langle\text{EVENT2}\rangle & \boxed{m} \\ \dots & \end{bmatrix} \rightsquigarrow \begin{array}{l} l_i : \textbf{pred}(x_j, x_k, \dots, h_l, h_m, \dots) \\ h_l \lhd^* l_l, h_m \lhd^* l_m \end{array}$$

  where *pred* is a subtype of *eventuality.*

## 5.2 Interaction with Quantificational NPs

By adding a quantifier to (4)/(5), we can provoke an additional semantic ambiguity due to an interaction with the adverb *again* as demonstrated in (7) and (8), with the new NP quantifier in object position and in subject position, respectively. In both cases, a reading where *again* outscopes the quantifier is possible: In (7), this reading signifies that it happened again that Bilbo opened more than half of the doors, but he did not necessarily open the same as before. In (8), wide scope of *again* signifies that it was again the case that there were two dwarfs (not necessarily the same as before) that walked to the hall.

(7) Bilbo opened more than half of the doors again.

(8) Two dwarfs walked to the hall again.

Let us consider the analysis of (8). The adjoining *again* picks the label of the result state as its minimal argument frame while the minimal argument of the quantifier *two dwarfs* is the label of the entire *walking* frame. As basis for this, the elementary construction of a directed motion with a goal PP provides the two different labels as MINS values at the VP node and the NP/PP nodes, respectively. We obtain the frame and the dominance constraints in Fig. 13.[5]

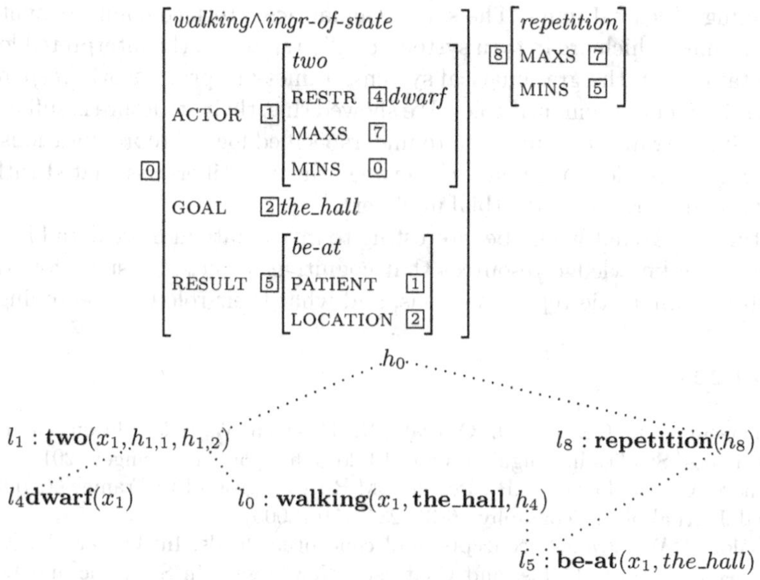


**Fig. 13.** Frame for (8)

The analysis of *again* proposed in this section has demonstrated that a combination of the detailed event decomposition of frames and the techniques of dominance constraints in underspecified scope representations paves the way to an elegant account of challenging scope phenomena. Our framework provides an analysis that avoids either lexical ambiguity or structural ambiguity, one of which was necessary in the two earlier types of accounts [32, 30, 28].

## 6 Conclusion

We added quantifier frames to Frame Semantics and defined a translation from frames to underspecified semantic representations that lead to resolved repre-

[5] The analysis of the definite article is left aside in this paper.

sentations with a conventional interpretation in models. With its integration in LTAG a new architecture emerged in which a Frame Semantics with fine-grained lexical decompositions of situations as frames supports a well-defined logical semantics with quantificational and intensional operators.

We distinguished a level of representation with frames as feature structures from a derived level of interpreted logical representations. Frames are taken to be cognitive representations of concepts, world knowledge and situational knowledge. When frame semantics represents a dog, it aims at the concept of a dog and an agent's knowledge of dogs including the range of their size or that they are carnivores etc. Type logical semantics on the other hand provides a predicate denoting the set of dogs in a model. Separating frames and logical semantics is therefore crucial to do justice to the fundamentally different nature of the two. As lexical representations, frames are supposed to reflect how speakers decompose and represent the meaning of lexical units. The syntax-semantics interface connects syntactic units to frames, which are in turn systematically related to the interpreted logical representations. In the grammatical system, frames are prominently responsible for the assignment of semantic roles. We showed that their structure is sufficiently rich to define a translation function to underspecified logical representations that explicate the possible scope relations between the quantifiers associated with the fillers of semantic roles or adverbial modifiers.

In future research it would be interesting to investigate in more detail how the comprehensive knowledge resources that cognitive scientists assume for frames are related to linguistic representations, and what their role is in reasoning.

## References

1. Gamerschlag, T., Gerland, D., Osswald, R., Petersen, W. (eds.): Frames and Concept Types. Studies in Linguistics and Philosophy, vol. 94. Springer (2014)
2. Fillmore, C.J., Johnson, C.R., Petruck, M.R.: Background to FrameNet. International Journal of Lexicography 16(3), 235–250 (2003)
3. Barsalou, L.W.: Frames, concepts, and conceptual fields. In: Lehrer, A., Kittay, E.F. (eds.) Frames, Fields, and Contrasts. New Essays in Semantic and Lexical Organization, pp. 21–74. Lawrence Erlbaum Associates, Hillsdale (1992)
4. Löbner, S.: Evidence for frames from human language. In: [1], pp. 23–67
5. Osswald, R., Van Valin, R.D.: FrameNet, frame structure, and the syntax-semantics interface. In: [1], pp. 125–156
6. Dowty, D.R.: Word Meaning and Montague Grammar. D. Reidel Publishing Company (1979)
7. Kallmeyer, L., Osswald, R.: Syntax-driven semantic frame composition in Lexicalized Tree Adjoining Grammars. Journal of Language Modelling 1(2), 267–330 (2013)
8. Joshi, A.K., Schabes, Y.: Tree-Adjoning Grammars. In: Rozenberg, G., Salomaa, A. (eds.) Handbook of Formal Languages, pp. 69–123. Springer, Berlin (1997)
9. Abeillé, A., Rambow, O.: Tree Adjoining Grammar: An Overview. In: Abeillé, A., Rambow, O. (eds.) Tree Adjoining Grammars: Formalisms, Linguistic Analysis and Processing, pp. 1–68. CSLI (2000)
10. Vijay-Shanker, K., Joshi, A.K.: Feature structures based tree adjoining grammar. In: Proceedings of COLING, Budapest, pp. 714–719 (1988)

11. Gardent, C., Kallmeyer, L.: Semantic Construction in FTAG. In: Proceedings of EACL 2003, Budapest, pp. 123–130 (2003)
12. Kallmeyer, L., Joshi, A.K.: Factoring Predicate Argument and Scope Semantics: Underspecified Semantics with LTAG. Research on Language and Computation 1(1-2), 3–58 (2003)
13. Kallmeyer, L., Romero, M.: Scope and situation binding in LTAG using semantic unification. Research on Language and Computation 6(1), 3–52 (2008)
14. Hegner, S.J.: Properties of Horn clauses in feature-structure logic. In: Rupp, C.J., Rosner, M.A., Johnson, R.L. (eds.) Constraints, Language and Computation, pp. 111–147. Academic Press, San Diego (1994)
15. Carpenter, B.: The Logic of Typed Feature Structures. Cambridge University Press (1992)
16. Van Valin, Jr., D.: Exploring the Syntax-Semantics Interface. Cambridge University Press (2005)
17. Bos, J.: Predicate logic unplugged. In: Dekker, P., Stokhof, M. (eds.) Proceedings of the 10th Amsterdam Colloquium, pp. 133–142 (1995)
18. Reyle, U.: Dealing with ambiguities by underspecification: Construction, representation and deduction. Journal of Semantics 10, 123–179 (1993)
19. Egg, M., Koller, A., Niehren, J.: The Constraint Language for Lambda Structures. Journal of Logic, Language and Information 10(4), 457–485 (2001)
20. Copestake, A., Flickinger, D., Pollard, C., Sag, I.A.: Minimal Recursion Semantics: An introduction. Research on Language and Computation 3, 281–332 (2005)
21. Althaus, E., Duchier, D., Koller, A., Mehlhorn, K., Niehren, J., Thiel, S.: An efficient graph algorithm for dominance constraints. Journal of Algorithms 48(1), 194–219 (2003)
22. Koller, A., Niehren, J., Treinen, R.: Dominance Constraints: Algorithms and Complexity. In: Moortgat, M. (ed.) LACL 1998. LNCS (LNAI), vol. 2014, pp. 106–125. Springer, Heidelberg (2001)
23. Heim, I., Kratzer, A.: Semantics in Generative Grammar. Blackwell (1998)
24. Barker, C.: Continuations and the Nature of Quantification. Natural Language Semantics 10, 167–210 (2002)
25. Sauerland, U.: DP is not a scope island. Linguistic Inquiry 36, 303–314 (2004)
26. Joshi, A.K., Kallmeyer, L., Romero, M.: Flexible Composition in LTAG: Quantifier Scope and Inverse Linking. In: Muskens, R., Bunt, H. (eds.) Computing Meaning Volume 3. Studies in Linguistics and Philosophy, vol. 83, pp. 233–256. Springer (2007)
27. Copestake, A., Flickinger, D., Pollard, C., Sag, I.A.: Minimal Recursion Semantics: An introduction. Research on Language and Computation 3, 281–332 (2005)
28. Beck, S.: There and back again: A semantic analysis. Journal of Semantics 22, 3–51 (2005)
29. Van Valin, Jr., D., LaPolla, R.: Syntax: Structure, meaning and function. Cambridge University Press (1997)
30. von Stechow, A.: The different readings of *wieder* "again": A structural account. Journal of Semantics 13, 87–138 (1996)
31. Egg, M.: Mismatches at the syntax-semantics interface. In: Müller, S. (ed.) Proceedings of the 11th International Conference on Head-Driven Phrase Structure Grammar, Center for Computational Linguistics, Katholieke Universiteit Leuven, Stanford, CA, pp. 119–139. CSLI Publications (2004)
32. Fabricius-Hansen, C.: Wieder ein *wieder*? Zur Semantik von *wieder*. In: Bäuerle, R., Schwarze, C., von Stechow, A. (eds.) Meaning, Use, and Interpretation of Language, pp. 97–120. Walter de Gruyter (1983)

# A Generalization of Linear Indexed Grammars Equivalent to Simple Context-Free Tree Grammars

Makoto Kanazawa

National Institute of Informatics, 2–1–2 Hitotsubashi,
Chiyoda-ku, Tokyo, 101–8430, Japan

**Abstract.** I define a generalization of linear indexed grammars that is equivalent to simple context-free tree grammars in the same way that linear indexed grammars are equivalent to tree-adjoining grammars.

## 1 Introduction

The equivalence in string generating power of *tree-adjoining grammars*, *head grammars*, and *linear indexed grammars* is one of the most celebrated results in the mathematics of grammar formalisms for natural language [11,27].[1] The title of Joshi et al.'s paper [11], "The convergence of mildly context-sensitive grammar formalisms", referred to this equivalence, but was somewhat misleading in that the relevant class of string languages—*tree-adjoining languages*—was properly included in a larger class, the class of *multiple context-free languages*, which has widely been regarded as a formal counterpart of the informal notion of mild context-sensitivity. In fact, this latter class has also been found to be characterized by a wide array of different formalisms [30,3,31,22,20,21,8,6,24].

Elsewhere [12], I have argued that a class of string languages that falls in between these two classes, namely, the class equivalently captured by *well-nested multiple context-free grammars* [13],[2] *coupled-context-free grammars* [10], *non-duplicating macro grammars* [25], *simple* (i.e., linear and non-deleting) *context-free tree grammars* [16], and *second-order abstract categorial grammars* of lexicon complexity 3 (see [14]), may be more attractive than the broader class as a formalization of mild context-sensitivity. I will not repeat the arguments here,[3] but one counterargument might be that this intermediate class (the class

[1] I exclude *combinatory categorial grammars*, another formalism that was shown to be equivalent, from the discussion here, for two reasons. First, the equivalence was proved with respect to a certain *restricted* version of combinatory categorial grammars, and is not known to hold for more general combinatory grammars that are actually used in practice [26]. Second, the definition of that version of combinatory categorial grammar is mathematically not as natural as the other three formalisms.

[2] The same formalism is called *well-nested linear context-free rewriting systems* by some people [5].

[3] Simple context-free tree grammars are also of interest because of their capacity to *lexicalize* tree-adjoining grammars preserving the set of derived trees [19].

G. Morrill et al. (Eds.): Formal Grammar 2014, LNCS 8612, pp. 86–103, 2014.

of well-nested multiple context-free languages) does not look as robust as the other two. The formalisms that capture it are all basically similar—they either define local sets of derivation trees that are evaluated bottom-up using "linear" functions (MCFGs and second-order ACGs) or present the same mechanism in a top-down rewriting perspective (coupled-context-free grammars, non-duplicating macro grammars, and simple context-free tree grammars). In contrast, at the level of tree-adjoining languages, linear indexed grammars have *non-local* (and non-regular) sets of derivation trees, and at the level of multiple context-free languages, *deterministic tree-walking transducers* [31] map trees to strings in a decidedly non-compositional way.

In this paper, I respond to this qualm by defining a natural generalization of linear indexed grammars, which generates the class of well-nested multiple context-free languages. This generalization, which I call *arboreal indexed grammars*, uses a "stack" attached to nonterminal symbols that stores tuples of trees, and is equivalent to simple context-free tree grammars in exactly the same way that linear indexed grammars are equivalent to tree-adjoining grammars (or more precisely, *monadic* simple context-free tree grammars [17]), in the following sense:

- For any simple context-free tree grammar, there is an arboreal indexed grammar such that the derived trees of the former may be obtained from the derivation trees of the latter by relabeling of nodes and deletion of some unary-branching nodes.
- For any arboreal indexed grammar, there is a simple context-free tree grammar whose derived trees are precisely the result of stripping the derivation trees of the arboreal indexed grammar of the "stack" part of their node labels.

The formalism of arboreal indexed grammar is closely related to the notion of Dyck tree language I introduced in [15]. This paper does not use this notion, however, and is completely self-contained.

Arboreal indexed grammars may be useful for devising new parsing algorithms for well-nested multiple context-free languages.

## 2 Indexed Grammars and Context-Free Tree Grammars

### 2.1 Indexed Grammars

An *indexed grammar* [1,9] is like a context-free grammar except that each occurrence of a nonterminal in a derivation tree has a string of indices attached to it, which acts as a pushdown stack. The stack is passed from a node to each of its nonterminal children, except that the production applied at that node may either push a symbol onto the stack or pop its topmost symbol.

A formal definition goes as follows. When $B$ is a nonterminal and $\chi$ is a string of indices, we write $B[\chi]$ for $B\chi$; thus, $B[]$ is just $B$. An *indexed grammar* is a quintuple $G = (N, \Sigma, I, P, S)$, where

**Table 1.** Standard interpretation of indexed grammar productions ($\chi \in I^*$)

| (TERM) | (DIST) | (PUSH) | (POP) |
|---|---|---|---|
| $A[] \to a$ | $A[] \to B_1[] \dots B_n[]$ | $A[] \to B[l]$ | $A[l] \to B[]$ |
| $A[\chi]$ | $A[\chi]$ | $A[\chi]$ | $A[l\chi]$ |
| $a$ | $B_1[\chi]$ $\cdots$ $B_n[\chi]$ | $B[l\chi]$ | $B[\chi]$ |

1. $N$ and $\Sigma$ are finite sets of nonterminals and terminals, respectively,
2. $I$ is a finite set of *indices*,
3. $S \in N$, and
4. $P$ is a finite set of productions, each having one of the following forms:[4]

$$A[] \to a, \qquad \text{(TERM)}$$
$$A[] \to B_1[] \dots B_n[], \qquad \text{(DIST)}$$
$$A[] \to B[l], \qquad \text{(PUSH)}$$
$$A[l] \to B[], \qquad \text{(POP)}$$

where $a \in \Sigma \cup \{\varepsilon\}$, $n \geq 1$, $A, B, B_1, \dots, B_n \in N$, and $l \in I$.

A *derivation tree* of $G$ is a finite labeled tree $\tau$ with node labels from $NI^* \cup \Sigma \cup \{\varepsilon\}$ such that

- each leaf node of $\tau$ is labeled by some $a \in \Sigma \cup \{\varepsilon\}$, and
- each internal node of $\tau$ is sanctioned by one of the productions of $G$,

where a node is said to be *sanctioned* by a production if it and its children are labeled as depicted in Table 1. For example, for a node to be sanctioned by a (TERM) production $A[] \to a$, it must be labeled by $A[\chi]$ for some $\chi \in I^*$, and its only child must be labeled by $a$. An internal node of a derivation tree is called a (TERM) node, (DIST) node, (PUSH) node, or (POP) node, depending on the type of production sanctioning it.

When a derivation tree has root label $A[\chi]$, we call it a derivation tree *from* $A[\chi]$. A *complete* derivation tree is a derivation tree from $S[]$. The *language* of $G$ is defined by

$$L(G) = \{\, \mathbf{y}(\tau) \mid \tau \text{ is a complete derivation tree of } G \,\},$$

where $\mathbf{y}(\tau)$ denotes the *yield* of $\tau$, the left-to-right concatenation of the labels of its leaf nodes.

[4] This is actually a normal form for indexed grammars which is more general than the normal form ("reduced form") given by Aho [1].

Note that in a derivation tree from $A[]$, each (POP) node must match exactly one (PUSH) node; a (PUSH) node may have zero, one, or more (POP) nodes matching it.[5]

### 2.2 Context-Free Tree Grammars

A *ranked alphabet* is a union $\Delta = \bigcup_{n \in \mathbb{N}} \Delta^{(n)}$ of disjoint finite sets of symbols. If $f \in \Delta^{(n)}$, then $n$ is the *rank* of $f$.

Let $\Sigma$ be an (unranked) alphabet and $\Delta$ be a ranked alphabet. We define the set $\mathbb{T}_{\Sigma,\Delta}$ of trees over $\Sigma, \Delta$ as follows:

1. If $f \in \Sigma \cup \Delta^{(0)}$, then $f \in \mathbb{T}_{\Sigma,\Delta}$.
2. If $f \in \Sigma \cup \Delta^{(n)}$ and $t_1, \dots, t_n \in \mathbb{T}_{\Sigma,\Delta}$ ($n \geq 1$), then $f(t_1, \dots, t_n) \in \mathbb{T}_{\Sigma,\Delta}$.

The notation $\mathbb{T}_\Sigma$ denotes the set of unranked trees over $\Sigma$; thus $\mathbb{T}_\Sigma = \mathbb{T}_{\Sigma,\varnothing}$.

We set aside special symbols $x_1, x_2, \dots$ called the *variables*. The set consisting of the first $n$ variables $x_1, \dots, x_n$ is denoted $X_n$. The set $\mathbb{T}_{\Sigma,\Delta}(X_n)$ is defined to be $\mathbb{T}_{\Sigma,\Delta \cup X_n}$, where $\Delta \cup X_n$ is the ranked alphabet where symbols in $X_n$ have rank 0. If $t[x_1, \dots, x_n] \in \mathbb{T}_{\Sigma,\Delta}(X_n)$ and $t_1, \dots, t_n \in \mathbb{T}_{\Sigma,\Delta}$, then $t[t_1, \dots, t_n]$ denotes the result of substituting $t_i$ for each occurrence of $x_i$ in $t$ ($i = 1, \dots, n$). Note that if $x_i$ does not occur in $t[x_1, \dots, x_n]$, then $t_i$ is deleted in $t[t_1, \dots, t_n]$, and if $x_i$ occurs more than once in $t[x_1, \dots, x_n]$, then $t_i$ is duplicated in $t[t_1, \dots, t_n]$. A tree $t[x_1, \dots, x_n] \in \mathbb{T}_{\Sigma,\Delta}(X_n)$ is called an *n-context* if each $x_i$ occurs exactly once in it. If $t[x_1, \dots, x_n]$ is an $n$-context, then each $t_i$ is neither deleted nor duplicated in $t[t_1, \dots, t_n]$.

We deviate from the standard practice and define context-free tree grammars using unranked alphabets of terminals. (This makes it easier to relate them to indexed grammars, but is not essential.) A *context-free tree grammar* [23,2] is a quadruple $G = (N, \Sigma, P, S)$, where

1. $N = \bigcup_{n \in \mathbb{N}} N^{(n)}$ is a finite ranked alphabet of nonterminals,
2. $\Sigma$ is a finite unranked alphabet of terminals,
3. $S$ is a nonterminal of rank 0, and
4. $P$ is a finite set of productions of the form

$$A(x_1, \dots, x_n) \to t[x_1, \dots, x_n],$$

where $A \in N^{(n)}$ and $t[x_1, \dots, x_n] \in \mathbb{T}_{\Sigma,N}(X_n)$.

We say that $G$ is of rank $m$ if the rank of nonterminals of $G$ does not exceed $m$.

The one-step rewriting relation $\Rightarrow_G$ on $\mathbb{T}_{\Sigma,N}$ is defined as follows: $u_1 \Rightarrow_G u_2$ if there is a 1-context $u[x_1] \in \mathbb{T}_{\Sigma,N}[X_1]$, a nonterminal $A \in N^{(n)}$, a production $A(x_1, \dots, x_n) \to t[x_1, \dots, x_n]$, and trees $t_1, \dots, t_n \in \mathbb{T}_{\Sigma,N}$ such that

[5] I leave to the reader a formal definition of the intuitively clear notion of a (POP) node *matching* a (PUSH) node.

$u_1 = u[A(t_1, \dots, t_n)]$ and $u_2 = u[t[t_1, \dots, t_n]]$. The *language* of a context-free tree grammar $G$ is[6]

$$L(G) = \{\, t \in \mathbb{T}_\Sigma \mid S \Rightarrow_G^* t \,\}.$$

We call elements of $L(G)$ *derived trees* of $G$.

We assume that when $\Sigma$ contains a special symbol $\varepsilon$, it is always used to label a leaf node and is interpreted as the empty string. The *string language* of a context-free tree grammar $G$ is defined to be

$$\{\, \mathbf{y}(t) \mid t \in L(G) \,\}.$$

*Example 1.* Here is a very simple example of a context-free tree grammar. Let $G = (N, \Sigma, P, S)$, where

$$\begin{aligned} N &= N^{(0)} \cup N^{(1)} = \{S, C\} \cup \{A, B\}, \\ \Sigma &= \{a, f, g\}, \end{aligned}$$

and $P$ consists of the following productions:

$$\begin{aligned} S &\to A(a), \\ A(x_1) &\to A(g(C, x_1)), \\ A(x_1) &\to B(x_1), \\ C &\to a, \\ B(x_1) &\to x_1, \\ B(x_1) &\to B(f(a, x_1, x_1)). \end{aligned}$$

Some elements of $L(G)$ are:

$$\begin{aligned} &a, \\ &g(a, a), \\ &g(a, g(a, a)), \\ &f(a, a, a), \\ &f(a, g(a, a), g(a, a)), \\ &f(a, g(a, g(a, a)), g(a, g(a, a))), \\ &f(a, f(a, a, a), f(a, a, a)), \\ &f(a, f(a, g(a, a), g(a, a)), f(a, g(a, a), g(a, a))), \\ &f(a, f(a, g(a, g(a, a)), g(a, g(a, a))), f(a, g(a, g(a, a)), g(a, g(a, a)))). \end{aligned}$$

A context-free tree grammar is *simple* if for each production $A(x_1, \dots, x_n) \to t[x_1, \dots, x_n]$, the right-hand side $t[x_1, \dots, x_n]$ is an $n$-context. *Monadic* simple context-free tree grammars, i.e., simple context-free tree grammars of rank 1, are, inessential details aside, the same as tree-adjoining grammars [17].

[6] This is the *OI*, as opposed to *IO*, interpretation of the grammar [2].

### 2.3 From Context-Free Tree Grammars to Indexed Grammars

In this section, we review Guessarian's [7] method (in slightly adapted form) of converting a context-free tree grammar $G$ to an indexed grammar $\mathrm{Ind}(G)$ that generates the same string language.

We refer to a node of a tree by a "Dewey decimal notation" [18] or "Gorn address", which is a string of positive integers separated by dots ".". If $t$ is a tree, define the *domain* of $t$, $\mathrm{dom}(t)$, by

$$\begin{aligned}\mathrm{dom}(a) &= \{\varepsilon\},\\ \mathrm{dom}(f(t_1,\ldots,t_n)) &= \{\varepsilon\} \cup \{\, i.p \mid 1 \le i \le n, p \in \mathrm{dom}(t_i)\,\}.\end{aligned}$$

If $p \in \mathrm{dom}(t)$, the label of the node at $p$, written $\mathrm{lab}(t,p)$ is defined by

$$\mathrm{lab}(a,\varepsilon) = a, \qquad \begin{aligned}\mathrm{lab}(f(t_1,\ldots,t_n),\varepsilon) &= f,\\ \mathrm{lab}(f(t_1,\ldots,t_n),i.p) &= \mathrm{lab}(t_i,p).\end{aligned}$$

Let $G = (N, \Sigma, P, S)$ be a context-free tree grammar. Let $t_i$ be the right-hand side tree of the $i$th production in $P$. Let

$$\begin{aligned}N' &= \{S'\} \cup \{\, (i,p) \mid 1 \le i \le |P| \text{ and } p \in \mathrm{dom}(t_i)\,\},\\ I &= \{\, (i,p) \in N' \mid \mathrm{lab}(t_i,p) \in N\,\}.\end{aligned}$$

Define the indexed grammar $\mathrm{Ind}(G) = (N', \Sigma, I, P', S')$, where $P'$ consists of the following productions:

- If the left-hand side of the $i$th production is $S$, then $P'$ contains the production
$$S'[] \to (i,\varepsilon)[]. \qquad (\mathrm{DIST}_1)$$
- If $\mathrm{lab}(t_i,p) \in \Sigma$ and $n = \max\{\, j \mid p.j \in \mathrm{dom}(t_i)\,\}$, then $P'$ contains the production
$$(i,p)[] \to (i,p.1)[] \ldots (i,p.n)[]. \qquad (\mathrm{DIST}_2)$$
- If $p$ is a leaf of $t_i$ and $\mathrm{lab}(t_i,p) = a \in \Sigma$, then $P'$ contains the production
$$(i,p)[] \to a. \qquad (\mathrm{TERM})$$
- If $\mathrm{lab}(t_i,p) = A \in N$ and the left-hand side nonterminal of the $j$th production is $A$, then $P'$ contains the production
$$(i,p)[] \to (j,\varepsilon)[(i,p)]. \qquad (\mathrm{PUSH})$$
and the production
$$(j,q)[(i,p)] \to (i,p.k)[] \qquad (\mathrm{POP})$$
for each $q \in \mathrm{dom}(t_j)$ such that $\mathrm{lab}(t_j,q) = x_k$.

If $\tau$ is a derivation tree of $\mathrm{Ind}(G)$, then let $h(\tau)$ be the result of removing all unary-branching nodes sanctioned by (TERM), ($\mathrm{DIST}_1$), (PUSH), or (POP) productions, and then changing the label of each remaining internal node from $(i,p)[\chi]$ to $\mathrm{lab}(t_i,p)$. Clearly, $\mathbf{y}(\tau) = \mathbf{y}(h(\tau))$. We can prove the following:

**Proposition 2.** *If $\tau$ is a complete derivation tree of* $\mathrm{Ind}(G)$, *then $h(\tau)$ is a derived tree of $G$. Conversely, if $t$ is a derived tree of $G$, then there exists a complete derivation tree $\tau$ of* $\mathrm{Ind}(G)$ *such that $t = h(\tau)$.*

**Corollary 3.** *For every context-free tree grammar $G$,* $\{\, \mathbf{y}(t) \mid t \in L(G) \,\} = L(\mathrm{Ind}(G))$.

We need two new notions to prove this proposition.[7] For an indexed grammar, a *derivation tree fragment* is defined like a derivation tree except that labels of the form $A[\chi]$ are allowed on leaf nodes. For a context-free tree grammar $G = (N, \Sigma, P, S)$, we extend the rewriting relation $\Rightarrow_G^*$ to $\mathbb{T}_{N \cup \Sigma}(X_n)$ in an obvious way.

*Example 4.* The result of applying the method to the context-free tree grammar $G$ of Example 1 is the following indexed grammar $\mathrm{Ind}(G)$:

$$\begin{aligned}
S'[] &\to (1,\varepsilon)[] \\
(1,\varepsilon)[] &\to (i,\varepsilon)[(1,\varepsilon)] \quad (i = 2,3) \\
(1,1)[] &\to a, \\
(2,\varepsilon)[] &\to (i,\varepsilon)[(2,\varepsilon)] \quad (i = 2,3) \\
(2,1)[] &\to (2,1.1)[]\,(2,1.2)[] \\
(2,1.1)[] &\to (4,\varepsilon)[(2,1.1)] \\
(2,1.2)[(i,\varepsilon)] &\to (i,1)[] \quad (i = 1,2) \\
(3,\varepsilon)[] &\to (i,\varepsilon)[(3,\varepsilon)] \quad (i = 5,6) \\
(3,1)[(i,\varepsilon)] &\to (i,1)[] \quad (i = 1,2) \\
(4,\varepsilon)[] &\to a \\
(5,\varepsilon)[(i,\varepsilon)] &\to (i,1)[] \quad (i = 3,6) \\
(6,\varepsilon)[] &\to (i,\varepsilon)[(6,\varepsilon)] \quad (i = 3,6) \\
(6,1)[] &\to (6,1.1)[]\,(6,1.2)[]\,(6,1.3)[] \\
(6,1.1)[] &\to a \\
(6,1.2)[(i,\varepsilon)] &\to (i,1)[] \quad (i = 3,6) \\
(6,1.3)[(i,\varepsilon)] &\to (i,1)[] \quad (i = 3,6)
\end{aligned}$$

Fig. 1 shows two derivation trees of $\mathrm{Ind}(G)$ corresponding to the derived trees $g(a,a)$ and $f(a, g(a,a), g(a,a))$ of $G$.

## 3 Linear Indexed Grammars and an Alternative Conception of Indexed Grammars

### 3.1 Linear Indexed Grammars

Gazdar [4] introduced *linear indexed grammars*,[8] which are a variant of indexed grammars where the stack attached to an internal node is passed to *exactly*

[7] Due to space limitations, I had to leave out all proofs.

[8] "Linear indexed grammar" seems to be a coinage of Vijay-Shanker [29].

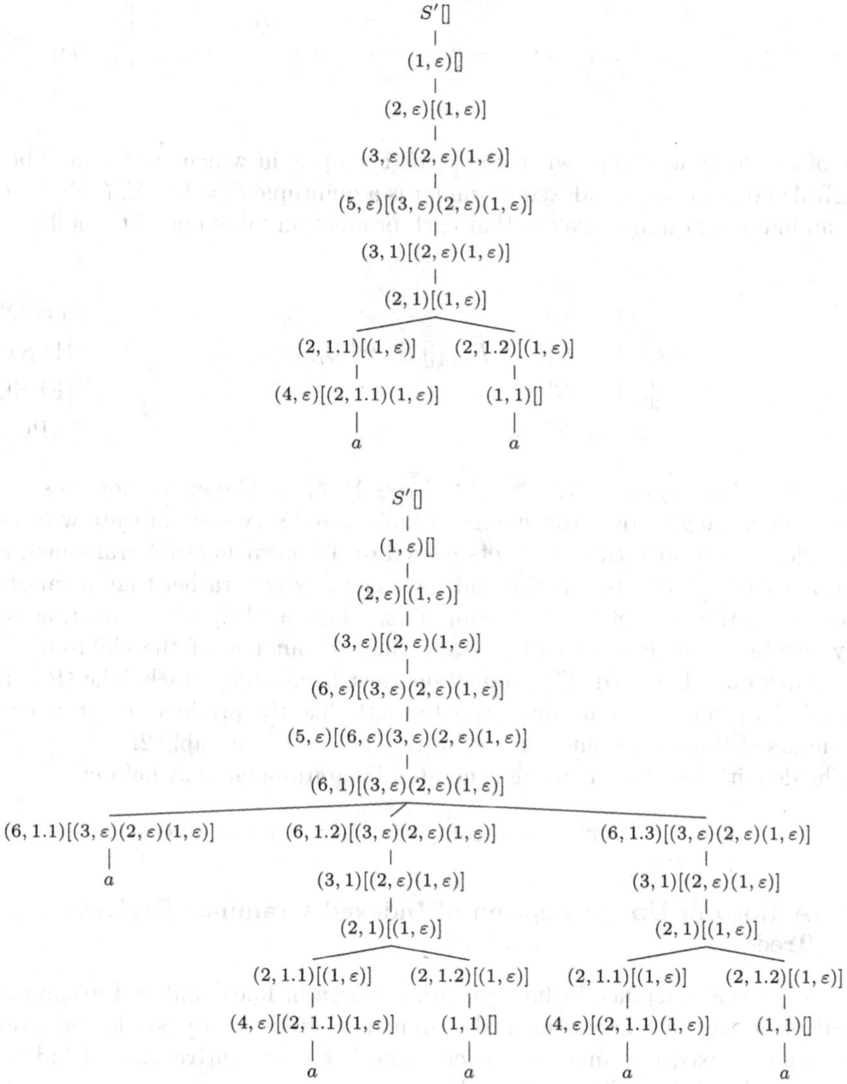


**Fig. 1.** Derivation trees of an indexed grammar obtained from a context-free tree grammar

**Table 2.** Interpretation of linear indexed grammar productions ($\chi \in I^*$)

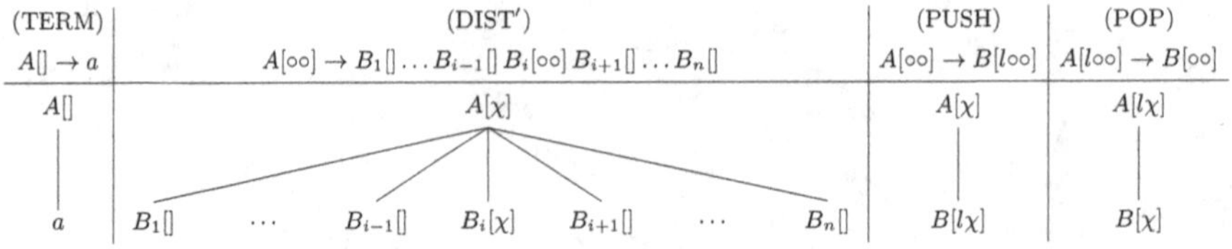


*one* of its children, except when the stack is empty, in which case it may be a (TERM) node. A linear indexed grammar is a quintuple $G = (N, \Sigma, I, P, S)$ just like an indexed grammar except that each production takes one of the following forms:[9]

$$\begin{aligned} A[] &\to a, && \text{(TERM)} \\ A[\circ\circ] &\to B_1[] \dots B_{i-1}[]\, B_i[\circ\circ]\, B_{i+1}[] \dots B_n[], && \text{(DIST}'\text{)} \\ A[\circ\circ] &\to B[l\circ\circ], && \text{(PUSH)} \\ A[l\circ\circ] &\to B[\circ\circ], && \text{(POP)} \end{aligned}$$

where $A, B, B_1, \dots, B_n \in N, n \geq 1, l \in I, a \in \Sigma \cup \{\varepsilon\}$. The expression $\circ\circ$ serves as a variable ranging over the strings of indices and serves to indicate which of the children of a node the stack gets passed to. In linear indexed grammars, an occurrence of "[]" in a production indicates empty stack, rather than a variable stack, as in the case of indexed grammars. Thus, a (TERM) production can only sanction a node with empty stack, and all but one of the children of a node sanctioned by a (DIST′) production must have empty stack. The (PUSH) and (POP) productions are interpreted exactly like the productions of indexed grammars of the same name, except for the notation. See Table 2.

The definition of the language generated by a grammar is as before:

$$L(G) = \{\, \mathbf{y}(\tau) \mid \tau \text{ is a complete derivation tree of } G \,\}.$$

### 3.2 A Bottom-Up Conception of Indexed Grammar Derivation Trees

Because of the difference in how the stack works in linear indexed grammars, a derivation tree of a linear indexed grammar is often *not* a possible derivation tree of an indexed grammar. However, there is an alternative view of indexed grammars that brings the two formalisms closer together.

Note that the labels of nodes in a derivation tree of a linear indexed grammar may be determined both top-down and bottom-up; once you know which

[9] Again, this is a normal form for linear indexed grammars, which we adopt here for convenience. Note that some authors, e.g., [27], place the top of the stack at the right end, contrary to the original convention of Aho [1].

**Table 3.** Bottom-up interpretation of indexed grammar productions ($\chi, \chi_1, \ldots, \chi_n \in I^*$). In (DIST), $\chi_1, \ldots, \chi_n$ must be pairwise compatible, and $i = \mathrm{argmax}_j\, |\chi_j|$.

| (TERM) | (DIST) | (PUSH) | (POP) |
|---|---|---|---|
| $A[] \to a$ | $A[] \to B_1[] \ldots B_n[]$ | $A[] \to B[l]$ | $A[l] \to B[]$ |
| $A[]$ | $A[\chi_i]$ | $A[\chi]$ or $A[]$ | $A[l\chi]$ |
| $a$ | $B_1[\chi_1]$ $\cdots$ $B_n[\chi_n]$ | $B[l\chi]$ $B[]$ | $B[\chi]$ |

production an internal node is sanctioned by, knowing its label uniquely determines its children's labels, and vice versa. Derivation trees of indexed grammars are deterministic only in the top-down direction, because when an internal node is sanctioned by a (TERM) production, the label of its unique child does not determine the stack portion of its label.

We can, however, adopt an alternative interpretation of indexed grammar productions and construct derivation trees bottom-up. With this alternative conception, (TERM) productions of an indexed grammar are interpreted in exactly the same way as in linear indexed grammars: they sanction a node only when its stack is empty. (POP) productions are interpreted in the same way as before, but (DIST) and (PUSH) productions are reinterpreted, as indicated in Table 3 (we call two strings *compatible* if one of them is a prefix of the other). Note that there are two ways in which a node may be sanctioned by a (PUSH) production; the second case is for a (PUSH) node with no matching (POP) node.

*Example 5.* Fig. 2 shows the same derivation trees in Fig. 1 relabeled in the new bottom-up way.

Indexed grammar derivation trees in the new sense correspond one-to-one with derivation trees in the original sense. The derivation trees under the two conceptions differ only in the stack portion of the labels of internal nodes, and they give rise to the same notion of the generated language. Let us adopt this new, bottom-up conception from now on, since it allows us to view derivation trees of linear indexed grammars as derivation trees of indexed grammars of a special kind.

Let $G$ be a linear indexed grammar, and let $G'$ be the indexed grammar that is the result of erasing all occurrences of $\infty$ from productions of $G$. Then every derivation tree of $G$ (from $A[]$ for some nonterminal $A$) is a derivation tree of $G'$ in which each (PUSH) node has exactly one matching (POP) node.

### 3.3 Monadic Indexed Grammars

Suppose we apply the method of Guessarian [7] reviewed in Section 2.3 to a monadic simple context-free tree grammar $G$. The resulting indexed grammar

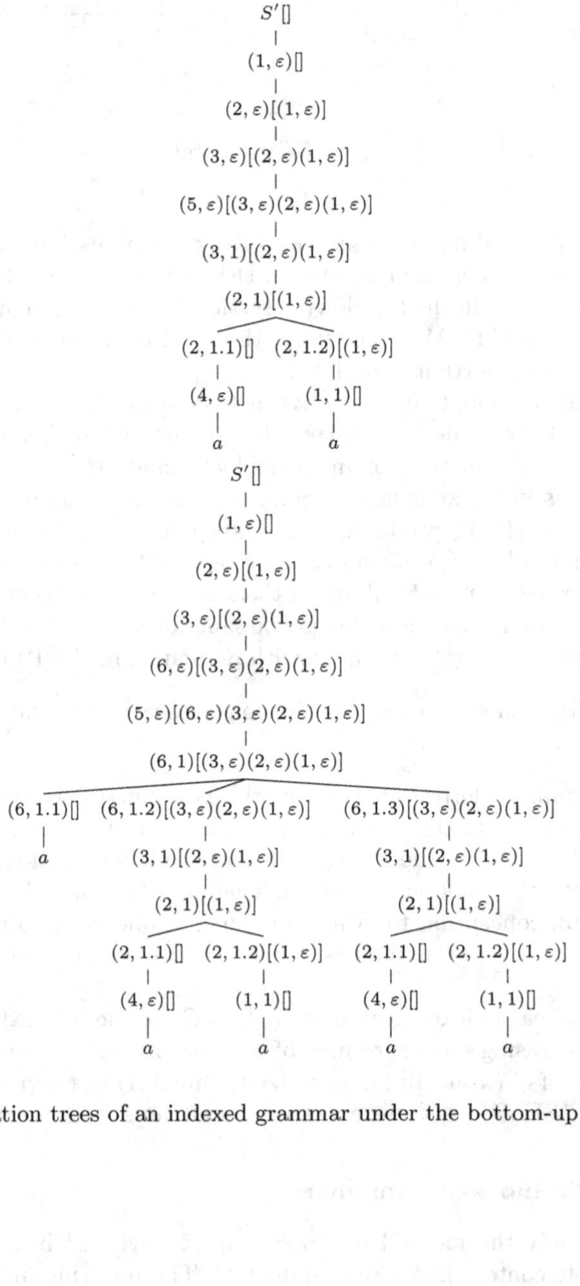


**Fig. 2.** Derivation trees of an indexed grammar under the bottom-up conception

Ind($G$) is very close to a linear indexed grammar. (For example, the grammar consisting of the first five productions of the grammar in Example 1 is a monadic simple context-free tree grammar, and the productions of the corresponding indexed grammar are the first 11 lines of the grammar in Example 4, with $i \neq 6$.) In any complete derivation tree of Ind($G$), every (PUSH) node has *at most one* matching (POP) node. More precisely, a (PUSH) node has no matching (POP) node when the (PUSH) production sanctioning it is related to a nonterminal of $G$ of rank 0; it has exactly one matching (POP) node when the production is related to a nonterminal of rank 1. Actually, it is easy to modify Guessarian's method and convert a monadic simple context-free tree grammar into a linear indexed grammar, instead of an indexed grammar.[10] However, the kind of indexed grammar that Ind($G$) exemplifies is interesting in its own right.

Let us call a derivation tree of an indexed grammar *monadic* if every (PUSH) node in it has at most one matching (POP) node. If $G$ is an indexed grammar, let us write $D(G)$ for the set of complete derivation trees of $G$ and $D_1(G)$ for the set of monadic complete derivation trees of $G$. Define

$$L_1(G) = \{\, \mathbf{y}(\tau) \mid \tau \in D_1(G) \,\}.$$

(Recall $L(G) = \{\, \mathbf{y}(\tau) \mid \tau \in D(G) \,\}$.) We can prove the following:

**Proposition 6.** *For every indexed grammar $G$, there is a linear indexed grammar $G'$ such that $L_1(G) = L(G')$.*

**Proposition 7.** *For every linear indexed grammar $G$, there is an indexed grammar $G'$ such that $L(G) = L_1(G') = L(G')$.*

Consider an indexed grammar $G$ together with $D_1(G)$ and $L_1(G)$. This can be thought of as another variant of indexed grammar where (DIST) production $A[] \rightarrow B_1[] \dots B_n[]$ sanctions a node only if all but one of its children has empty stack. Let us call an indexed grammar with this interpretation a *monadic indexed grammar.* (See Table 4.) The indexed grammar obtained from a monadic simple context-free tree grammar by Guessarian's method can be regarded equivalently as a monadic indexed grammar. As Propositions 6 and 7 show, monadic indexed grammars are equivalent to linear indexed grammars, and the derivation trees of the two formalisms are also almost identical.

Vijay-Shanker and Weir [27] give a method of converting a linear indexed grammar to an equivalent head grammar. Combined with a conversion from head grammars to tree-adjoining grammars, it gives a method of converting a linear indexed grammar into a tree-adjoining grammar that generates the same string language. This result can be strengthened. For a (linear/monadic/general) indexed grammar $G$, let us call the result $\widehat{\tau}$ of erasing all indices from a (complete) derivation tree $\tau$ of $G$ a *stripped (complete) derivation tree.* We can easily turn Vijay-Shanker and Weir's method into one that establishes the following:

[10] A variant of this modification is given by Vijay-Shanker and Weir [28] for tree-adjoining grammars.

**Table 4.** Interpretation of monadic indexed grammar productions ($\chi \in I^*, 1 \leq i \leq n$)

| (TERM) | (DIST) | (PUSH) | (POP) |
|---|---|---|---|
| $A[] \to a$ | $A[] \to B_1[] \dots B_n[]$ | $A[] \to B[l]$ | $A[l] \to B[]$ |
| $A[]$<br>\|<br>$a$ | $A[\chi]$<br>$B_1[] \;\cdots\; B_{i-1}[]\; B_i[\chi]\; B_{i+1}[] \;\cdots\; B_n[]$ | $A[\chi]$ or $A[]$<br>\| \|<br>$B[l\chi]$ $B[]$ | $A[l\chi]$<br>\|<br>$B[\chi]$ |

**Proposition 8.** *For every linear or monadic indexed grammar $G$, there is a monadic simple context-free tree grammar $G'$ that generates the set of all stripped complete derivation trees of $G$.*

I use monadic indexed grammars, rather than linear indexed grammars, as the point of departure for my generalization of linear indexed grammars. This is not strictly necessary, but will greatly simplify the definition of the generalized formalism.

## 4 Arboreal Indexed Grammars

Given Proposition 8, an obvious variant of indexed grammars corresponding to simple context-free tree grammars of rank $m$ suggests itself: indexed grammars interpreted in such a way that all (PUSH) nodes must have at most $m$ matching (POP) nodes.

### 4.1 From $m$-adic Indexed Grammars to Simple Context-Free Tree Grammars of Rank $m$

Let us call a derivation tree fragment of an indexed grammar *$m$-adic* if each (PUSH) node in it has at most $m$ matching (POP) nodes. Write $D_m(G)$ for the set of $m$-adic complete derivation trees of an indexed grammar $G$, and let $L_m(G) = \{\, \mathbf{y}(\tau) \mid \tau \in D_m(G) \,\}$. It is easy to check the following:

**Proposition 9.** *If $G$ is a simple context-free tree grammar of rank $m$, then $D(\mathrm{Ind}(G)) = D_m(\mathrm{Ind}(G))$, and consequently, $L(\mathrm{Ind}(G)) = L_m(\mathrm{Ind}(G))$.*

I now present a generalization of the construction underlying Proposition 8. Consider an indexed grammar $G$. A *path* in a derivation tree fragment of $G$ is a sequence of nodes, always passing from a parent node to one of its children. We say that a path $\rho$ is *clean* if every (PUSH) node on $\rho$ is matched by a (POP) node on $\rho$.

**Lemma 10.** *If a (POP) node matches a (PUSH) node in a derivation tree fragment, the path from the child of the (PUSH) node to the (POP) node is a clean path.*

**Lemma 11.** *Let $\tau$ be a derivation tree fragment from $A[]$ such that $\mathbf{y}(\tau) = w_0\, B_1[]\, w_1 \dots B_n[]\, w_n$ for some $w_0, \dots, w_n \in \Sigma^*$ and for every $i = 1, \dots, n$, the path $\rho_i$ from the root to the leaf node labeled by $B_i[]$ is a clean path. Let $\tau'$ be the result of changing the label of each node on $\rho_1, \dots, \rho_n$ from $C[\chi]$ to $C[\chi l]$. Then $\tau'$ is a derivation tree fragment from $A[l]$ with $\mathbf{y}(\tau') = w_0\, B_1[l]\, w_1 \dots B_n[l]\, w_n$.*

Let $m \geq 1$. Given an indexed grammar $G = (N, \Sigma, I, P, S)$, define a simple context-free tree grammar $\mathrm{CFT}^m_{\mathrm{sp}}(G) = (N', \Sigma', P', \langle S \rangle)$ where

$$N'^{(k)} = \begin{cases} \{\, \langle AB_1 \dots B_k \rangle \mid A, B_1, \dots, B_k \in N \,\} & \text{if } k \leq m, \\ \varnothing & \text{otherwise,} \end{cases}$$

$$\Sigma' = \Sigma \cup \{\varepsilon\} \cup N,$$

and $P'$ consists of the following productions:

(A) If $A[] \to a$ is a (TERM) production in $P$, $P'$ contains the production

$$\langle A \rangle \to A(a).$$

(B) For each nonterminal $A \in N$, $P'$ contains the production

$$\langle AA \rangle(x_1) \to x_1.$$

(C) For each nonterminal $\langle AB_1 \dots B_k \rangle \in N'$, if $A[] \to C_1[] \dots C_n[]$ is a (DIST) production in $P$ and $0 \leq k_1 \leq \dots \leq k_n = k$, then $P'$ contains the production

$$\begin{aligned} &\langle AB_1 \dots B_k \rangle(x_1, \dots, x_k) \to \\ &A(\langle C_1 B_1 \dots B_{k_1} \rangle(x_1, \dots, x_{k_1}), \dots, \langle C_n B_{k_{n-1}+1} \dots B_{k_n} \rangle(x_{k_{n-1}+1}, \dots, x_{k_n})). \end{aligned}$$

(D) For each nonterminal $\langle AB_1 \dots B_k \rangle \in N'$, if $A[] \to C[l]$ is a (PUSH) production in $P$, $D_1[l] \to E_1[], \dots, D_n[l] \to E_n[]$ are (POP) productions in $P$ $(0 \leq n \leq m)$, and $0 \leq k_1 \leq \dots \leq k_n = k$, then $P'$ contains the production

$$\begin{aligned} &\langle AB_1 \dots B_k \rangle(x_1, \dots, x_k) \to \\ &A(\langle CD_1 \dots D_n \rangle(D_1(\langle E_1 B_1 \dots B_{k_1} \rangle(x_1, \dots, x_{k_1})), \\ &\qquad \dots, \\ &\qquad D_n(\langle E_n B_{k_{n-1}+1} \dots B_{k_n} \rangle(x_{k_{n-1}+1}, \dots, x_{k_n})))). \end{aligned}$$

(When $n = 0$, this production is $\langle A \rangle \to A(\langle C \rangle)$.)

**Lemma 12.** *Let $\langle AB_1 \dots B_k \rangle$ be a nonterminal of $\mathrm{CFT}^m_{\mathrm{sp}}(G)$. For every $t[x_1, \dots, x_k] \in \mathbb{T}_{\Sigma \cup \{\varepsilon\} \cup N}[X_k]$, the following are equivalent:*

(i) $\langle AB_1 \dots B_k \rangle(x_1, \dots, x_k) \Rightarrow^*_{\mathrm{CFT}^m_{\mathrm{sp}}(G)} t[x_1, \dots, x_k]$.

(ii) *There is an $m$-adic derivation tree fragment $\tau$ of $G$ such that*
   - $t[B_1, \dots, B_k] = \widehat{\tau}$,
   - *the root of $\tau$ is labeled by $A[]$,*
   - $\mathbf{y}(\tau) = w_0\, B_1[]\, w_1 \dots B_k[]\, w_k$ *for some $w_0, w_1, \dots, w_k \in \Sigma^*$, and*

– *for each* $i = 1, \ldots, k$, *the path from the root of* $\tau$ *to the leaf node labeled by* $B_i[]$ *is a clean path.*

**Theorem 13.** *For every indexed grammar* $G$, $L(\mathrm{CFT}^m_{\mathrm{sp}}(G)) = \{\widehat{\tau} \mid \tau \in D_m(G)\}$.

Let us call indexed grammars with the restriction to $m$-adic complete derivation trees *$m$-adic indexed grammars.* Proposition 9 and Theorem 13 establish the equivalence between $m$-adic indexed grammars and simple context-free tree grammars of rank $m$.

### 4.2 Storing Tuples of Trees in the Stack

Our job is not done yet. The notion of an $m$-adic indexed grammar has not been defined in terms of how the productions are interpreted. In the case of monadic indexed grammars, the restriction on the way a (DIST) production may sanction a node carved out precisely the set of monadic derivation trees. We cannot obtain the $m$-adic derivation trees in a similar way, for $m \geq 2$.

In order to express the restriction to $m$-adic derivation trees, we have to somehow record at each node the number of (POP) nodes below the node that are to match a given (PUSH) node above the node. This means that a (PUSH) production should push $k$ copies of the same index ($k \leq m$) onto the stack, and a (DIST) production should distribute different copies of the same index to (possibly) different children.

Such stack actions cannot be realized with strings of indices acting as pushdown storage. The most natural solution is to store a *tuple of trees*, rather than a string, in the stack. We only need to store a tuple of trees $s_1, \ldots, s_k$ with very special properties. First, all the nodes of $s_1, \ldots, s_k$ of the same level (i.e., at the same distance from the root) must have the same label. Second, for $m$-adic indexed grammars, the number of nodes of $s_1, \ldots, s_k$ of the same level may not exceed $m$. (This implies that the trees are at most $m$-branching; it does *not* imply that the number of leaves is bounded.) The number of components of the tuple may vary from node to node, but of course cannot exceed $m$. Let us call such a tuple of trees *$m$-limited.*

We consider trees in which each leaf node has a label from $\Sigma \cup \{\varepsilon\}$ and each internal node has a label of the form $A[s_1, \ldots, s_k]$, where $s_1, \ldots, s_k$ is an $m$-limited tuple of trees over $I$. Such trees may be sanctioned by productions of an indexed grammar as indicated in Table 5, in which case we call them *$m$-adic arboreal derivation trees.*

If $s_1, \ldots, s_k$ is an $m$-limited tuple of trees over $I$, all the paths in $s_1, \ldots, s_k$ starting from the root and ending in some leaf node give mutually compatible strings of indices. Let $\overline{s_1, \ldots, s_k}$ be the string given by a maximal path. If $\upsilon$ is an $m$-adic arboreal derivation tree of an indexed grammar $G$, let $\overline{\upsilon}$ be the result of changing each label $C[s_1, \ldots, s_k]$ in $\upsilon$ to $C[\overline{s_1, \ldots, s_k}]$. Then $\overline{\upsilon}$ is always an ordinary derivation tree. It is easy to see the following:

**Table 5.** Interpretation of $m$-adic arboreal indexed grammar productions. Here, $\sigma$ and $\sigma_i$ range over ($m$-limited) tuples of trees over $I$. When $\sigma$ is empty, $l(\sigma)$ stands for $l$. The concatenation of tuples $\sigma_1, \ldots, \sigma_n$ is written as $\sigma_1 \ldots \sigma_n$.

| (TERM) | (DIST) | (PUSH) | (POP) |
|---|---|---|---|
| $A[] \to a$ | $A[] \to B_1[] \ldots B_n[]$ | $A[] \to B[l]$ | $A[l] \to B[]$ |
| $A[]$ | $A[\sigma_1 \ldots \sigma_n]$ | $A[\sigma_1 \ldots \sigma_k] \quad (k \leq m)$ | $A[l(\sigma)]$ |
| $a$ | $B_1[\sigma_1] \cdots B_n[\sigma_n]$ | $B[l(\sigma_1), \ldots, l(\sigma_k)]$ | $B[\sigma]$ |

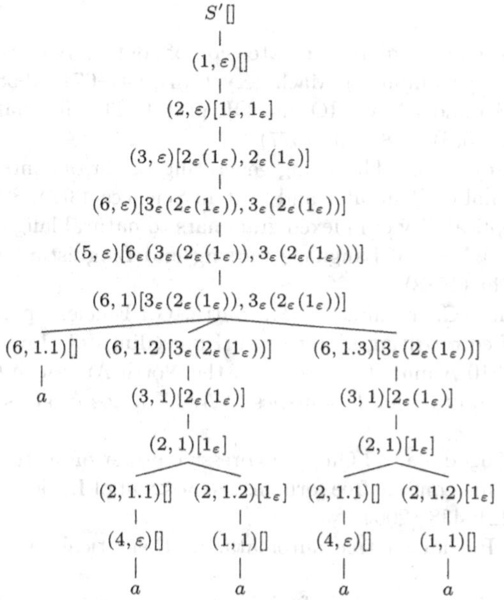

**Fig. 3.** An example of a 2-adic arboreal derivation tree

**Lemma 14.** *Let $G$ be an indexed grammar. An (ordinary) derivation tree $\tau$ of $G$ is $m$-adic if and only if there is an $m$-adic arboreal derivation tree $\upsilon$ of $G$ such that $\tau = \overline{\upsilon}$.*

*Example 15.* The second derivation tree in Fig. 2 was 2-adic. The 2-adic arboreal derivation tree corresponding to it is shown in Fig. 3, where we abbreviate indices $(i, p)$ by $i_p$.

We call an indexed grammar together with the interpretation of productions given in Table 5 an *$m$-adic arboreal indexed grammar*. We have established the following:

**Theorem 16.** (i) *For every $m$-adic arboreal indexed grammar $G$, there is a simple context-free tree grammar $G'$ of rank $m$ such that the derived trees of $G'$ are precisely the stripped complete derivation trees of $G$.*
(ii) *For every simple context-free tree grammar $G'$ of rank $m$, there is an $m$-adic arboreal indexed grammar $G$ such that the derived trees of $G'$ are obtained from the stripped complete derivation trees of $G$ by deleting some unary-branching nodes.*

**Corollary 17.** *Simple context-free tree grammars of rank $m$ and $m$-adic arboreal indexed grammars are equivalent in string-generating power.*

## References

1. Aho, A.V.: Indexed grammars—an extension of context-free grammars. Journal of the Association for Computing Machinery 15(4), 647–671 (1968)
2. Engelfriet, J., Schmidt, E.M.: IO and OI, part I. The Journal of Computer and System Sciences 15(3), 328–353 (1977)
3. Engelfriet, J., Heyker, L.: The string generating power of context-free hypergraph grammars. Journal of Computer and System Sciences 43(2), 328–360 (1991)
4. Gazdar, G.: Applicability of indexed grammars to natural languages. In: Reyle, U., Rohrer, C. (eds.) Natural Language Parsing and Linguistic Theories, pp. 69–94. Reidel, Dordrecht (1988)
5. Gómez-Rodríguez, C., Kuhlmann, M., Satta, G.: Efficient parsing of well-nested linear context-free rewriting systems. In: Proceedings of Human Language Technologies: The 2010 Annual Conference of the North American Chapter of the Association for Computational Linguistics (NAACL), Los Angeles, USA, pp. 276–284 (2010)
6. de Groote, P., Pogodalla, S.: On the expressive power of abstract categorial grammars: Representing context-free formalisms. Journal of Logic, Language and Information 13(4), 421–438 (2004)
7. Guessarian, I.: Pushdown tree automata. Mathematical Systems Theory 16(1), 237–263 (1983)
8. Harkema, H.: A characterization of minimalist languages. In: de Groote, P., Morrill, G., Retoré, C. (eds.) LACL 2001. LNCS (LNAI), vol. 2099, pp. 193–211. Springer, Heidelberg (2001)
9. Hopcroft, J.E., Ullman, J.D.: Introduction to Automata Theory, Languages, and Computation. Addison-Wesley, Reading (1979)
10. Hotz, G., Pitsch, G.: On parsing coupled-context-free languages. Thoretical Computer Science 161(1-2), 205–253 (1996)
11. Joshi, A.K., Vijay-Shanker, K., Weir, D.: The convergence of mildly context-sensitive grammar formalisms. In: Sells, P., Shieber, S., Wasow, T. (eds.) Processing of Linguistic Structure, pp. 31–81. MIT Press, Cambridge (1991)
12. Kanazawa, M.: The convergence of well-nested mildly context-sensitive grammar formalisms. An invited talk given at the 14th Conference on Formal Grammar, Bordeaux, France (July 2009), `http://research.nii.ac.jp/~kanazawa/talks/fg2009_talk.pdf`
13. Kanazawa, M.: The pumping lemma for well-nested multiple context-free languages. In: Diekert, V., Nowotka, D. (eds.) DLT 2009. LNCS, vol. 5583, pp. 312–325. Springer, Heidelberg (2009)

14. Kanazawa, M.: Second-order abstract categorial grammars. Lecture notes for a course taught at ESSLLI 2009 (2009), http://research.nii.ac.jp/~kanazawa/publications/esslli2009_lectures.pdf
15. Kanazawa, M.: Multi-dimensional trees and a Chomsky-Schützenberger-Weir representation theorem for simple context-free tree grammars. Journal of Logic and Computation (to appear)
16. Kepser, S., Mönnich, U.: Closure properties of linear context-free tree languages with an application to optimality theory. Theoretical Computer Science 354(1), 82–97 (2006)
17. Kepser, S., Rogers, J.: The equivalence of tree adjoining grammars and monadic linear context-free tree grammars. Journal of Logic, Language and Information 20(3), 361–384 (2011)
18. Knuth, D.E.: The Art of Computer Programming, Fundamental Algorithms, 3rd edn., vol. I. Addison-Wesley, Reading (1997)
19. Maletti, A., Engelfriet, J.: Strong lexicalization of tree adjoining grammars. In: Proceedings of the 50th Annual Meeting of the Association for Computational Linguistics, pp. 506–515. Association for Computational Linguistics (2012)
20. Michaelis, J.: Derivational minimalism is mildly context-sensitive. In: Moortgat, M. (ed.) LACL 1998. LNCS (LNAI), vol. 2014, pp. 179–198. Springer, Heidelberg (2001)
21. Michaelis, J.: Transforming linear context-free rewriting systems into minimalist grammars. In: de Groote, P., Morrill, G., Retoré, C. (eds.) LACL 2001. LNCS (LNAI), vol. 2099, pp. 228–244. Springer, Heidelberg (2001)
22. Rambow, O., Satta, G.: Independent parallelism in finite copying parallel rewriting systems. Theoretical Computer Science 223(1-2), 87–120 (1999)
23. Rounds, W.: Mappings and grammars on trees. Mathematical Systems Theory 4(3), 257–287 (1970)
24. Salvati, S.: Encoding second order string ACG with deterministic tree walking transducers. In: Wintner, S. (ed.) Proceedings of FG 2006: The 11th Conference on Formal Grammar, pp. 143–156. FG Online Proceedings, CSLI Publications (2007)
25. Seki, H., Kato, Y.: On the generative power of multiple context-free grammars and macro grammars. IEICE Transactions on Information and Systems E91-D(2), 209–221 (2008)
26. Steedman, M.: The Syntactic Process. MIT Press, Cambridge (2000)
27. Vijay-Shanker, K., Weir, D.J.: The equivalence of four extensions of context-free grammars. Mathematical Systems Theory 27(6), 511–546 (1994)
28. Vijay-Shanker, K., Weir, D.J.: Parsing some constrained grammar formalisms. Computational Linguistics 19(4), 591–636 (1993)
29. Vijayashanker, K.: A Study of Tree Adjoining Grammars. Ph.D. thesis, University of Pennsylvania (1987)
30. Weir, D.J.: Characterizing Mildly Context-Sensitive Grammar Formalisms. Ph.D. thesis, University of Pennsylvania (1988)
31. Weir, D.J.: Linear context-free rewriting systems and deterministic tree-walking transducers. In: Proceedings of the 30th Annual Meeting of the Association for Computational Linguistics, pp. 136–143 (1992)

# Unifying Local and Nonlocal Modelling of Respective and Symmetrical Predicates

Yusuke Kubota[1] and Robert Levine[2]

[1] JSPS, Ohio State University, USA
kubota.7@osu.edu
[2] Ohio State University, USA
levine@ling.ohio-state.edu

**Abstract.** We propose a unified analysis of 'respective' readings of plural and conjoined expressions and the internal readings of symmetrical predicates such as *same* and *different*. The two problems have both been recognized as significant challenges in the literature of syntax and semantics, but so far there is no analysis which captures their close parallel via some uniform mechanism. In fact, the representative compositional analyses of the two phenomena in the current literature (Gawron and Kehler (2004) (G&K) on 'respective' readings and Barker (2007) on symmetrical predicates) look superficially quite different from each other, where one (Barker) employs a movement-like nonlocal mechanism for mediating the dependency between the relevant terms whereas the other (G&K) achieves a similar effect via a chain of local composition operations.

In this paper, we first point out the parallels and interactions between the two phenomena that motivate a unified analysis. We then briefly review G&K's and Barker's analyses and show that the G&K-style analysis can be modelled by the Barker-style analysis once we formulate the relevant rules within an explicit syntax-semantics interface couched in a variant of Type-Logical Categorial Grammar called Hybrid TLCG. After clarifying the hitherto unnoticed formal relations between the Barker-style nonlocal modelling and the G&K-style local modelling by focusing on the analysis of 'respective' readings, we present our unified analysis of 'respective' readings and symmetrical predicates and show how their parallel behaviors and interactions can be systematically accounted for.

**Keywords:** 'respective' reading, symmetrical predicate, categorial grammar, Hybrid Type-Logical Categorial Grammar, coordination, parasitic scope.

## 1 Introduction

The so-called 'respective' readings of plural and conjoined expressions and the internal readings of symmetrical predicates such as *same* and *different* as in (1) have posed difficult challenges to theories of the syntax-semantics interface.

(1) a. John and Bill married Mary and Sue, (respectively).
(= 'John married Mary and Bill married Sue.')
b. John and Bill bought the same book.
(= 'There is a single identical book which both John and Bill bought.')

G. Morrill et al. (Eds.): Formal Grammar 2014, LNCS 8612, pp. 104–120, 2014.

These phenomena interact with coordination, including the 'noncanonical' types of coordination (both Right-Node Raising and Dependent Cluster Coordination):

(2) a. John read, and Bill reviewed, *Barriers* and *LGB*, (respectively).
b. John introduced the same girl to Chris on Thursday and (to) Peter on Friday.

Moreover, these expressions can themselves be iterated and interact with one another to induce multiple dependencies:

(3) a. John and Bill introduced Mary and Sue to Chris and Pat (respectively).
b. John and Bill gave the same book to the same man.
c. John and Bill gave the same book to Mary and Sue (respectively).

Any adequate analysis of these phenomena needs to account for these empirical facts. In particular, the parallel between the phenomena in the multiple dependency cases in (3), especially, the interdependency between 'respective' and symmetrical predicates in (3c), raises the interesting possibility that the same (or a similar) mechanism is at the core of the semantics of these two phenomena.

The present paper has two inter-related goals, one empirical and the other theoretical. The empirical goal is to develop an explicit analysis of 'respective' and symmetrical predicates that systematically accounts for the empirical facts just reviewed above. In particular, we argue that the core mechanism underlying both 'respective' and symmetrical predicates is a pairwise predication that establishes a one-to-one correspondence between elements of two ordered sets of denotata each associated with plural, conjoined or symmetrical terms (i.e. expressions like *the same man*). Formally, we treat such 'ordered sets' by means of tuples, enriching the semantic ontology slightly by introducing product-type elements for semantic objects of any arbitrary type. This enables us to formulate a unified analysis of these phenomena that immediately accounts for the complex yet systematic empirical facts noted above.

The theoretical goal of the paper is to explicitly establish a (hidden) connection between two representative compositional analyses of these phenomena proposed by previous authors: Gawron and Kehler (2004) (G&K) on 'respective' readings and Barker (2007) on symmetrical predicates. G&K's analysis builds on the idea of recursively assigning a tuple-like object as the denotation of phrase containing a plural or a conjoined term at each step of local semantic composition, so that the ordering inherent in the original conjoined or plural term is retained in the larger structure which undergoes pairwise predication. By contrast, Barker (2007) proposes to analyze the semantics of symmetrical predicates in terms of a nonlocal, movement-like process of 'parasitic scope' whereby the symmetrical term (*the same book*) and the plural NP (*John and Bill*) that is related to it are scoped out of their local positions and are treated essentially as an interdependent complex quantifier.

While the strictly local approach by G&K and the nonlocal approach by Barker look superficially quite different, the effects of the two types of operations (or series of operations) that they respectively invoke are rather similar: they both allow one to establish some correspondence between the internal structures of two terms that do not necessarily appear adjacent to each other in the surface form of the sentence. The main difference is *how* this correspondence is established: G&K opt for a series of local composition operations (somewhat reminiscent of the way long-distance dependencies are handled in lexicalist frameworks such as CCG and G/HPSG), whereas Barker does it by a single step of nonlocal mechanism (in a way analogous to a movement-based analysis of long-distance dependencies). But then, is it just an accident that G&K and Barker proposed their respective solutions for the phenomena they were dealing with, or do we need both types of approach, but for different phenomena, or can we unify the two approaches somehow?

We attempt to shed some light on these questions by simulating G&K's and Barker's approaches in Hybrid Type-Logical Categorial Grammar (Hybrid TLCG), a variant of categorial grammar that is notable for its flexible and systematic syntax-semantics interface (Kubota and Levine, 2012, 2013; Kubota, to appear). A comparison of the two approaches in this setting reveals that the G&K-style local modelling of 'respective' predication can be modelled by the Barker-style approach once we recognize one independently needed mechanism for dealing with (non-'respective') distributive predication. We take this result to be highly illuminating, as it once again shows that the logic-based setup of TLCG enables us to gain a deeper insight into the underlying connections between two related empirical phenomena and two apparently different but deeply related approaches to each, which, without such a perspective, may have gone unnoticed.

## 2 Modelling 'Respectively' Readings Locally and Nonlocally

We start by briefly reviewing the key components of G&K's and Barker's analyses. In order to facilitate the comparison (both to each other and to the unified analysis that we present below), we replace sums in their analyses that model complex structured objects with the notion of tuples (which has inherent ordering of elements), but nothing essential in their respective analyses are lost by this adjustment.

### 2.1 Local Modelling of 'Respective' Readings by Gawron and Kehler (2004)

G&K propose to analyze 'respective' readings of sentences like the following via a recursive application of 'respective' and distributive operators:

(4) John and Bill married Mary and Sue, (respectively).

Since they assume a simple phrase structure grammar for syntax, we model it via a simple AB grammar, with the following two rules of /E and \E alone:

(5) a. **Forward Slash Elimination** b. **Backward Slash Elimination**

$$\frac{a;\ \mathcal{F};\ A/B \quad b;\ \mathcal{G};\ B}{a \circ b;\ \mathcal{F}(\mathcal{G});\ A}/\mathrm{E} \qquad \frac{b;\ \mathcal{G};\ B \quad a;\ \mathcal{F};\ B\backslash A}{b \circ a;\ \mathcal{F}(\mathcal{G});\ A}\backslash\mathrm{E}$$

As noted above, we replace their sum-based treatment with a tuple-based treatment, where the two NPs *John and Bill* and *Mary and Sue* both denote tuples (or pairs) of individuals $\langle \mathbf{j}, \mathbf{b} \rangle$ and $\langle \mathbf{m}, \mathbf{s} \rangle$.[1]

The core (empty) semantic operators that G&K exploit are the following dist(ributive) and resp(ective) operators:

(6) $\varepsilon;\ \lambda P \lambda g. \prod_i^n P(\pi_i(g));\ \mathrm{X/X}$

(7) $\varepsilon;\ \lambda F \lambda x. \prod_i^n \pi_i(F)(\pi_i(x));\ \mathrm{X/X}$

There is in addition the following 'boolean reduction' operator which takes a tuple of propositions and returns their boolean conjunction at the S level:

(8) $\varepsilon;\ \lambda p. \bigwedge_i \pi_i(p);\ \mathrm{S|S}$

We can analyze (4) as follows:

(9)

$$\frac{\varepsilon;\ \lambda P \lambda g. \prod_i^n P(\pi_i(g));\ \mathrm{X/X} \quad \mathsf{married};\ \mathbf{marry};\ (\mathrm{NP\backslash S})/\mathrm{NP}}{\mathsf{married};\ \lambda g. \prod_i^n \mathbf{marry}(\pi_i(g));\ (\mathrm{NP\backslash S})/\mathrm{NP}}$$

$$\frac{\mathsf{married};\ \lambda g. \prod_i^n \mathbf{marry}(\pi_i(g));\ (\mathrm{NP\backslash S})/\mathrm{NP} \quad \mathsf{mary \circ and \circ sue};\ \langle \mathbf{m}, \mathbf{s} \rangle;\ \mathrm{NP}}{\mathsf{married \circ mary \circ and \circ sue};\ \prod_i^n \mathbf{marry}(\pi_i(\langle \mathbf{m}, \mathbf{s} \rangle));\ \mathrm{NP\backslash S}}$$

$$\cdots\cdots\cdots$$

$$\mathsf{married \circ mary \circ and \circ sue};\ \langle \mathbf{marry}(\mathbf{m}), \mathbf{marry}(\mathbf{s}) \rangle;\ \mathrm{NP\backslash S}$$

$$\frac{\varepsilon;\ \lambda F \lambda x. \prod_i^n \pi_i(F)(\pi_i(x));\ \mathrm{X/X} \quad \mathsf{married \circ mary \circ and \circ sue};\ \langle \mathbf{marry}(\mathbf{m}), \mathbf{marry}(\mathbf{s}) \rangle;\ \mathrm{NP\backslash S}}{\mathsf{married \circ mary \circ and \circ sue};\ \lambda x. \prod_i^n \pi_i(\langle \mathbf{marry}(\mathbf{m}), \mathbf{marry}(\mathbf{s}) \rangle)(\pi_i(x));\ \mathrm{NP\backslash S}}$$

$$\frac{\mathsf{john \circ and \circ bill};\ \langle \mathbf{j}, \mathbf{b} \rangle;\ \mathrm{NP} \quad \mathsf{married \circ mary \circ and \circ sue};\ \lambda x. \prod_i^n \pi_i(\langle \mathbf{marry}(\mathbf{m}), \mathbf{marry}(\mathbf{s}) \rangle)(\pi_i(x));\ \mathrm{NP\backslash S}}{\mathsf{john \circ and \circ bill \circ married \circ mary \circ and \circ sue};\ \prod_i^n \pi_i(\langle \mathbf{marry}(\mathbf{m}), \mathbf{marry}(\mathbf{s}) \rangle)(\pi_i(\langle \mathbf{j}, \mathbf{b} \rangle));\ \mathrm{S}}$$

$$\cdots\cdots\cdots$$

$$\mathsf{john \circ and \circ bill \circ married \circ mary \circ and \circ sue};\ \langle \mathbf{marry}(\mathbf{m})(\mathbf{j}), \mathbf{marry}(\mathbf{s})(\mathbf{b}) \rangle;\ \mathrm{S}$$

$$\frac{\varepsilon;\ \lambda p. \bigwedge_i \pi_i(p);\ \mathrm{S|S} \quad \mathsf{john \circ and \circ bill \circ married \circ mary \circ and \circ sue};\ \langle \mathbf{marry}(\mathbf{m})(\mathbf{j}), \mathbf{marry}(\mathbf{s})(\mathbf{b}) \rangle;\ \mathrm{S}}{\mathsf{john \circ and \circ bill \circ married \circ mary \circ and \circ sue};\ \mathbf{marry}(\mathbf{m})(\mathbf{j}) \wedge \mathbf{marry}(\mathbf{s})(\mathbf{b});\ \mathrm{S}}$$

The derivation in (11) for (10) illustrates a more complex case involving a recursive application of the 'respective' operator (here, **dist** and **resp** abbreviate the semantic translations of the two operators in (6) and (7)).

---

[1] This also removes G&K's ontological commitment of taking propositions rather than worlds as primitives (which is necessary for them since sums of two extensionally identical properties in the Montagovian setup collapse to a single property). While such a position is not necessarily implausible, we do not think that the semantics of respective readings should be taken to form a basis for this ontological choice.

(10)  John and Mary drove to Berkeley and Santa Cruz on Monday and Tuesday.

(11)

$\varepsilon$; on;
dist; on;
X/X    (VP\VP)/NP    mon $\circ$ and $\circ$ tue;
$\langle$m, t$\rangle$; NP

$\varepsilon$;  drove;
dist;  drive;
X/X   VP/PP    to $\circ$ bkl $\circ$ and $\circ$ sc;
$\langle$b, s$\rangle$; PP

drove;
$\lambda g. \prod_i^n$ drive$(\pi_i(g))$; VP/PP

$\varepsilon$;
resp; X/X

on;
$\lambda g. \prod_i^n$ on$(\pi_i(g))$; (VP\VP)/NP

on $\circ$ mon $\circ$ and $\circ$ tue;
$\langle$on(m), on(t)$\rangle$; VP\VP

drove $\circ$ to $\circ$ bkl $\circ$ and $\circ$ sc;
$\langle$drive(b), drive(s)$\rangle$; VP

on $\circ$ mon $\circ$ and $\circ$ tue;
$\lambda g. \prod_i^n \pi_i(\langle$on(m), on(t)$\rangle)(\pi_i(g))$; VP\VP

john $\circ$ and $\circ$ mary;
$\langle$j, m$\rangle$; NP

$\varepsilon$;
resp; X/X

drove $\circ$ to $\circ$ bkl $\circ$ and $\circ$ sc $\circ$ on $\circ$ mon $\circ$ and $\circ$ tue;
$\langle$on(m)(drive(b)), on(t)(drive(s))$\rangle$; VP

drove $\circ$ to $\circ$ bkl $\circ$ and $\circ$ sc $\circ$ on $\circ$ mon $\circ$ and $\circ$ tue;
$\lambda x. \prod_i^n \pi_i(\langle$on(m)(drive(b)), on(t)(drive(s))$\rangle)(\pi_i(x))$; VP

john $\circ$ and $\circ$ mary $\circ$ drove $\circ$ to $\circ$ bkl $\circ$ and $\circ$ sc $\circ$ on $\circ$ mon $\circ$ and $\circ$ tue;
$\prod_i^n \pi_i(\langle$on(m)(drive(b)), on(t)(drive(s))$\rangle)(\pi_i(\langle$j, m$\rangle))$; S

Note that at each step where a functor takes a product-type term as an argument, the dist operator is first applied to the functor so that the functor is distributively applied to each member of the tuple and the result is 'summed up' as a tuple (rather than conjoined by a generalized conjunction operator as in the standard definition of the distributive operator in the plurality literature). Thus, the larger constituent inherits the ordering of elements in its subconstituent.

Another notable property of G&K's analysis is that after the application of the resp operator, the larger constituent still denotes a tuple (of two properties of type $e \to t$, in the case of (11)), rather than boolean conjunction. This is crucial for making the recursive application of the resp operator straightforward. Since the tuple structure is preserved after the application of the first resp operator, the result can simply be taken up by another resp operator which relates it to another tuple in a 'respective' manner.

Although G&K does not discuss this point explicitly, in order to generalize this analysis to cases like the following in which the tuple structure is percolated from the functor rather than the argument, one either needs to assume that type-raising is generally available in the grammar so that the functor-argument relation of any arbitrary pair of functor and argument types can be flipped, or else needs to introduce another version of the dist operator, call it dist', which distributes a single argument meaning to a tuple of functor meanings.[2]

(12)   a.  John and Bill read and reviewed the book, respectively.

---

[2] G&K speculate on a possibility of unifying their dist and resp operators toward the end of their paper; if this unification is successfully done, both the argument-distributing dist operator in (6) above and the functor-distributing dist' operator under discussion here might be thought of as special cases of a single unified 'predication' operator. But this part of their proposal remains somewhat obscure and not worked out in full detail.

b. John and Bill sent the bomb and the letter to the president yesterday, respectively.

Essentially, at the expense of applying either the dist or resp operator at each step of local composition, G&K does away with hypothetical reasoning entirely and their fragment can be modelled by a simple AB grammar.

## 2.2 Nonlocal Modeling of 'Respective' Readings Building on Barker (2007)

In contrast to G&K, Barker (2007) extensively relies on hypothetical reasoning for characterizing the semantics of symmetrical predicates. In order to facilitate a comparison with G&K's analysis, we first discuss an extension of Barker's approach to 'respective' readings (it should be noted that Barker himself confines his analysis to the case of symmetrical predicates, mostly focusing on the analysis of *same*), and come back to the case of symmetrical predicates in the next section.

The key idea behind Barker's proposal is that the interdependency between the relevant two complex terms (i.e. the two plural or conjoined terms in the case of 'respective' readings) can be straightforwardly mediated by abstracting over the positions in the sentence that such terms occupy and then directly giving the relevant terms (and the abstracted proposition) as arguments to the operator that mediates their interdependency.

For modelling this 'covert' movement treatment of 'respective'/symmetrical predicates, we introduce here a new connective |, called 'vertical slash', together with the Elimination and Introduction rules for it formulated in (21) (just like /, we write the argument to the right for this slash; thus, in $A|B$, $B$ is the argument).[3]

(13) a. **Vertical Slash Introduction** b. **Vertical Slash Elimination**

$$\dfrac{\dfrac{\begin{matrix} \vdots & \vdots & \underline{[\varphi; x; A]^n} & \vdots & \vdots \\ \vdots & \vdots & \vdots \quad \vdots & \vdots & \vdots \end{matrix}}{b; \mathcal{F}; B}}{\lambda\varphi.b; \lambda x.\mathcal{F}; B|A}\,|\mathrm{I}^n \qquad\qquad \dfrac{a; \mathcal{F}; A|B \quad b; \mathcal{G}; B}{a(b); \mathcal{F}(\mathcal{G}); A}\,|\mathrm{E}$$

These rules are essentially the same as the rules for the linear implication connective ($\multimap$) posited in the family of 'Linear Categorial Grammars' (Oehrle, 1994; de Groote, 2001; Muskens, 2003; Mihaliček and Pollard, 2012).

[3] These rules introduce functional prosodic objects. One might wonder how the grammar (or the prosodic calculus that is part of it) is constrained in such a way that it does not admit of uninterpretable prosodic objects such as $\mathsf{john} \circ \lambda\varphi.\varphi$ (i.e., 'concatenation' of a string and a function from strings to strings). In fact, Hybrid TLCG does not admit of any such ill-formed prosodic objects. Such an expression would be obtained only by applying a functor that has a syntactic type of the form X/(Y|Z) to its argument Y|Z, but a syntactic type of the form X/(Y|Z), with the vertical slash 'under' a directional slash, are explicitly excluded from the grammar. For the details of the syntax-prosody mapping which ensures this, see Kubota and Levine (2014).

With this vertical slash, extending Barker's 'parasitic scope' analysis to 'respective' readings is in fact mostly straightforward, with one extra complication discussed below. Assuming (as above) that plural and conjoined terms denote tuples (of the relevant type of semantic objects), we just need the following three-place 'respective' operator which semantically takes a relation (denoted by the sentence containing the two 'gap' positions for the two product-type terms) and two tuples as arguments and returns a tuple as an output (this is so that, as above, multiple 'respective' readings can be handled by recursive application of this operator).

(14) $\lambda\sigma_0\lambda\varphi_1\lambda\varphi_2.\sigma_0(\varphi_1)(\varphi_2)$; **resp3**; $(Z|X|Y)|(Z|X|Y)$

As can be seen in (14), the resp operator is a (polymorphic) identity function both syntactically and prosodically. The semantics is unpacked in (15).

(15) $\mathbf{resp3} = \lambda\mathcal{R}\lambda\mathcal{T}_{\times}\lambda\mathcal{U}_{\times}.\prod_i \mathcal{R}(\pi_i(\mathcal{T}_{\times}))(\pi_i(\mathcal{U}_{\times}))$

Semantically, this operator relates the elements of the two tuples in a pairwise manner with respect to the relation in question. Note that this three place **resp3** operator is distinct from the two place **resp** operator posited in G&K's approach though their semantic effects are similar. We come back to the relationship between these two operators immediately below (see (19)).

The analysis of the simple 'respective' sentence is then straightforward:

(16)

$$\dfrac{\mathsf{married};\ \mathbf{marry};\ (\mathrm{NP}\backslash\mathrm{S})/\mathrm{NP}}{\vdots\quad\vdots}$$

$$\dfrac{\lambda\sigma_0\lambda\varphi_1\lambda\varphi_2.\sigma_0(\varphi_1)(\varphi_2);\ \mathbf{resp3};\ (Z|X|Y)|(Z|X|Y) \qquad \lambda\varphi_3\lambda\varphi_4.\varphi_4 \circ \mathsf{married} \circ \varphi_3;\ \mathbf{marry};\ (\mathrm{S}|\mathrm{NP})|\mathrm{NP}}{\lambda\varphi_1\lambda\varphi_2.\varphi_2 \circ \mathsf{married} \circ \varphi_1;\ \mathbf{resp3(marry)};\ \mathrm{S}|\mathrm{NP}|\mathrm{NP}}$$

$$\dfrac{\mathsf{john} \circ \mathsf{and} \circ \mathsf{bill};\ \langle \mathbf{j}, \mathbf{b}\rangle;\ \mathrm{NP} \qquad \lambda\varphi_1\lambda\varphi_2.\varphi_2 \circ \mathsf{married} \circ \varphi_1;\ \mathbf{resp3(marry)};\ \mathrm{S}|\mathrm{NP}|\mathrm{NP}}{\lambda\varphi_2.\varphi_2 \circ \mathsf{married} \circ \mathsf{john} \circ \mathsf{and} \circ \mathsf{bill};\ \mathbf{resp3(marry)}(\langle \mathbf{j}, \mathbf{b}\rangle);\ \mathrm{S}|\mathrm{NP}}$$

$$\dfrac{\mathsf{mary} \circ \mathsf{and} \circ \mathsf{sue};\ \langle \mathbf{m}, \mathbf{s}\rangle;\ \mathrm{NP} \qquad \lambda\varphi_2.\varphi_2 \circ \mathsf{married} \circ \mathsf{john} \circ \mathsf{and} \circ \mathsf{bill};\ \mathbf{resp3(marry)}(\langle \mathbf{j}, \mathbf{b}\rangle);\ \mathrm{S}|\mathrm{NP}}{\mathsf{mary} \circ \mathsf{and} \circ \mathsf{sue} \circ \mathsf{married} \circ \mathsf{john} \circ \mathsf{and} \circ \mathsf{bill};\ \mathbf{resp3(marry)}(\langle \mathbf{j}, \mathbf{b}\rangle)(\langle \mathbf{m}, \mathbf{s}\rangle);\ \mathrm{S}}$$

$$\cdots\cdots\cdots\cdots\cdots\cdots\cdots\cdots\cdots$$

$$\mathsf{mary} \circ \mathsf{and} \circ \mathsf{sue} \circ \mathsf{married} \circ \mathsf{john} \circ \mathsf{and} \circ \mathsf{bill};\ \langle \mathbf{marry(j)(m)}, \mathbf{marry(b)(s)}\rangle;\ \mathrm{S}$$

$$\dfrac{\lambda\varphi_1.\varphi_1;\ \lambda p.\bigwedge_i \pi_i(p);\ \mathrm{S}|\mathrm{S} \qquad \mathsf{mary} \circ \mathsf{and} \circ \mathsf{sue} \circ \mathsf{married} \circ \mathsf{john} \circ \mathsf{and} \circ \mathsf{bill};\ \langle \mathbf{marry(j)(m)}, \mathbf{marry(b)(s)}\rangle;\ \mathrm{S}}{\mathsf{mary} \circ \mathsf{and} \circ \mathsf{sue} \circ \mathsf{married} \circ \mathsf{john} \circ \mathsf{and} \circ \mathsf{bill};\ \mathbf{marry(j)(m)} \wedge \mathbf{marry(b)(s)};\ \mathrm{S}}$$

Just as in G&K's analysis, multiple 'respective' readings in examples like the following are obtained via recursive application of the **resp** operator:

(17) Tolstoy and Dostoevsky sent Anna Karenina and the Idiot to Dickens and Thackeray (respectively).

The analysis is in fact straightforward. After two of the tuple-denoting terms are related to each other with respect to the predicate denoted by the verb, the resultant S|NP expression is a tuple of two properties.

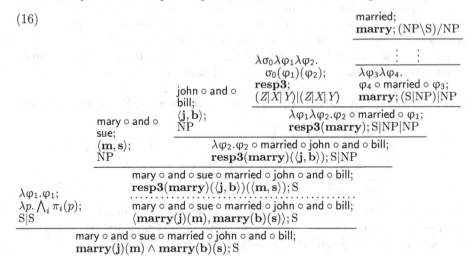

(18) $\Longrightarrow$

$$\dfrac{\dfrac{\dfrac{\begin{array}{l}\lambda\sigma_0\lambda\varphi_1\lambda\varphi_2.\sigma_0(\varphi_1)(\varphi_2);\\ \mathbf{resp3};\\ (Z|X|Y)|(Z|X|Y)\end{array} \qquad \begin{array}{l}\vdots\\ \lambda\varphi_1\lambda\varphi_2\lambda\varphi_3.\\ \varphi_3 \circ \mathsf{sent} \circ \varphi_1 \circ \mathsf{to} \circ \varphi_2;\\ \mathbf{send}; \mathrm{S|NP|NP|NP}\end{array}}{\begin{array}{l}\mathsf{AK} \circ \mathsf{and} \circ \mathsf{Id};\\ \langle \mathbf{ak}, \mathbf{id}\rangle; \mathrm{NP}\end{array} \quad \begin{array}{l}\lambda\varphi_1\lambda\varphi_2\lambda\varphi_3.\varphi_3 \circ \mathsf{sent} \circ \varphi_1 \circ \mathsf{to} \circ \varphi_2;\\ \mathbf{resp3}(\mathbf{send}); \mathrm{S|NP|NP|NP}\end{array}}}{\begin{array}{l}\mathsf{Di} \circ \mathsf{and} \circ \mathsf{Th};\\ \langle \mathbf{di}, \mathbf{th}\rangle; \mathrm{NP}\end{array} \quad \begin{array}{l}\lambda\varphi_2\lambda\varphi_3.\varphi_3 \circ \mathsf{sent} \circ \mathsf{AK} \circ \mathsf{and} \circ \mathsf{Id} \circ \mathsf{to} \circ \varphi_2;\\ \mathbf{resp3}(\mathbf{send})(\langle \mathbf{ak}, \mathbf{id}\rangle); \mathrm{S|NP|NP}\end{array}}}{\begin{array}{l}\lambda\varphi_3.\varphi_3 \circ \mathsf{sent} \circ \mathsf{AK} \circ \mathsf{and} \circ \mathsf{Id} \circ \mathsf{to} \circ \mathsf{Di} \circ \mathsf{and} \circ \mathsf{Th};\\ \mathbf{resp3}(\mathbf{send})(\langle \mathbf{ak}, \mathbf{id}\rangle)(\langle \mathbf{di}, \mathbf{th}\rangle); \mathrm{S|NP}\\ \cdots\cdots\cdots\cdots\cdots\cdots\cdots\cdots\cdots\cdots\\ \lambda\varphi_3.\varphi_3 \circ \mathsf{sent} \circ \mathsf{AK} \circ \mathsf{and} \circ \mathsf{Id} \circ \mathsf{to} \circ \mathsf{Di} \circ \mathsf{and} \circ \mathsf{Th};\\ \langle \mathbf{send}(\mathbf{ak})(\mathbf{di}), \mathbf{send}(\mathbf{id})(\mathbf{th})\rangle; \mathrm{S|NP}\end{array}}$$

And the remaining conjoined term $\langle \mathbf{To}, \mathbf{Do}\rangle$ is related to this product-type property in the following way:

(19)

$$\dfrac{\begin{array}{l}\lambda\rho\lambda\sigma\lambda\varphi.\rho(\sigma)(\varphi);\\ \lambda\mathcal{R}\lambda\mathcal{T}_\times\lambda\mathcal{U}_\times.\prod_i \mathcal{R}(\pi_i(\mathcal{T}_\times))(\pi_i(\mathcal{U}_\times));\\ (Z|X|Y)|(Z|X|Y)\end{array} \qquad \dfrac{\dfrac{\dfrac{[\sigma; f; \mathrm{S|NP}]^1 \quad [\varphi; x; \mathrm{NP}]^2}{\sigma(\varphi);\ f(x);\ \mathrm{S}}|\mathrm{E}}{\lambda\varphi.\sigma(\varphi);\ \lambda x.f(x);\ \mathrm{S|NP}}|\mathrm{I}^2}{\lambda\sigma\lambda\varphi.\sigma(\varphi);\ \lambda f\lambda x.f(x);\ (\mathrm{S|NP})|(\mathrm{S|NP})}|\mathrm{I}^1}{\lambda\sigma_1\lambda\varphi_1.\sigma_1(\varphi_1);\ \lambda P_\times\lambda X_\times.\prod_i \pi_i(P_\times)(\pi_i(X_\times));\ (\mathrm{S|NP})|(\mathrm{S|NP})}|\mathrm{E}$$

$$\dfrac{\begin{array}{l}\mathsf{To} \circ \mathsf{and} \circ \mathsf{Do};\\ \langle \mathbf{to}, \mathbf{do}\rangle;\\ \mathrm{NP}\end{array} \quad \dfrac{\begin{array}{l}\vdots\\ \lambda\sigma_2\lambda\varphi_2.\sigma_2(\varphi_2);\\ \lambda P_\times\lambda X_\times.\\ \quad\prod_i \pi_i(P_\times)(\pi_i(X_\times));\\ (\mathrm{S|NP})|(\mathrm{S|NP})\end{array} \qquad \begin{array}{l}\vdots\\ \lambda\varphi_3.\varphi_3 \circ \mathsf{sent} \circ \mathsf{AK} \circ \mathsf{and} \circ \mathsf{Id} \circ\\ \mathsf{to} \circ \mathsf{Di} \circ \mathsf{and} \circ \mathsf{Th};\\ \langle \mathbf{send}(\mathbf{ak})(\mathbf{di}), \mathbf{send}(\mathbf{id})(\mathbf{th})\rangle; \mathrm{S|NP}\end{array}}{\begin{array}{l}\lambda\varphi_2.\varphi_2 \circ \mathsf{sent} \circ \mathsf{AK} \circ \mathsf{and} \circ \mathsf{Id} \circ \mathsf{to} \circ \mathsf{Di} \circ \mathsf{and} \circ \mathsf{Th};\\ \lambda Y_\times.\prod_i \pi_i(\langle \mathbf{send}(\mathbf{ak})(\mathbf{di}), \mathbf{send}(\mathbf{id})(\mathbf{th})\rangle)(\pi_i(Y_\times)); \mathrm{S|NP}\end{array}}|\mathrm{E}}{\begin{array}{l}\mathsf{To} \circ \mathsf{and} \circ \mathsf{Do} \circ \mathsf{sent} \circ \mathsf{AK} \circ \mathsf{and} \circ \mathsf{Id} \circ \mathsf{to} \circ \mathsf{Di} \circ \mathsf{and} \circ \mathsf{Th};\\ \prod_i \pi_i(\langle \mathbf{send}(\mathbf{ak})(\mathbf{di}), \mathbf{send}(\mathbf{i})(\mathbf{th})\rangle)(\pi_i(\langle \mathbf{to}, \mathbf{do}\rangle)); \mathrm{S}\\ \cdots\cdots\cdots\cdots\cdots\cdots\cdots\cdots\cdots\cdots\\ \mathsf{To} \circ \mathsf{and} \circ \mathsf{Do} \circ \mathsf{sent} \circ \mathsf{AK} \circ \mathsf{and} \circ \mathsf{Id} \circ \mathsf{to} \circ \mathsf{Di} \circ \mathsf{and} \circ \mathsf{Th};\\ \langle \mathbf{send}(\mathbf{ak})(\mathbf{di})(\mathbf{to}), \mathbf{send}(\mathbf{id})(\mathbf{th})(\mathbf{do})\rangle; \mathrm{S}\end{array}}|\mathrm{E}$$

The first chunk of derivation in (19) (where $X$ is instantiated as NP, $Y$ as S|NP, and $Z$ as S), the point of which may not be immediately clear, can be thought of as a way of deriving the two-place **resp** operator (identical to the one that G&K posit) from the lexically specified three-place **resp3** operator introduced above. As in G&K's analysis discussed above, the two place **resp** operator directly relates the product-type property (of type S|NP) derived in (18) with the product-type NP occupying the subject position via function application of the corresponding elements.

## 3 Comparison of Local Modelling and Nonlocal Modelling

We now show that the G&K-style 'local' modelling of 'respective' predication can be simulated by the Barker-style 'nonlocal' approach. Consider first a case which contains only two product-type terms to be related in a 'respective' manner. The structure of the derivation for a sentence containing two product-type terms in the G&K-style analysis can be schematically shown in (20), where $i, j, n, m \geq 0, n \geq i, m \geq j$ and $l \geq 2$ and for any $k$, $\gamma_k$ or $\boldsymbol{\delta}_k$ is some linguistic sign.[4] Note here that both $\Psi$ and $\Phi$, which are meanings of expressions that contain exactly one tuple-denoting (lexical) term inside themselves, denote tuples, and they are then related by the two-place **resp** operator with each other.

(20)

$$\dfrac{\dfrac{\varepsilon;\ \mathbf{resp};\ \mathrm{Z/Z} \qquad \dfrac{\gamma_1 \ldots \gamma_i \quad \dfrac{\mathsf{a};\ \langle a_1, \ldots, a_l \rangle;\ \mathrm{A}}{\vdots\ \vdots} \quad \gamma_{i+1} \ldots \gamma_n}{\mathsf{c};\ \Psi;\mathrm{X/Y}}}{\mathsf{c};\ \mathbf{resp}(\Psi);\ \mathrm{X/Y}} \qquad \dfrac{\boldsymbol{\delta}_1 \ldots \boldsymbol{\delta}_j \quad \dfrac{\mathsf{b};\ \langle b_1, \ldots, b_l \rangle;\ \mathrm{B}}{\vdots\ \vdots} \quad \boldsymbol{\delta}_{j+1} \ldots \boldsymbol{\delta}_m}{\mathsf{d};\ \Phi;\mathrm{Y}}}{\mathsf{c} \circ \mathsf{d};\ \mathbf{resp}(\Psi)(\Phi);\ \mathrm{X}}$$

We derive two auxiliary rules in G&K's system to facilitate the comparison to the Barker-style analysis.[5]

(21) a. **Rule 1**

$$\dfrac{\mathsf{a};\ f;\ \mathrm{A/B} \quad \mathsf{b};\ \langle a_1 \ldots a_l \rangle;\ \mathrm{B}}{\mathsf{a} \circ \mathsf{b};\ \langle f(a_1) \ldots f(a_l) \rangle;\ \mathrm{A}}$$

b. **Rule 2**

$$\dfrac{\mathsf{a};\ \langle f_1 \ldots f_n \rangle;\ \mathrm{A/B} \quad \mathsf{b};\ a;\ \mathrm{B}}{\mathsf{a} \circ \mathsf{b};\ \langle f_1(a) \ldots f_n(a) \rangle;\ \mathrm{A}}$$

Rule 1 is obtained by applying the dist operator (6) to the functor $f$ and then applying it to its tuple argument. Rule 2 is obtained by applying the dist' operator discussed above (see the discussion pertaining to (12)) to the argument $a$ and applying it to the tuple functor. (We remain agnostic about how dist' is obtained in G&K's setup.) These two rules are introduced here just for expository ease. We show below how they can be derived from the more general **resp3** operator in the present setup with the use of hypothetical reasoning by introducing one auxiliary rule converting an atomic object to an $n$-tuple of identical objects.

By assumption, among the signs $\gamma_1 \ldots \gamma_n$, $\boldsymbol{\delta}_1 \ldots \boldsymbol{\delta}_n$, and a and b constituting the leaves of (20), only a and b have product-type meanings. Thus, at each step of local composition inside c and d, either the functor or the argument (but not both) has a product-type meaning. From this it further follows that each local step of composition inside c and d instantiates either Rule 1 or Rule 2.

Now, consider a structure in which we replace the two product-type terms in (20) by the variables $x$ and $y$, both fresh in $\Psi$ and $\Phi$.

[4] We assume here that the lefthand substructure is the functor. The same result obtains for a structure where the righthand substructure is the functor, by merely replacing the linear order between c and d in (20).

[5] Here we are inspired by Bekki's (2006) reformulation of G&K's analysis in terms of product-types.

(22)
$$
\dfrac{\dfrac{\dfrac{\gamma_1 \ldots \gamma_i \quad \underline{\varphi_1; x; \mathrm{A}} \quad \gamma_{i+1} \ldots \gamma_n}{\vdots}}{\mathsf{c}';\ \Gamma; \mathrm{X/Y}} \qquad \dfrac{\dfrac{\delta_1 \ldots \delta_j \quad \underline{\varphi_2; y; \mathrm{B}} \quad \delta_{i+1} \ldots \delta_m}{\vdots}}{\mathsf{d}';\ \Delta; \mathrm{Y}}}{\mathsf{c}' \circ \mathsf{d}';\ \Gamma(\Delta);\ \mathrm{X}}
$$

The relation between the internal structures of (22) and (20) is such that each step of function application in (22) is replaced by an application of either Rule 1 or Rule 2 in (20). Thus, by induction,[6]

(23) $\Psi = \langle \Gamma[x/a_1], \ldots, \Gamma[x/a_l] \rangle$

(where $\Gamma[x/a_k]$ is a term identical to $\Gamma$ except that all occurrences of $x$ in $\Gamma$ are replaced by $a_k$). Similarly,

(24) $\Phi = \langle \Delta[y/b_1], \ldots, \Delta[y/b_l] \rangle$

Thus,

(25)
$$
\begin{aligned}
&\mathbf{resp}(\Phi)(\Psi) \\
&= \mathbf{resp}(\langle \Gamma[x/a_1], \ldots, \Gamma[x/a_l] \rangle)(\langle \Delta[y/b_1], \ldots, \Delta[y/b_l] \rangle) \\
&= \langle \Gamma[x/a_1](\Delta[y/b_1]), \ldots, \Gamma[x/a_l](\Delta[y/b_l]) \rangle
\end{aligned}
$$

This is exactly the same interpretation that we obtain in the following Barker-style analysis of the same string of words:

(26)
$$
\dfrac{\dfrac{\dfrac{\dfrac{\dfrac{\gamma_1 \ldots \gamma_i \quad \underline{\begin{bmatrix}\varphi_1; \\ x; \\ \mathrm{A}\end{bmatrix}^1} \quad \gamma_{i+1} \ldots \gamma_n}{\vdots}}{\mathsf{c}';\ \Gamma; \mathrm{X/Y}} \qquad \dfrac{\dfrac{\delta_1 \ldots \delta_j \quad \underline{\begin{bmatrix}\varphi_2; \\ y; \\ \mathrm{B}\end{bmatrix}^2} \quad \delta_{j+1} \ldots \delta_m}{\vdots}}{\mathsf{d}';\ \Delta; \mathrm{Y}}}{\mathsf{c}' \circ \mathsf{d}';\ \Gamma(\Delta);\ \mathrm{X}}}{\lambda\varphi_2.\mathsf{c}' \circ \mathsf{d}';\ \lambda y.\Gamma(\Delta);\ \mathrm{X|B}}\ |\mathrm{I}^2}{\lambda\varphi_1\lambda\varphi_2.\mathsf{c}' \circ \mathsf{d}';\ \lambda x\lambda y.\Gamma(\Delta);\ \mathrm{X|B|A}}\ |\mathrm{I}^1
$$

$$
\dfrac{\dfrac{\dfrac{\begin{matrix}\vdots \\ \lambda\varphi_1\lambda\varphi_2.\mathsf{c}' \circ \mathsf{d}';\ \lambda x\lambda y.\Gamma(\Delta);\ \mathrm{X|B|A}\end{matrix} \qquad \begin{matrix}\lambda\varphi.\varphi; \\ \mathbf{resp3}; \\ (\mathrm{Z|X|Y})|(\mathrm{Z|X|Y})\end{matrix}}{\lambda\varphi_1\lambda\varphi_2.\mathsf{c}' \circ \mathsf{d}';\ \mathbf{resp3}(\lambda x\lambda y.\Gamma(\Delta));\ \mathrm{X|B|A}} \qquad \begin{matrix}\mathsf{a}; \\ \langle a_1, \ldots, a_l \rangle; \\ \mathrm{A}\end{matrix}}{\lambda\varphi_2.\mathsf{c} \circ \mathsf{d}';\ \mathbf{resp3}(\lambda x\lambda y.\Gamma(\Delta))(\langle a_1, \ldots, a_l \rangle);\ \mathrm{X|B}} \qquad \begin{matrix}\mathsf{b}; \\ \langle b_1, \ldots, b_l \rangle; \\ \mathrm{B}\end{matrix}}{\mathsf{c} \circ \mathsf{d};\ \mathbf{resp3}(\lambda x\lambda y.\Gamma(\Delta))(\langle a_1, \ldots, a_l \rangle)(\langle b_1, \ldots, b_l \rangle);\ \mathrm{X}}
$$

The final translation we obtain in this derivation is

(27) $\mathbf{resp3}(\lambda x\lambda y.\Gamma(\Delta))(\langle a_1, \ldots, a_l \rangle)(\langle b_1, \ldots, b_l \rangle)$

[6] See Appendix for a formal proof.

Since $|$ is linear, $x$ is fresh in $\Delta$ and $y$ in $\Gamma$. Thus, for any $k$, $\lambda x \lambda y.[\Gamma(\Delta)](a_k)(b_k) = \Gamma[x/a_k](\Delta[y/b_k])$. From this it follows that

(28) $$\begin{aligned}&\mathbf{resp3}(\lambda x\lambda y.\Gamma(\Delta))(\langle a_1,\ldots,a_l\rangle)(\langle b_1,\ldots,b_l\rangle)\\ &= \langle \Gamma[x/a_1](\Delta[y/b_1]),\ldots,\Gamma[x/a_l](\Delta[y/b_l])\rangle\end{aligned}$$

For cases containing more than two respective terms, the correspondence between the G&K-style analysis and the Barker style analysis can be established recursively by taking the whole structure (20)/(26) to instantiate either a or b and relating it to the next 'adjacent' product-type term one by one.

It now remains to show how Rule 1 and Rule 2 can be derived in the Barker-style setup. For this, we need a mechanism that derives the two dist operators in the G&K setup from the three-place resp operator posited in the Barker system in (15). Following Bekki (2006), we assume that the following 'product duplicator' is responsible for this operation, which takes some term $x$ and returns an $n$-tuple consisting of $x$: $\langle x,\ldots,x\rangle$:

(29) $\lambda\varphi.\varphi;\ \lambda x.\prod_i^n x;\ \mathrm{X|X}$

With this operator and the three-place resp operator in (15), Rule 1 and Rule 2 can be derived as follows:

(30)
$$\dfrac{\dfrac{\begin{matrix}\lambda\varphi.\varphi;\\ \lambda x.\prod_i^l x;\\ \mathrm{X|X}\end{matrix}\quad \begin{matrix}\mathsf{a};\\ f;\mathrm{A/B}\end{matrix}}{\mathsf{a};\ \langle f,\ldots,f\rangle;\ \mathrm{A/B}}\qquad \dfrac{\begin{matrix}\mathsf{b};\\ \langle a_1,\ldots,a_l\rangle;\\ \mathrm{B}\end{matrix}\quad \dfrac{\begin{matrix}\lambda\sigma_0\lambda\varphi_1\lambda\varphi_2.\\ \sigma_0(\varphi_1)(\varphi_2);\\ \mathbf{resp3};\\ (\mathrm{Z|X|Y})|(\mathrm{Z|X|Y})\end{matrix}\quad \dfrac{\dfrac{\dfrac{\begin{bmatrix}\varphi_1;\\ f;\mathrm{A/B}\end{bmatrix}^1\ \begin{bmatrix}\varphi_2;\\ x;\mathrm{B}\end{bmatrix}^2}{\varphi_1\circ\varphi_2;\ f(x);\ \mathrm{A}}}{\begin{matrix}\lambda\varphi_1.\varphi_1\circ\varphi_2;\\ \lambda f.f(x);\mathrm{A|(A/B)}\end{matrix}}|\mathrm{I}^1}{\begin{matrix}\lambda\varphi_2\lambda\varphi_1.\varphi_1\circ\varphi_2;\\ \lambda x\lambda f.f(x);\mathrm{A|(A/B)|B}\end{matrix}}|\mathrm{I}^2}{\lambda\varphi_1\lambda\varphi_2.\varphi_2\circ\varphi_1;\ \mathbf{resp3}(\lambda x\lambda f.f(x));\ \mathrm{A|(A/B)|B}}}{\lambda\varphi_2.\varphi_2\circ\mathsf{b};\ \mathbf{resp3}(\lambda x\lambda f.f(x))(\langle a_1,\ldots,a_l\rangle);\ \mathrm{A|(A/B)}}}{\dfrac{\mathsf{a}\circ\mathsf{b};\ \mathbf{resp3}(\lambda x\lambda f.f(x))(\langle a_1,\ldots,a_l\rangle)(\langle f,\ldots,f\rangle);\ \mathrm{A}}{\mathsf{a}\circ\mathsf{b};\ \langle f(a_1),\ldots,f(a_l)\rangle;\ \mathrm{A}}\ (\text{dotted})}$$

(31)
$$\dfrac{\begin{matrix}\mathsf{a};\\ \langle f_1,\ldots,f_l\rangle;\\ \mathrm{A/B}\end{matrix}\quad \dfrac{\dfrac{\begin{matrix}\lambda\varphi.\varphi;\\ \lambda x.\prod_i^l x;\\ \mathrm{X|X}\end{matrix}\quad \begin{matrix}\mathsf{b};\\ a;\mathrm{B}\end{matrix}}{\mathsf{b};\ \langle a,\ldots,a\rangle;\ \mathrm{B}}\quad \begin{matrix}\vdots\ \ \vdots\\ \lambda\varphi_1\lambda\varphi_2.\varphi_2\circ\varphi_1;\\ \mathbf{resp3}(\lambda x\lambda f.f(x));\mathrm{A|(A/B)|B}\end{matrix}}{\lambda\varphi_2.\varphi_2\circ\mathsf{b};\ \mathbf{resp3}(\lambda x\lambda f.f(x))(\langle a,\ldots,a\rangle);\ \mathrm{A|(A/B)}}}{\dfrac{\mathsf{a}\circ\mathsf{b};\ \mathbf{resp3}(\lambda x\lambda f.f(x))(\langle a,\ldots,a\rangle)(\langle f_1,\ldots,f_l\rangle);\ \mathrm{A}}{\mathsf{a}\circ\mathsf{b};\ \langle f_1(a),\ldots,f_l(a)\rangle;\ \mathrm{A}}\ (\text{dotted})}$$

## 4 Extension of the Analysis

In this section, we extend the above analysis in two ways. We first show that, by enriching the calculus with rules for hypothetical reasoning for directional slashes / and \, the interaction between 'respective' readings and nonconstituent coordination exemplified by the data such as (2) straightforwardly falls out. We then extend the tuple-based analysis to symmetrical predicates and show that this analysis immediately extends to multiple dependencies among symmetrical and 'respective' predicates observed in (3).

For the analysis of NCC, we add the following Introduction rules for directional slashes / and \:

(32) a. **Forward Slash Introduction**

$$\dfrac{\dfrac{\begin{array}{ccc} \vdots\ \vdots & \underline{[\varphi;\, x;\, A]^n} & \vdots\ \vdots \\ \vdots\ \vdots & \vdots\ \vdots & \vdots\ \vdots \end{array}}{b \circ \varphi;\, \mathcal{F};\, B}}{b;\, \lambda x.\mathcal{F};\, B/A}\,/\mathrm{I}^n$$

b. **Backward Slash Introduction**

$$\dfrac{\dfrac{\begin{array}{ccc} \vdots\ \vdots & \underline{[\varphi;\, x;\, A]^n} & \vdots\ \vdots \\ \vdots\ \vdots & \vdots\ \vdots & \vdots\ \vdots \end{array}}{\varphi \circ b;\, \mathcal{F};\, B}}{b;\, \lambda x.\mathcal{F};\, A\backslash B}\,\backslash\mathrm{I}^n$$

In TLCG, dependent cluster coordination is analyzed by directly analyzing the apparent nonconstituents that are coordinated in examples like (33) to be a (higher-order) derived constituent, via hypothetical reasoning.

(33) I lent *Syntactic Structures* and *Barriers* to Robin on Thursday and to Mary on Friday, respectively.

Specifically, by hypothesizing the verb and the direct object and withdrawing them after a whole VP is formed, the string *to Robin on Thursday* can be analyzed as a constituent of type NP\(VP/PP/NP)\VP:

(34)

$$\dfrac{\dfrac{\dfrac{\dfrac{\dfrac{[\varphi_1; P; \mathrm{VP/PP/NP}]^1 \quad [\varphi_2; x; \mathrm{NP}]^2}{\varphi_1 \circ \varphi_2;\, P(x);\, \mathrm{VP/PP}}/\mathrm{E} \quad \begin{array}{c}\mathsf{to} \circ \mathsf{robin};\\ \mathbf{r}; \mathrm{PP}\end{array}}{\varphi_1 \circ \varphi_2 \circ \mathsf{to} \circ \mathsf{robin};\, P(x)(\mathbf{r});\, \mathrm{VP}}/\mathrm{E} \quad \begin{array}{c}\mathsf{on} \circ \mathsf{thursday};\\ \mathbf{onTh}; \mathrm{VP\backslash VP}\end{array}}{\varphi_1 \circ \varphi_2 \circ \mathsf{to} \circ \mathsf{robin} \circ \mathsf{on} \circ \mathsf{thursday};\, \mathbf{onTh}(P(x)(\mathbf{r}));\, \mathrm{VP}}\backslash\mathrm{E}}{\varphi_2 \circ \mathsf{to} \circ \mathsf{robin} \circ \mathsf{on} \circ \mathsf{thursday};\, \lambda P.\mathbf{onTh}(P(x)(\mathbf{r}));\, \mathrm{(VP/PP/NP)\backslash VP}}\backslash\mathrm{I}^1}{\mathsf{to} \circ \mathsf{robin} \circ \mathsf{on} \circ \mathsf{thursday};\, \lambda x \lambda P.\mathbf{onTh}(P(x)(\mathbf{r}));\, \mathrm{NP\backslash(VP/PP/NP)\backslash VP}}\backslash\mathrm{I}^2$$

We then derive a sentence containing gap positions corresponding to this derived constituent and the object NP, that is, an expression that has the syntactic type S|(NP\(VP/PP/NP)\VP)|NP, to be given as an argument to the three-place **resp3** operator introduced above. Since the relevant steps are the same as in the previous examples, we omit the details and just reproduce the derived sign:

(35) $\lambda\varphi_1\lambda\varphi_2.\mathsf{I} \circ \mathsf{lent} \circ \varphi_1 \circ \varphi_2;$
$\lambda x \lambda f.f(x)(\mathbf{lend})(\mathbf{I}); \mathrm{S|(NP\backslash(VP/PP/NP)\backslash VP)|NP}$

The rest of the derivation just involves giving this relation and the two product-type arguments of types NP and NP\(VP/PP/NP)\VP respectively as arguments to the **resp3** operator. The final translation obtained:

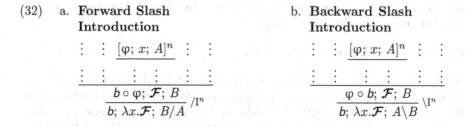

(36) $\mathbf{onTh(lend(s)(r))(I)} \wedge \mathbf{onFr(lend(b)(l))(I)}$

corresponds exactly to the relevant reading of the sentence.

We now turn to an extension of the analysis to symmetrical predicates. The key intuition behind our proposal here is that the NP containing *same*, *different*, etc. (we call such NPs 'symmetrical terms' below) in examples like (37) denotes a tuple (linked to the other tuple denoted by the plural *John and Bill* in the same way as in the 'respective' readings above) but that it imposes a specifial condition on each member of the tuple.

(37) John and Bill read the same book.

Specifically, to assign the right meaning to (37), John and Bill need to be each paired with an identical book, and in the case of *different*, they need to be paired with distinct books. To capture this additional constraint on the tuples denoted by symmetrical terms, we assign to them GQ-type meanings of type S|(S|NP), where the abstracted NP in their arguments are product-type expressions:[7]

(38) a. $\lambda\varphi_0\lambda\sigma_0.\sigma_0(\mathsf{the} \circ \mathsf{same} \circ \varphi_0)$;
$\lambda P \lambda Q.\exists X_\times \forall i\, P(\pi_i(X_\times)) \wedge \forall i \forall j [\pi_i(X_\times) = \pi_j(X_\times)] \wedge Q(X_\times)$;
S|(S|NP)|N

b. $\lambda\varphi_0\lambda\sigma_0.\sigma_0(\mathsf{different} \circ \varphi_0)$;
$\lambda P \lambda Q.\exists X_\times \forall i\, P(\pi_i(X_\times)) \wedge \forall i \forall j [i \neq j \rightarrow \pi_i(X_\times) \neq \pi_j(X_\times)] \wedge Q(X_\times)$;
S|(S|NP)|N

---

[7] So far as we can tell, the lexical meanings given in (38) capture the truth conditions for the internal readings of *same* and *different* correctly. A reviewer raises a concern that (38a) may be too weak as the meaning of *same* since the existentially bound $X_\times$ would merely be 'a common subset of the books read by John and by [Bill], while the actual sets of read books may still differ'. We do not agree with this reviewer. We believe that (37) is true and felicitous as long as one can identify (at least) one book commonly read by John and Bill. They may have read other books in addition, but that doesn't make (37) false or infelicitous. Such an implication, if felt to be present, is presumably a conversational implicature since it's clearly cancellable:

(i) John and Bill read the same book, although they both read several different books in addition.

Likewise, the same reviewer says that in (38b), as it stand, 'it suffices if $X_\times$ is taken, say, as a pair of different books read by both, all other books still being the same', on the basis of which s/he claims that the truth conditions need to be strengthed in such a way that $X_\times$ satisfies some maximality condition. Here again, we disagree. The following (ii) shows that the maximality implication excluding the existence of common books read by the two (if present at all) is not part of the entailment of the sentence.

(ii) John and Bill read different books, although they read the same books too.

For both *the same N* and *different Ns*, the relevant tuple (which enters into the 'respective' relation with another tuple via the **resp3** operator) consists of objects that satisfy the description provided by the *N*. The difference is that in the case of *same*, the elements of the tuple are all constrained to be identical, whereas in the case of *different*, they are constrained to differ from one another.

The analysis for (37) now goes as follows:

(39)

$$\dfrac{\begin{array}{l}\lambda\sigma_0\lambda\varphi_1\lambda\varphi_2.\\ \sigma_0(\varphi_1)(\varphi_2);\\ \mathbf{resp3};\\ (Z|X|Y)|(Z|X|Y)\end{array} \qquad \begin{array}{l}\vdots\quad\vdots\\ \lambda\varphi_3\lambda\varphi_4.\\ \varphi_4 \circ \mathsf{read} \circ\\ \varphi_3;\\ \mathbf{read};\\ \mathrm{S|NP|NP}\end{array}}{\begin{array}{l}\lambda\varphi_1\lambda\varphi_2.\varphi_2 \circ \mathsf{read} \circ \varphi_1;\\ \mathbf{resp3(read)};\\ \mathrm{S|NP|NP}\end{array}}$$

$$\dfrac{\begin{bmatrix}\varphi;\\ X_\times;\\ \mathrm{NP}\end{bmatrix}^1 \qquad \begin{array}{l}\lambda\varphi_1\lambda\varphi_2.\varphi_2 \circ \mathsf{read} \circ \varphi_1;\\ \mathbf{resp3(read)};\\ \mathrm{S|NP|NP}\end{array}}{\begin{array}{l}\lambda\varphi_2.\varphi_2 \circ \mathsf{read} \circ \varphi;\\ \mathbf{resp3(read)}(X_\times);\\ \mathrm{S|NP}\end{array}}$$

$$\dfrac{\begin{array}{l}\mathsf{john} \circ\\ \mathsf{and} \circ\\ \mathsf{bill};\\ \langle \mathbf{j}, \mathbf{b}\rangle;\\ \mathrm{NP}\end{array} \qquad \begin{array}{l}\lambda\varphi_2.\varphi_2 \circ \mathsf{read} \circ \varphi;\\ \mathbf{resp3(read)}(X_\times);\\ \mathrm{S|NP}\end{array}}{\mathsf{john} \circ \mathsf{and} \circ \mathsf{bill} \circ \mathsf{read} \circ \varphi;\ \mathbf{resp3(read)}(X_\times)(\langle \mathbf{j}, \mathbf{b}\rangle);\ \mathrm{S}}$$

$$\dfrac{\dfrac{\begin{array}{l}\lambda\varphi_1.\varphi_1;\\ \lambda p.\bigwedge_i\\ \pi_i(p);\\ \mathrm{S|S}\end{array} \qquad \mathsf{john} \circ \mathsf{and} \circ \mathsf{bill} \circ \mathsf{read} \circ \varphi;\ \mathbf{resp3(read)}(X_\times)(\langle \mathbf{j}, \mathbf{b}\rangle);\ \mathrm{S}}{\begin{array}{l}\mathsf{john} \circ \mathsf{and} \circ \mathsf{bill} \circ \mathsf{read} \circ \varphi;\\ \bigwedge_i \pi_i(\mathbf{resp3(read)}(X_\times)(\langle \mathbf{j}, \mathbf{b}\rangle));\mathrm{S}\end{array}}}{\begin{array}{l}\lambda\varphi.\mathsf{john} \circ \mathsf{and} \circ \mathsf{bill} \circ \mathsf{read} \circ \varphi;\\ \lambda X_\times . \bigwedge_i \pi_i(\mathbf{resp3(read)}(X_\times)(\langle \mathbf{j}, \mathbf{b}\rangle));\mathrm{S|NP}\end{array}}\ |\mathrm{I}^1$$

$$\dfrac{\begin{array}{l}\lambda\varphi_0\lambda\sigma_0.\\ \quad\sigma_0(\mathsf{the} \circ\\ \quad\mathsf{same} \circ \varphi_0);\\ \mathbf{same};\\ \mathrm{S|(S|NP)|N}\end{array} \qquad \begin{array}{l}\mathsf{book};\\ \mathbf{book};\\ \mathrm{N}\end{array}}{\begin{array}{l}\lambda\sigma_0.\sigma_0(\mathsf{the} \circ\\ \quad\mathsf{same} \circ \mathsf{book});\\ \mathbf{same(book)};\\ \mathrm{S|(S|NP)}\end{array}}$$

$$\dfrac{\begin{array}{l}\lambda\sigma_0.\sigma_0(\mathsf{the} \circ \mathsf{same} \circ \mathsf{book});\\ \mathbf{same(book)};\\ \mathrm{S|(S|NP)}\end{array} \qquad \begin{array}{l}\lambda\varphi.\mathsf{john} \circ \mathsf{and} \circ \mathsf{bill} \circ \mathsf{read} \circ \varphi;\\ \lambda X_\times . \bigwedge_i \pi_i(\mathbf{resp3(read)}(X_\times)(\langle \mathbf{j}, \mathbf{b}\rangle));\mathrm{S|NP}\end{array}}{\begin{array}{l}\mathsf{john} \circ \mathsf{and} \circ \mathsf{bill} \circ \mathsf{read} \circ \mathsf{the} \circ \mathsf{same} \circ \mathsf{book};\\ \mathbf{same(book)}(\lambda X_\times . \bigwedge_i \pi_i(\mathbf{resp3(read)}(X_\times)(\langle \mathbf{j}, \mathbf{b}\rangle)));\mathrm{S}\end{array}}$$

The derivation proceeds by first positing a product-type variable $X_\times$, which is related to the other product-type term denoted by *John and Bill* via the resp operator. Then, after the boolean reduction operator reduces the pair of propositions to their conjunction, the variable $X_\times$ is abstracted over to yield a property of product-type objects (of syntactic type S|NP). Since *the same book* is a GQ over product-type terms, it takes this property as an argument to return a proposition.

The final translation is unpacked in (40):

(40)
$$\begin{array}{l}\mathbf{same(book)}(\lambda X_\times . \bigwedge_i \pi_i(\mathbf{resp3(read)}(X_\times)(\langle \mathbf{j}, \mathbf{b}\rangle)))\\ = \exists X_\times \forall i\, \mathbf{book}(\pi_i(X_\times)) \wedge \forall i \forall j[\pi_i(X_\times) = \pi_j(X_\times)] \wedge \bigwedge_i \pi_i(\mathbf{resp3(read)}(X_\times)(\langle \mathbf{j}, \mathbf{b}\rangle))\\ = \exists X_\times \forall i\, \mathbf{book}(\pi_i(X_\times)) \wedge \forall i \forall j[\pi_i(X_\times) = \pi_j(X_\times)] \wedge \mathbf{read}(\pi_1(X_\times))(\mathbf{j}) \wedge\\ \quad \mathbf{read}(\pi_2(X_\times))(\mathbf{b})\end{array}$$

Since, by definition, $\pi_1(X_\times) = \pi_2(X_\times)$, this correctly ensures that the book that John read and the one that Bill read are identical.

Importantly, since the same 'respective' operator is at the core of the analysis as in the case of 'respective' readings, this analysis immediately predicts that symmetrical predicates can enter into multiple dependencies both among themselves and with respect to 'respective' predication, as exemplified by the data in (3). Since the relevant derivations can be reconstructed easily by taking

(18)–(19) as a model, we omit the details and reproduce here only the derived meanings for (3b) and (3c) in (41) and (42), respectively.

(41) $\mathbf{same(book)}(\lambda X_{\times}.\mathbf{give(m)}(\pi_1(X_{\times}))(\mathbf{j}) \wedge \mathbf{give(s)}(\pi_2(X_{\times}))(\mathbf{b}))$
$= \exists X_{\times} \forall i\, \mathbf{book}(\pi_i(X_{\times})) \wedge \forall i \forall j[\pi_i(X_{\times}) = \pi_j(X_{\times})] \wedge \mathbf{give(m)}(\pi_1(X_{\times}))(\mathbf{j}) \wedge \mathbf{give(s)}(\pi_2(X_{\times}))(\mathbf{b})$

(42) $\mathbf{same(book)}(\lambda X_{\times}.\mathbf{same(man)}(\lambda Y_{\times}.\mathbf{give}(\pi_1(Y_{\times}))(\pi_1(X_{\times}))(\mathbf{j}) \wedge \mathbf{give}(\pi_2(Y_{\times}))(\pi_2(X_{\times}))(\mathbf{b})))$
$= \exists X_{\times} \forall i\, \mathbf{book}(\pi_i(X_{\times})) \wedge \forall i \forall j[\pi_i(X_{\times}) = \pi_j(X_{\times})] \wedge \exists Y_{\times} \forall i\, \mathbf{man}(\pi_i(Y_{\times})) \wedge \forall i \forall j[\pi_i(Y_{\times}) = \pi_j(Y_{\times})] \wedge \mathbf{give}(\pi_1(Y_{\times}))(\pi_1(X_{\times}))(\mathbf{j}) \wedge \mathbf{give}(\pi_2(Y_{\times}))(\pi_2(X_{\times}))(\mathbf{b})$

## 5 Conclusion

In this paper, we have proposed a unified analysis of 'respective' readings and symmetrical predicates, building on the previous accounts of the two phenomena by Gawron and Kehler (2004) and Barker (2007). While these two previous proposals look apparently quite different from each other, in that one involves a nonlocal mechanism for obtaining the right meaning of the sentence whereas the other involves a chain of local operations, we showed that the underlying mechanisms that they rely on are not so different from each other, and that, by recasting the two analyses in a general calculus of the syntax-semantics interface, one (G&K) can essentially be seen as a 'lexicalized' version of the other (Barker), in the sense that it involves only local composition rules but these local composition rules themselves can be derived from the general rules for 'pairwise' predication posited in the latter. We argued that this enables us to unify the analyses of 'respective' readings and symmetrical predicates, and that such a unified analysis is empirically desirable; it immediately accounts for the close parallels and interactions between 'respective' and symmetrical predication via a single uniform mechanism of pairwise predication that is at the core of the semantics of both phenomena. We have demonstrated this point by working out an explicit analysis that captures these parallels and interactions between the two phenomena systematically.

We would like to comment on one technical (and conceptual) point (albeit briefly) before concluding the paper. As noted by two reviewers, the present system relies heavily on empty operators manipulating tuple-denoting objects to yield 'respective' readings and these operators do not affect the syntactic types of the expressions that they apply to. So, for example, a perfectly well-formed syntactic derivation may nonetheless yield an incongruent semantic translation because there is a type mismatch in the semantics. One could alternatively explicitly distinguish tuple-denoting expressions from expressions denoting non-tuple objects by enriching the syntactic typing system with product connectives (so that, for example, *John and Bill* denoting the tuple $\langle \mathbf{j}, \mathbf{b} \rangle$ has syntactic type NP×NP rather than NP). This will enable us to retain the straightforward functional mapping from syntactic types to semantic types standard in the categorial grammar syntax-semantics interface.[8] Moreover, such an approach may enable us to do away with the empty operators that we posit as lexical assumptions in the current system by letting the deductive rules for the product types do the work that these operators undertake in the current fragment. Thus, this seems to be a promising possibility to explore, which may elucidate the 'logic' underlying 'respective' and symmetrical predication even more. We do not see any obstacle in principle for refining the analysis presented above along these lines, and would like to explore this possibility in a future study.

**Acknowledgments.** We thank reviewers for FG 2014 for their comments. The first author acknowledges the financial support from the Japan Society for the Promotion of Science (Postdoctoral Fellowship for Research Abroad).

## References

Barker, C.: Parasitic scope. Linguistics and Philosophy 30, 407–444 (2007)

Bekki, D.: Heikooteki-kaishaku-niokeru yoosokan-junjo-to bunmyaku-izon-sei (The order of elements and context dependence in the 'respective' interpretation). In: Nihon Gengo-Gakkai Dai 132-kai Taikai Yokooshuu (Proceedings from the 132nd Meeting of the Linguistic Society of Japan), pp. 47–52 (2006)

de Groote, P.: Towards abstract categorial grammars. In: Association for Computational Linguistics, 39th Annual Meeting and 10th Conference of the European Chapter, pp. 148–155 (2001)

Gawron, J.M., Kehler, A.: The semantics of respective readings, conjunction, and filler-gap dependencies. Linguistics and Philosophy 27(2), 169–207 (2004)

Kubota, Y.: Nonconstituent coordination in Japanese as constituent coordination: An analysis in Hybrid Type-Logical Categorial Grammar. To appear in Linguistic Inquiry (to appear)

Kubota, Y., Levine, R.: Gapping as like-category coordination. In: Béchet, D., Dikovsky, A. (eds.) Logical Aspects of Computational Linguistics. LNCS, vol. 7351, pp. 135–150. Springer, Heidelberg (2012)

Kubota, Y., Levine, R.: Determiner gapping as higher-order discontinuous constituency. In: Morrill, G., Nederhof, M.-J. (eds.) FG 2012/2013. LNCS, vol. 8036, pp. 225–241. Springer, Heidelberg (2013)

Kubota, Y., Levine, R.: Pseudogapping as pseudo-VP ellipsis. In: Asher, N., Soloviev, S. (eds.) LACL 2014. LNCS, vol. 8535, pp. 122–137. Springer, Heidelberg (2014)

Mihaliček, V., Pollard, C.: Distinguishing phenogrammar from tectogrammar simplifies the analysis of interrogatives. In: de Groote, P., Nederhof, M.-J. (eds.) FG 2010/2011. LNCS, vol. 7395, pp. 130–145. Springer, Heidelberg (2012)

Muskens, R.: Language, lambdas, and logic. In: Kruijff, G.-J., Oehrle, R. (eds.) Resource Sensitivity in Binding and Anaphora, pp. 23–54. Kluwer, Dordrecht (2003)

Oehrle, R.T.: Term-labeled categorial type systems. Linguistics and Philosophy 17(6), 633–678 (1994)

[8] But note that, though standard, it's not clear whether this assumption is empirically motivated. See for example Linear Grammar (Mihaliček and Pollard, 2012), which explicitly rejects the functional mapping from syntactic types to semantic types (see in particular Worth (2014, section 1.1) for an explicit statement of this point).

Worth, C.: The phenogrammar of coordination. In: Proceedings of the EACL 2014 Workshop on Type Theory and Natural Language Semantics (TTNLS), pp. 28–36. Association for Computational Linguistics, Gothenburg (2014)

# Appendix

**Lemma:** For any arbitrary complex structure $S$ licensed by the G&K fragment with semantic translation $\Gamma$ and which contains exactly one occurrence of a term $t$ whose semantic translation is $x$, we obtain a structure $S'$ by replacing $t$ in $S$ with a term whose translation is $\langle a_1, \ldots, a_l \rangle$. Then for the semantic translation of $S'$ $\Psi$, the following holds:

$(\star)\quad \Psi = \langle \Gamma[x/a_1], \ldots, \Gamma[x/a_l] \rangle$

**Proof:** The proof is by induction.

**Base Case:** Since $\Gamma = x$ and $\Psi = \langle a_1, \ldots, a_l \rangle$, it trivially follows that $(\star)$ holds.

**Inductive Step:** We have two cases to consider: (i) $S$ consists of a function $f$ and a structure $T$ (with translation $\Omega$, which is an argument of $f$) that satisfies $(\star)$; (ii) $S$ consists of a structure $T$ (with translation $\Omega$) that satisfies $(\star)$ and a term $c$ that is an argument of $\Omega$. We consider (i) first.

(i)

$$\dfrac{\varphi_2;\ f;\ \mathrm{X/Y} \qquad \dfrac{\vdots\vdots\vdots \quad \dfrac{}{\varphi_0;\ x;\ \mathrm{A}} \quad \vdots\vdots\vdots}{\varphi_1;\ \Omega;\ \mathrm{Y}}}{\varphi_2 \circ \varphi_1;\ f(\Omega);\ \mathrm{X}}$$

Since $T$ satisfies $(\star)$, there is a structure $T'$ in which $x$ in $T$ is replaced by $\langle a_1, ..., a_l \rangle$ such that the following holds between $\Omega$ and $\Omega'$, the translations of $T$ and $T'$: $\Omega' = \langle \Omega[x/a_1], ..., \Omega[x/a_n] \rangle$.

We are interested in the translation $\Gamma'$ of a structure $S'$, which can be obtained by replacing $t$ with a term whose translation is $\langle a_1, ..., a_l \rangle$. By replacing $T$ in $S$ with $T'$, we obtain just such a structure:

(i')

$$\dfrac{\varphi_2;\ f;\ \mathrm{X/Y} \qquad \dfrac{\vdots\vdots\vdots \quad \dfrac{}{\varphi_0';\ \langle a_1, ..., a_l \rangle;\ \mathrm{A}} \quad \vdots\vdots\vdots}{\varphi_1';\ \Omega';\ \mathrm{Y}}}{\varphi_2 \circ \varphi_1';\ \Gamma';\ \mathrm{X}}\ \text{Rule 1}$$

Thus,

$$\begin{aligned} \Gamma' &= \langle f(\Omega[x/a_1]), ..., f(\Omega[x/a_n]) \rangle && \text{(via Rule 1)} \\ &= \langle (f(\Omega))[x/a_1], ..., (f(\Omega))[x/a_n] \rangle && \text{(since } x \text{ is fresh in } f) \\ &= \langle \Gamma[x/a_1], ..., \Gamma[x/a_n] \rangle && \text{(since } \Gamma = f(\Omega)) \end{aligned}$$

Case (ii) can be proven similarly to case (i).

# Learning Context Free Grammars with the Finite Context Property: A Correction of A. Clark's Algorithm

Hans Leiß

Centrum für Informations- und Sprachverarbeitung
Universität München
Oettingenstr.67, 80538 München, Germany
leiss@cis.uni-muenchen.de

**Abstract.** A. Clark[2] has shown that the class of languages which have a context-free grammar whose nonterminals can be defined by a finite set of contexts can be identified in the limit, given an enumeration of the language and a test for membership. We show by example that Clark's algorithm may converge to a grammar that does not define the input language. We review the theoretical background, provide a non-obvious modification of the algorithm and prove its correctness.

## 1 Introduction

An important goal of structural linguistics was to analyse and describe a language in terms of distributions. Given an alphabet $\Sigma$, the *distribution* of a word $v \in \Sigma^*$ with respect to a language $L \subseteq \Sigma^*$ is the set

$$D(v) = \{(u, w) \in \Sigma^* \times \Sigma^* \mid uvw \in L\}$$

of all contexts where $v$ appears in $L$. Having the same distribution with respect to $L$ is a congruence relation $\equiv_L$ on $\Sigma^*$, the *syntactic congruence.* It partitions $\Sigma^*$ into disjoint distribution classes $[v] := \{v' \in \Sigma^* \mid v \equiv_L v'\}$. By the Myhill/Nerode theorem (c.f.[6]), $\equiv_L$ has finitely many distribution classes if and only if $L$ is a regular language.

When the monoid operations of $(\Sigma^*, \cdot, \epsilon)$ are lifted to word sets by $A \cdot B := \{a \cdot b \mid a \in A, b \in B\}$ and $1 = \{\epsilon\}$, one obtains a monoid $(\mathcal{P}(\Sigma^*), \cdot, 1)$, which is partially ordered by $\subseteq$. The operation $(u, w) \odot v := uvw$ of *filling* a context $(u, w)$ with a word $v$ is lifted to context sets $C$ and word sets $A$ by

$$C \odot A := \{(u, w) \odot v \mid (u, w) \in C, v \in A\}.$$

With respect to $L$, each set of contexts $C$ has a largest set of fillers, $C^{\triangleleft} = \{v \in \Sigma^* \mid C \odot \{v\} \subseteq L\}$, and each set $A$ of words has a largest set of contexts, $A^{\triangleright} = \{(u, w) \in \Sigma^* \times \Sigma^* \mid \{(u, w)\} \odot A \subseteq L\}$. Notice that $D(v) = \{v\}^{\triangleright} =: v^{\triangleright}$ and

$$\{v\}^{\triangleright\triangleleft} = \{u \mid v^{\triangleright} \subseteq u^{\triangleright}\} \supseteq \{u \mid v^{\triangleright} = u^{\triangleright}\} = [v].$$

G. Morrill et al. (Eds.): Formal Grammar 2014, LNCS 8612, pp. 121–137, 2014.

The maps $A \mapsto A^{\triangleright\triangleleft}$ and $C \mapsto C^{\triangleleft\triangleright}$ are closure operators. Via a Galois-connection between sets of words and sets of contexts, the partial orders of closed sets of words and closed sets of contexts are anti-isomorphic. Clark[2] defines a *(syntactic) concept of* $L$ to be a pair $\langle A, C\rangle$ such that $A^{\triangleright} = C$ and $C^{\triangleleft} = A$. As here each component is closed and determines the other one, one can use the component which is the better representation for a given purpose.

Note that $L$ is itself a concept, namely $L = \{(\epsilon,\epsilon)\}^{\triangleleft}$, and can be represented by a finite set of contexts. Suppose $L$ has a context-free grammar $G$ (in Chomsky normal form, CNF) whose nonterminals $N$ are concepts of $L$, i.e. $N = N^{\triangleright\triangleleft}$ when $N$ is identified with $\{w \in \Sigma^* \mid N \Rightarrow_G^* w\}$. A branching rule $(N \to AB)$ of $G$ then corresponds to $N \supseteq AB$, which is equivalent to $N \supseteq (AB)^{\triangleright\triangleleft}$. If $N, A, B$ are represented by context sets $C, D, E$, this means $C^{\triangleleft} \supseteq (D^{\triangleleft}E^{\triangleleft})^{\triangleright\triangleleft}$.

A. Clark[2] developed an algorithm to identify in the limit a CNF-grammar for $L$ from membership queries and an enumeration of $L$, provided $L$ has a CNF-grammar whose nonterminals can be defined by finite sets of contexts. The basic idea is to extract from a finite subset $E \subseteq L$ a finite set $F$ of contexts and a finite set $K$ of subwords of $L$ and relativize the operations $\cdot^{\triangleright}$ and $\cdot^{\triangleleft}$ of taking all contexts resp. fillers with respect to $L$ to $\cdot^F : \mathcal{P}(\Sigma^*) \to \mathcal{P}(F)$ and $\cdot^K : \mathcal{P}(F) \to \mathcal{P}(K)$ by $A^F := A^{\triangleright} \cap F$ and $C^K := C^{\triangleleft} \cap K$. Then there are only finitely many relativized concepts of $L$, the pairs $\langle A, C\rangle \in \mathcal{P}(K) \times \mathcal{P}(F)$ where $A = C^K$ and $C = A^F$; using relativized closed sets $C = C^{KF}$ of contexts to represent them, Clark builds a CNF-grammar $G(K, L, F)$ by taking as branching rules those triples $(C \to DE)$ where $C^K \supseteq (D^K E^K)^{FK}$. It is claimed that as $K$ and $F$ increase, the grammars $G(K, L, F)$ converge to a grammar for $L$.

But there is a technical problem: Clark's criterion for $C, D, E$ to form a grammar rule is right when working with infinite filler sets, i.e. $C^{\triangleleft} \supseteq (D^{\triangleleft}E^{\triangleleft})^{\triangleright\triangleleft}$, since $\cdot^{\triangleright\triangleleft}$ is a closure operator on $\mathcal{P}(\Sigma^*)$ and hence $D^{\triangleleft}E^{\triangleleft} \subseteq (D^{\triangleleft}E^{\triangleleft})^{\triangleright\triangleleft}$. But the criterion is not correct with finite filler sets, i.e. $C^K \supseteq (D^K E^K)^{FK}$ is not equivalent to $C^K \supseteq D^K E^K$: although $\cdot^{FK}$ is a closure operator on $\mathcal{P}(K)$, it is generally not the case that $D^K E^K \subseteq (D^K E^K)^{FK}$, as the left hand side need not be a subset of $K$. We give an example where Clark's algorithm does not converge to a grammar of the intended language.

Clark's algorithm can be fixed by three modifications: (i) the criterion for when three concepts $C, D, E$ constitute a grammar rule has to be changed from $C^K \supseteq (D^K E^K)^{FK}$ to $C \subseteq (D^K E^K)^F$. Since this works directly with context sets, it emphasizes the importance of the Galois correspondence between word sets and context sets. (ii) the criterion in the learning algorithm that makes the hypothesis grammar shrink is weakened; the effect is that the learner cannot converge to a grammar that defines a strict superset of the intended language. (iii) since for our modified definition, $L(G(K, L, F))$ is neither antitone in $K$ nor monotone in $F$ (as pointed out by R.Yoshinaka), we need a different line of reasoning to show the convergence of the grammar inference process.

We prove that the modified algorithm indeed identifies in the limit the class of context free languages with the finite context property. If we admit only concepts whose context sets are closures of bounded context sets, the algorithm has polynomial update time. Ryo Yoshinaka [8] has a different modification of Clark's algorithm where nonterminals need not be closed sets. This makes the reasoning simpler, but results in a larger grammar.

## 2 Correspondence between Word and Context Sets

Let $\Sigma$ be a finite alphabet and $L \subseteq \Sigma^*$ a set of words. An *L-context of* $A \subseteq \Sigma^*$ is a word pair or *context* $(u, w)$ such that $uAw \subseteq L$. The *largest set of L-contexts of* $A \subseteq \Sigma^*$ is

$$A^{\rhd} := \{(u,w) \mid \{(u,w)\} \odot A \subseteq L\} = \{(u,w) \mid uAw \subseteq L\}. \qquad (1)$$

An *L-filler* of a set $C \subseteq \Sigma^* \times \Sigma^*$ of word pairs is a word $w \in \Sigma^*$ such that $uwv \in L$ for each $(u, v) \in C$. Let $C \odot A := \{uvw \mid (u,w) \in C, v \in A\}$. The *largest set of L-fillers of* $C \subseteq \Sigma^* \times \Sigma^*$ is

$$C^{\lhd} := \{v \mid C \odot \{v\} \subseteq L\}. \qquad (2)$$

If $C \neq \emptyset$, $C^{\lhd} \subseteq Sub(L)$, where $Sub(L) := \{v \in \Sigma^* \mid \exists u, w \in \Sigma^*\ uvw \in L\}$ is the set of all subwords of $L$. The following equivalence (3) is easy to check:

**Proposition 1.** *The functions* $\cdot^{\rhd} : (\mathcal{P}(\Sigma^*), \supseteq) \rightleftarrows (\mathcal{P}(\Sigma^* \times \Sigma^*), \subseteq) : \cdot^{\lhd}$ *form a Galois-connection, i.e., for all* $A \in \mathcal{P}(\Sigma^*)$ *and* $C \in \mathcal{P}(\Sigma^* \times \Sigma^*)$ *we have*

$$C^{\lhd} \supseteq A \iff C \subseteq A^{\rhd}. \qquad (3)$$

*In particular, we have:*

(i) $\cdot^{\rhd}$ *and* $\cdot^{\lhd}$ *are antitone:* $A^{\rhd} \supseteq B^{\rhd}$ *for all word sets* $A \subseteq B$ *and* $C^{\lhd} \supseteq D^{\lhd}$ *for all context sets* $C \subseteq D$.
(ii) $A^{\rhd\lhd\rhd} = A^{\rhd}$ *and* $C^{\lhd\rhd\lhd} = C^{\lhd}$ *for all word sets* $A$ *and context sets* $C$.
(iii) $\cdot^{\rhd\lhd}$ *is a closure operator on* $(\mathcal{P}(\Sigma^*), \subseteq)$ *and* $\cdot^{\lhd\rhd}$ *is a closure operator on* $(\mathcal{P}(\Sigma^* \times \Sigma^*), \subseteq)$.
(iv) $\cdot^{\rhd} : (\{C^{\lhd} \mid C \subseteq \Sigma^* \times \Sigma^*\}, \supseteq) \rightleftarrows (\{A^{\rhd} \mid A \subseteq \Sigma^*\}, \subseteq) : \cdot^{\lhd}$ *form an order isomorphism between the closed word sets and the closed context sets.*

Claims (i)-(iv) are standard consequences of a Galois-connection. Since $A \subseteq L$ gives $(\epsilon, \epsilon) \in A^{\rhd}$ , hence $A^{\rhd\lhd} \subseteq \{(\epsilon, \epsilon)\}^{\lhd} = L$, we have

$$A \subseteq L \iff A^{\rhd\lhd} \subseteq L. \qquad (4)$$

In particular, $L$ itself is closed: $L^{\rhd\lhd} = L$.

**Proposition 2.** *For all* $A, B \subseteq \Sigma^*$, $(A^{\rhd\lhd} B^{\rhd\lhd})^{\rhd\lhd} = (AB)^{\rhd\lhd}$.

*Proof.* Since $\cdot^{\rhd\lhd}$ is a closure operator on $\mathcal{P}(\Sigma^*)$, it is sufficient to show that $A^{\rhd\lhd} B^{\rhd\lhd} \subseteq (AB)^{\rhd\lhd}$. Let $a \in A^{\rhd\lhd}$, $b \in B^{\rhd\lhd}$, hence $A^{\rhd} \subseteq a^{\rhd}$ and $B^{\rhd} \subseteq b^{\rhd}$. Moreover, let $(u,w) \in (AB)^{\rhd}$, so $uABw \subseteq L$. Since $\{u\} \times Bw \subseteq A^{\rhd} \subseteq a^{\rhd}$ we have $uaBw \subseteq L$, so $(ua, w) \in B^{\rhd} \subseteq b^{\rhd}$, and $uabw \in L$. Thus, $ab \in (AB)^{\rhd\lhd}$.

For the *residuation* $A/B := \{v \in \Sigma^* \mid \{v\}B \subseteq A\}$ of word sets $A, B$, we have:

**Proposition 3.** *For all* $A, B \subseteq \Sigma^*$, $(A/B)^{\rhd\lhd} \subseteq A^{\rhd\lhd}/B$. *If* $A = A^{\rhd\lhd}$, *then* $(A/B)^{\rhd\lhd} = A/B$.

### 2.1 The Residuated Lattice $\mathcal{B}(L)$ of All Concepts of $L$

A syntactic *concept of* $L$ is a pair $\langle A, C\rangle$ of a word set $A$ and context set $C$ such that $A^{\triangleright} = C$ and $C^{\triangleleft} = A$. Note that here $A = A^{\triangleright\triangleleft}$ and $C = C^{\triangleleft\triangleright}$ are closed sets, and by the order isomorphism of Proposition 1 (iv) and (ii), one can represent a concept $\langle A, C\rangle$ by its closed word set $A^{\triangleright\triangleleft}$ or by its closed context set $C^{\triangleleft\triangleright}$.

**Theorem 1.** *(Clark[1]) The set* $B(L)$ *of all concepts of* $L$ *forms a monoidal, residuated and complete lattice,*

$$\mathcal{B}(L) := (B(L), \circ, \mathbb{1}, \backslash, /, \vee, \wedge, \top, \bot, \leq)$$

*where the operations are, in terms of closed word sets, given by*

$$\begin{aligned} A \circ B &:= (A \cdot B)^{\triangleright\triangleleft}, \\ \mathbb{1} &:= \{\epsilon\}^{\triangleright\triangleleft}, \\ B\backslash A &:= \{v \in \Sigma^* \mid B \cdot \{v\} \subseteq A\}, \\ A/B &:= \{v \in \Sigma^* \mid \{v\} \cdot B \subseteq A\}, \end{aligned} \qquad \begin{aligned} A \vee B &:= (A \cup B)^{\triangleright\triangleleft}, \\ A \wedge B &:= A \cap B, \\ \top &:= \Sigma^*, \\ \bot &:= \emptyset^{\triangleright\triangleleft}, \\ A \leq B &:\iff A \subseteq B. \end{aligned}$$

*Proof.* (Sketch) Monoid properties for $\circ$ and $\mathbb{1}$ follow from Proposition 2:

$$(A \circ B) \circ C = ((AB)^{\triangleright\triangleleft} C)^{\triangleright\triangleleft} = ((AB)^{\triangleright\triangleleft} C^{\triangleright\triangleleft})^{\triangleright\triangleleft} = (ABC)^{\triangleright\triangleleft}\,,$$
$$A \circ \mathbb{1} = (A \cdot \{\epsilon\}^{\triangleright\triangleleft})^{\triangleright\triangleleft} = (A \cdot \{\epsilon\})^{\triangleright\triangleleft} = A.$$

By proposition 3, we know that $A/B$ is closed when $A$ is. Since the residual laws hold in $\mathcal{P}(\Sigma^*)$ they hold in $B(L)$, because for concepts $A, B, C$,

$$A \circ B \subseteq C \iff (A \cdot B)^{\triangleright\triangleleft} \subseteq C \iff A \cdot B \subseteq C. \tag{5}$$

(See also Jipsen e.a.[4], Lemma 7.1.)

The residuations make the syntactic concept lattices complete for the full Lambek calculus, see [7]. Clark emphasizes the lattice structure "as a good search space" for grammatical inference. To learn CFGs, it is sufficient that $(B(L), \vee, \bot, \circ, \mathbb{1})$ is a complete idempotent semiring, in which context-free grammars can be interpreted.

**Proposition 4.** *The syntactic concepts of* $L$ *form a complete idempotent semiring*

$$\mathcal{B}(L) = (B(L), +^{\mathcal{B}(L)}, 0^{\mathcal{B}(L)}, \cdot^{\mathcal{B}(L)}, 1^{\mathcal{B}(L)}) := (B(L), \vee, \bot, \circ, \mathbb{1}),$$

*and the mapping* $h : \mathcal{P}(\Sigma^*) \to \mathcal{B}(L)$ *given by* $h(A) := \langle A^{\triangleright\triangleleft}, A^{\triangleright}\rangle$*, is a continuous semiring-homomorphism.*

A context-free grammar $G$ with constants from $\Sigma$ is a system of polynomial equations $X_i = p_i(X_1, \ldots, X_n)$. Its least solution in $\mathcal{P}(\Sigma^*)$, the tuple of languages $L(G, A)$ for nonterminals $A$, is componentwise mapped by $h$ to its least solution in $\mathcal{B}(L)$, the tuple of closed sets $L(G, A)^{\triangleright\triangleleft}$. For the main component, we have $h(L) = h(L(G, S)) = L(G, S)^{\triangleright\triangleleft} = L^{\triangleright\triangleleft} = L$.

*Remark 1.* Since the Kleene-closure $A^*$ is the least solution of $AX + 1 \leq X$ in $\mathcal{P}(\Sigma^*)$ and $h$ preserves least fixed-points, the semiring of syntactic concepts of $L$ can be expanded to a Kleene algebra $(B(L), \vee, \bot, \circ, \mathbb{1}, ^{\circledast})$, using $\langle A, C\rangle^{\circledast} := \langle (A^*)^{\triangleright\triangleleft}, ((C^{\triangleleft})^*)^{\triangleright}\rangle$.

## 2.2 The Lattice $\mathcal{B}(L, F)$ of Concepts of $L$ Relative to $F$

We restrict ourselves to context-free grammars in *weak Chomsky Normal Form* (CNF), where rules may have the forms $(C \to \epsilon), (C \to a), (C \to DE)$, where $\epsilon$ is the empty word, $a$ is a terminal and $C, D, E$ are nonterminals of the grammar. A motivating idea of Clark was that a finite set $V$ of concepts in the monoidal lattice $B(L)$ gives rise to a grammar $G(L, V)$ that defines a sublanguage of $L$.

**Proposition 5.** *(c.f. Lemma 1 in [2]) Let $L \subseteq \Sigma^*$ and $V \subseteq B(L)$ be a finite set of concepts, here viewed as context sets $C \subseteq \Sigma^* \times \Sigma^*$ that are closed, i.e. $C = C^{\lhd\rhd}$. Let $G(L,V) = (\Sigma, V, P, S)$ be the grammar with*

$$\begin{aligned} S &:= \{(\epsilon, \epsilon)\}^{\lhd\rhd}, \\ P &:= \{(C \to w) \mid w \in \Sigma \cup \{\epsilon\}, C \in V, w \in C^{\lhd}\} \\ &\quad \cup \{(C \to DE) \mid C, D, E \in V, (D^{\lhd}E^{\lhd})^{\rhd\lhd} \subseteq C^{\lhd}\}. \end{aligned}$$

*Then $L(G(L,V)) \subseteq L$.*

*Proof.* By induction, one shows that if $C \Rightarrow^* w$, then $w \in C^{\lhd}$, and $L = S^{\lhd}$.

In suitable cases, each concept $C = C^{\lhd\rhd}$ of $V$ may be generated by a finite subset $C_f \subseteq C$ (c.f. the *diagnostic contexts* in 3.4 of Harris[3]), i.e. $C = C_f^{\lhd\rhd}$, and then $V$ is determined by a collection of subsets of a finite set $F \subseteq \Sigma^* \times \Sigma^*$ of contexts. In particular, when $L$ does have a grammar $G$ whose nonterminals $A$ define word sets $L(G, A)$ that are the filler sets $C_A^{\lhd}$ of a finite set $C_A \subseteq F$ of contexts, we can hope to find such a grammar $G$ from a finite fragment $E \subseteq L$ that provides each context $(u, v)$ of $F$ through some word $uwv \in E$.

The idea now is to construct from a finite set $F$ of contexts a finite "approximation" $B(L, F)$ of $B(L)$ and define a variant $G(L, F)$ of $G(L, V)$ such that $L(G(L, F)) \subseteq L$. In suitable cases, $G(L, F)$ defines $L$.

**Proposition 6.** *For sets $F \subseteq \Sigma^* \times \Sigma^*$ of contexts and $A \subseteq \Sigma^*$ of words, let $A^F := A^{\rhd} \cap F$. The mappings $\cdot^F : (\mathcal{P}(\Sigma^*), \supseteq) \rightleftarrows (\mathcal{P}(F), \subseteq) : \cdot^{\lhd}$ form a Galois-connection, i.e. for all $A \in \mathcal{P}(\Sigma^*)$ and $C \in \mathcal{P}(F)$ we have*

$$C^{\lhd} \supseteq A \iff C \subseteq A^F.$$

*In particular, properties (i) – (iv) of proposition 1 hold with $\cdot^F$ instead of $\cdot^{\rhd}$.*

A *concept of $L$ relative to $F$* is a pair $\langle A, C\rangle$ of $\mathcal{P}(\Sigma^*) \times \mathcal{P}(F)$ such that $A^F = C$ and $C^{\lhd} = A$. In this case, $A = C^{\lhd} = A^{F\lhd}$ is closed with respect to $\cdot^{F\lhd}$ and $C = A^F = C^{\lhd F}$ is closed with respect to $\cdot^{\lhd F}$.

**Proposition 7.** *Let $F \subseteq \Sigma^* \times \Sigma^*$. The set of all concepts of $L$ relative to $F$ forms a complete lattice $\mathcal{B}(L, F) = (B(L, F), \vee, \wedge, \top, \bot, \leq)$, where*

$$\begin{aligned} \top &:= \langle \Sigma^*, \Sigma^{*F}\rangle \qquad \bot := \langle F^{\lhd}, F\rangle \\ \langle A_1, C_1\rangle \vee \langle A_2, C_2\rangle &:= \langle (A_1 \cup A_2)^{F\lhd}, C_1 \cap C_2\rangle \\ \langle A_1, C_1\rangle \wedge \langle A_2, C_2\rangle &:= \langle A_1 \cap A_2, (C_1 \cup C_2)^{\lhd F}\rangle \\ \langle A_1, C_1\rangle \leq \langle A_2, C_2\rangle &:\iff A_1 \subseteq A_2 \wedge C_1 \supseteq C_2. \end{aligned}$$

(Notice that $\emptyset^F = F$ and $\emptyset^\triangleleft = \Sigma^*$.) Define an operation $\circ : B(L,F) \times B(L,F) \to B(L,F)$ by

$$\langle A_1, C_1\rangle \circ \langle A_2, C_2\rangle := \langle (A_1A_2)^{F\triangleleft}, (C_1^\triangleleft C_2^\triangleleft)^F\rangle.$$

Indeed, $(A_1A_2)^{F\triangleleft F} = (A_1A_2)^F = (C_1^\triangleleft C_2^\triangleleft)^F$ and $(C_1^\triangleleft C_2^\triangleleft)^{F\triangleleft} = (A_1A_2)^{F\triangleleft}$. Moreover, $\circ$ is monotone with respect to $\leq$. However, proposition 2 does not extend from $\cdot^{\triangleright\triangleleft}$ to $\cdot^{F\triangleleft}$, and $\circ$ is not a monoid operation on $B(L,F)$; nor is $\mathbb{1} = \langle \{\epsilon\}^{F\triangleleft}, \{\epsilon\}^F\rangle$ neutral with respect to $\circ$.

Let $\mathcal{P}^{F\triangleleft}(\Sigma^*)$ be the set of $\cdot^{F\triangleleft}$-closed word sets and $\mathcal{P}^{\triangleleft F}(F)$ the set of $\cdot^{\triangleleft F}$-closed sets of contexts from $F$. We sometimes use the component functions $\circ^{F\triangleleft}$ on $\mathcal{P}^{F\triangleleft}(\Sigma^*)$ and $\circ^{\triangleleft F}$ on $\mathcal{P}^{\triangleleft F}(F)$ of $\circ$, defined by

$$A_1 \circ^{F\triangleleft} A_2 := (A_1A_2)^{F\triangleleft} \quad\text{and}\quad C_1 \circ^{\triangleleft F} C_2 := (C_1^\triangleleft C_2^\triangleleft)^F.$$

If $F$ is finite, a concept $\langle A, C\rangle$ of $L$ relative to $F$ has a finite representation by its closed context set $C$.

**Lemma 1.** *If $F \subseteq \Sigma^* \times \Sigma^*$ is finite and $(\epsilon,\epsilon) \in F$, then $L(G(L,F)) \subseteq L$, where $G(L,F)$ is the CNF-grammar $(V, \Sigma, P, S)$ with*

$$\begin{aligned} V &:= \{C \mid \emptyset \neq C \subseteq F,\ C^{\triangleleft F} = C\},\\ S &:= \{(\epsilon,\epsilon)\}^{\triangleleft F},\\ P &:= \{(C \to w) \mid C \in V, w \in \Sigma \cup \{\epsilon\}, C \subseteq w^F\},\\ &\ \cup \{(C \to DE) \mid C, D, E \in V,\ C \subseteq (D^\triangleleft E^\triangleleft)^F\}. \end{aligned}$$

*Proof.* As for proposition 5, we show by induction on the length of derivations that if $C \Rightarrow^* w$, then $w \in C^\triangleleft$. The claim follows from $S^\triangleleft = \{(\epsilon,\epsilon)\}^{\triangleleft F\triangleleft} = \{(\epsilon,\epsilon)\}^\triangleleft = L$. For a derivation $C \Rightarrow DE \Rightarrow^* uE \Rightarrow^* uv$ with $u, v \in \Sigma^*$, we have, by induction,

$$uv \in D^\triangleleft E^\triangleleft \subseteq (D^\triangleleft E^\triangleleft)^{F\triangleleft} \subseteq C^\triangleleft,$$

using that $\cdot^{F\triangleleft}$ is a closure operator on $\mathcal{P}(\Sigma^*)$, that $C \subseteq (D^\triangleleft E^\triangleleft)^F$, and that $\cdot^\triangleleft$ is antitone.

If $G_1 = (V_1, \Sigma, P_1, S_1)$ and $G_2 = (V_2, \Sigma, P_2, S_2)$ are CNF-grammars over $\Sigma$, the mapping $\tilde{\cdot} : V_1 \to V_2$ induces a *grammar homomorphism* from $G_1$ to $G_2$, if $\tilde{S}_1 = S_2$ and $(\tilde{C} \to a), (\tilde{C} \to \tilde{D}\tilde{E}) \in P_2$ for all rules $(C \to a), (C \to DE) \in P_1$. In this case, clearly $L(G_1) \subseteq L(G_2)$.

We remark that the mapping $C \mapsto C^{\triangleleft\triangleright}$ induces a grammar isomorphism from $G(L,F)$ to $G(L,V)$, where $V = \{C^{\triangleleft\triangleright} \mid C \in V_F\}$, and $V_F$ is the set of nonterminals of $G(L,F)$. Moreover, if $(\epsilon,\epsilon) \in F_1 \subseteq F_2$, then $C \mapsto C^{\triangleleft F_2}$ induces a grammar homomorphism from $G(L,F_1)$ to $G(L,F_2)$. It follows that $L(G(L,F))$ is monotone in $F$, but we will not exploit this.

### 2.3 Grammars with the Finite Context Property

A context-free grammar $G$ *has the finite context property* (FCP), if the $\cdot^{\triangleright\triangleleft}$-closure of the word set $L(G,A) := \{v \in \Sigma^* \mid A \Rightarrow_G^* v\}$ of every nonterminal $A$ of $G$ can be defined by a finite non-empty[1] set $C_A \subseteq \Sigma^* \times \Sigma^*$ of contexts, i.e.

[1] Non-emptyness of $C_A$ is not demanded in [2], [8], but is needed for $E \subseteq L$ in lemma 3. The stronger condition $L(G,A) = C_A^\triangleleft$ is used in [2], [5].

$$L(G, A)^{\triangleright\triangleleft} = C_A^{\triangleleft}.$$

The grammar has the finite context property *with respect to the context set* $F$, if all the above $C_A$ are subsets of $F$. Clearly, this is monotone in $F$, and we can replace the $C_A \subseteq F$ by their closures $C_A^{\triangleleft F}$, because $C_A^{\triangleleft F \triangleleft} = C_A^{\triangleleft}$,

**Lemma 2.** *If $L$ has a CNF-grammar $G$ with the FCP with respect to the finite set $F$ of contexts, then $G(L, F)$ contains a homomorphic image of $G$ and $L = L(G(L, F))$.*

*Proof.* Suppose $G = (V, \Sigma, P, S)$, and for each $A \in V$, suppose $L(G, A)^{\triangleright\triangleleft} = C_A^{\triangleleft}$ for some $\emptyset \neq C_A \subseteq F$. We can assume that $C_A \in \mathcal{P}^{\triangleleft F}(F)$. Let $G(L, F) = (V(L, F), \Sigma, P(L, F), S(L, F))$. Under $A \mapsto C_A$, each rule of $G$ is mapped to a rule of $G(L, F)$. If $(A \to BD) \in P$, then $L(G, B)L(G, D) \subseteq L(G, A)$, hence $C_B^{\triangleleft} C_D^{\triangleleft} \subseteq C_A^{\triangleleft}$ using proposition 2, hence $C_A = C_A^{\triangleleft F} \subseteq (C_B^{\triangleleft} C_D^{\triangleleft})^F$, and therefore $(C_A \to C_B C_D) \in P(L, F)$. If $(A \to a) \in P$ for $a \in \Sigma \cup \{\epsilon\}$, then $a \in L(G, A) \subseteq C_A^{\triangleleft}$, so $C_A = C_A^{\triangleleft F} \subseteq a^F$, hence $(C_A \to a) \in P(L, F)$. Moreover, $S$ is mapped to $S(L, F) = \{(\epsilon, \epsilon)\}^{\triangleleft F}$: since $L = L^{\triangleright\triangleleft} = L(G, S)^{\triangleright\triangleleft} = C_S^{\triangleleft} = \{(\epsilon, \epsilon)\}^{\triangleleft}$, we have $C_S = C_S^{\triangleleft F} = \{(\epsilon, \epsilon)\}^{\triangleleft F} = S(L, F)$. Thus $A \mapsto C_A$ induces a grammar homomorphism from $G$ to $G(L, F)$. It follows that $L = L(G) \subseteq L(G(L, F))$. By lemma 1, $L(G(L, F)) \subseteq L$.

The *set of contexts derived from $v \in L$ resp. $E \subseteq \Sigma^*$* is

$$\begin{aligned} Con(v) &:= \{(u, w) \in \Sigma^* \times \Sigma^* \mid \exists \tilde{v} \in \Sigma^* \, v = u\tilde{v}w\}, \\ Con(E) &:= \bigcup \{Con(v) \mid v \in E\}. \end{aligned}$$

**Lemma 3.** *Suppose $G = (V, \Sigma, P, S)$ is a CNF-grammar without unnecessary nonterminals, and $L = L(G) \neq \emptyset$. If $G$ has the FCP, there are finite sets $E \subseteq L$ and $F \subseteq Con(E)$ such that $G$ has the FCP with respect to $F$.*

*Proof.* For each nonterminal $A$ of $G$, there is a finite set of contexts $C_A \neq \emptyset$ such that $L(G, A)^{\triangleright\triangleleft} = C_A^{\triangleleft}$. Let $F = \bigcup \{C_A \mid A \in V\}$ be the union of all $C_A$. Since each $A \in V$ is necessary, $C_A^{\triangleleft} \neq \emptyset$, so there is $v_A \in \Sigma^*$ such that $v_A \in C_A^{\triangleleft}$. Then $C_A \subseteq Con(C_A \odot v_A)$, and $C_A \odot v_A \subseteq L$ is finite. It follows that $F \subseteq Con(E)$ for $E = \bigcup \{C_A \odot v_A \mid A \in V\} \subseteq L$.

It follows that if $L$ has a grammar with the FCP, then in order to find one, we can search finite subsets $E$ of $L$ and consider $G(L, F)$ with $F = Con(E)$. We know that $L(G(L, F))$ is a subset of $L$, and equals $L$ when $F$ is large. However, to construct $G(L, F)$ we must avoid computing infinite filler sets $C^{\triangleleft}$ in order to find the closed sets $C = C^{\triangleleft F}$. We need a truely finite representation of $G(L, F)$.

## 3 Lattices $\mathcal{B}(K, L, F)$ of Relativized Concepts of $L$

Let $K \subseteq \Sigma^*$ and $F \subseteq \Sigma^* \times \Sigma^*$ be word- and context sets. Put

$$A^F := A^{\triangleright} \cap F, \quad C^K := C^{\triangleleft} \cap K, \quad A^{FK} := (A^F)^K, \quad C^{KF} := (C^K)^F.$$

Then $A^F$ and $C^K$ are monotone in $F$ and $K$, but antitone in $A$ und $C$.

**Proposition 8.** *The mappings* $\cdot^F : (\mathcal{P}(K), \supseteq) \rightleftarrows (\mathcal{P}(F), \subseteq) : \cdot^K$ *form a Galois-connection, i.e. for all* $A \in \mathcal{P}(K)$ *and* $C \in \mathcal{P}(F)$ *we have*

$$C^K \supseteq A \iff C \subseteq A^F.$$

*In particular,*

(i) $\cdot^F$ *and* $\cdot^K$ *are antitone:* $A^F \supseteq B^F$ *for all* $A \subseteq B \subseteq K$ *and* $C^K \supseteq D^K$ *for all* $C \subseteq D \subseteq F$,
(ii) $A^{FKF} = A^F$ *and* $C^{KFK} = C^K$ *for all* $A \subseteq K$ *and* $C \subseteq F$.
(iii) $\cdot^{FK}$ *resp.* $\cdot^{KF}$ *is a closure operator on* $(\mathcal{P}(K), \subseteq)$ *resp.* $(\mathcal{P}(F), \subseteq)$.
(iv) $\cdot^F : (\{C^K \mid C \subseteq F\}, \supseteq) \rightleftarrows (\{A^F \mid A \subseteq K\}, \subseteq) : \cdot^K$ *form an order isomorphism.*

We call a pair $\langle A, C\rangle$ such that $A^F = C$ and $C^K = A$ a *relativized concept of* $L$. Let $B(K, L, F)$ be the set of all relativized concepts of $L$. The components of a relativized concept $\langle A, C\rangle$ are closed with respect to $\cdot^{FK}$ and $\cdot^{KF}$, respectively, as $A = A^{FK}$ and $C = C^{KF}$, and via (ii) and (iv) one can represent a relativized concept of $L$ by its closed word set or its closed context set. Writing

$$\mathcal{P}^{FK}(K) := \{A \subseteq K \mid A = A^{FK}\} \quad \text{and} \quad \mathcal{P}^{KF}(F) := \{C \subseteq F \mid C = C^{KF}\}$$

for the set of closed members of $\mathcal{P}(K)$ and $\mathcal{P}(F)$, respectively, (iv) gives an order isomorphism

$$\cdot^F : (\mathcal{P}^{FK}(K), \supseteq) \rightleftarrows (\mathcal{P}^{KF}(F), \subseteq) : \cdot^K .$$

**Proposition 9.** *Let* $K \subseteq \Sigma^*$ *and* $F \subseteq \Sigma^* \times \Sigma^*$. *The set of all relativized concepts of* $L$ *forms a complete lattice* $\mathcal{B}(K, L, F) = (B(K, L, F), \vee, \wedge, \top, \bot, \leq)$ *with the following operations:*

$$\begin{aligned}
\top := \langle K, K^F\rangle \qquad & \bot := \langle \emptyset^{FK}, F\rangle \\
\langle A_1, C_1\rangle \vee \langle A_2, C_2\rangle \ &:= \ \langle (A_1 \cup A_2)^{FK}, C_1 \cap C_2\rangle \\
\langle A_1, C_1\rangle \wedge \langle A_2, C_2\rangle \ &:= \ \langle A_1 \cap A_2, (C_1 \cup C_2)^{KF}\rangle \\
\langle A_1, C_1\rangle \leq \langle A_2, C_2\rangle :&\iff A_1 \subseteq A_2 \wedge C_1 \supseteq C_2.
\end{aligned}$$

When $K = \Sigma^*$, we have $C^K = C^{\lhd}$ for all $C \subseteq F$, and $\mathcal{B}(\Sigma^*, L, F) = \mathcal{B}(L, F)$. When $F$ is finite, we want to use $\mathcal{B}(L, F)$ as a finite "approximation" of the generally infinite $\mathcal{B}(L)$, and when $K$ is also finite, $\mathcal{B}(K, L, F)$ is a finite, effective approximation of $\mathcal{B}(L, F)$. To relativize to $B(K, L, F)$ the monoid operation $\circ$ of $\mathcal{B}(L)$ with its component functions $\circ^{\rhd\lhd}$ and $\circ^{\lhd\rhd}$,

$$\langle A_1, C_1\rangle \circ \langle A_2, C_2\rangle = \langle A_1 \circ^{\rhd\lhd} A_2, C_1 \circ^{\lhd\rhd} C_2\rangle := \langle (A_1 A_2)^{\rhd\lhd}, (C_1^{\lhd} C_2^{\lhd})^{\rhd}\rangle,$$

Clark [1] (Def. 7) defines a *partial* operation $\circ$ on $\mathcal{B}(K, L, F)$ by

$$\langle A_1, C_1\rangle \circ \langle A_2, C_2\rangle := \langle A_1 \circ^{FK} A_2, C_1 \circ^{KF} C_2\rangle := \langle (A_1 A_2)^{FK}, (C_1^K C_2^K)^F\rangle,$$

which need not be a monoid operation. (Proposition 2 for $\cdot^{\rhd\lhd}$ does not extend to all $\cdot^{FK}$.) It is only a partial operation on $\mathcal{B}(K, L, F)$, because although $(A_1 A_2)^{FK}$

is closed in $\mathcal{P}(K)$ and the $\cdot^K$-image of $(C_1^K C_2^K)^F$, the latter need not belong to $\mathcal{P}^{KF}(F)$: $C_1^K C_2^K$ need not be a subset of $K$, whence its $\cdot^F$-image need not be $\cdot^{KF}$-closed in $\mathcal{P}(F)$.

Moreover, the embedding $\langle A, C\rangle \mapsto \langle C^\triangleleft, C\rangle$ from $B(K, L, F)$ to $B(L, F)$ need not preserve $\circ$. However, if $K$ is large enough, everything is fine:

**Lemma 4.** *(cf. Clark[1], Lemma 6): For any $L \subseteq \Sigma^*$ and finite $F \subseteq \Sigma^* \times \Sigma^*$, there is a finite $K \subseteq \Sigma^*$ such that*

$$(B(K, L, F), \vee, \wedge, \bot, \top, \leq, \circ) \simeq (B(L, F), \vee, \wedge, \top, \bot, \leq, \circ).$$

We detail the four line proof sketch from [1] for later reference.

*Proof.* First note that if $C \subseteq F$ is closed in $\mathcal{P}^{KF}(F)$, it is closed in $\mathcal{P}^{\triangleleft F}(F)$ as well: if $C = C^{KF}$, then since $C^\triangleleft \supseteq C^K$ gives $C^{\triangleleft F} \subseteq C^{KF} = C$, so $C = C^{\triangleleft F}$ as $\cdot^{\triangleleft F}$ is a closure operator on $\mathcal{P}(F)$. Therefore $\langle A, C\rangle \mapsto \langle A^{F\triangleleft}, C\rangle = \langle C^\triangleleft, C\rangle$ embeds $B(K, L, F)$ into $B(L, F)$.

To make this embedding be onto $B(L, F)$, we have to choose $K$ sufficiently large. Recall that $C^{KF}$ is antitone in $K$, and $C^{\triangleleft F} \subseteq C^{KF}$ for any $C \subseteq F$ and any $K$. If $C^{KF} \not\subseteq C^{\triangleleft F}$, there is $(u, v) \in C^{KF} \backslash C^{\triangleleft F}$, and hence some $w \in C^\triangleleft \backslash C^K$ such that $uwv \not\subseteq L$, hence $(u, v) \notin C^{(K\cup\{w\})F}$. By adding at most $|C^{KF} \setminus C^{\triangleleft F}|$ many elements from $C^\triangleleft$ to $K$ we obtain $K' \supseteq K$ such that $C^{\triangleleft F} = C^{K'F} \subseteq C^{KF}$. Since $\mathcal{P}(F)$ is finite, there is thus a finite set $K \subseteq \Sigma^*$ such that

$$C^{KF} = C^{\triangleleft F}, \qquad \text{for all } C \subseteq F. \tag{6}$$

It follows that $\mapsto$: $(B(K, L, F), \vee, \wedge, \bot, \top, \leq) \simeq (B(L, F), \vee, \wedge, \bot, \top, \leq)$, and that $A^{FKF} = A^{F\triangleleft F} = A^F$ for any $A \subseteq \Sigma^*$ (not just for $A \subseteq K$). In particular, $\circ$ is total on $B(K, L, F)$, as for each $\langle A_1, C_1\rangle, \langle A_2, C_2\rangle \in B(K, L, F)$ we get

$$\langle A_1, C_1\rangle \circ \langle A_2, C_2\rangle = \langle (A_1A_2)^{FK}, (C_1^K C_2^K)^F\rangle = \langle (A_1A_2)^{FK}, (C_1^K C_2^K)^{FKF}\rangle.$$

Moreover, by similar means we can achieve that

$$(C_1^K C_2^K)^F = (C_1^\triangleleft C_2^\triangleleft)^F, \qquad \text{for all } C_1, C_2 \subseteq F, \tag{7}$$

so that $\mapsto$ preserves $\circ$ since for each $\langle A_1, C_1\rangle, \langle A_2, C_2\rangle \in B(K, L, F)$, we get

$$\begin{aligned} \langle A_1, C_1\rangle \circ^{\mathcal{B}(K,L,F)} \langle A_2, C_2\rangle &= \langle (A_1A_2)^{FK}, (C_1^K C_2^K)^F\rangle \\ \mapsto \quad \langle (C_1^\triangleleft C_2^\triangleleft)^{F\triangleleft}, (C_1^\triangleleft C_2^\triangleleft)^F\rangle &= \langle C_1^\triangleleft, C_1\rangle \circ^{\mathcal{B}(L,F)} \langle C_2^\triangleleft, C_2\rangle. \end{aligned}$$

To see (7), notice that $(C_1^K C_2^K)^F \supseteq (C_1^\triangleleft C_2^\triangleleft)^F$ for any $K$, and if we have $\neq$ here, there are $u \in C_1^\triangleleft \setminus C_1^K$ and $v \in C_2^\triangleleft \setminus C_2^K$. Adding these to $K$ lets $(C_1^K C_2^K)^F$ shrink stricty. By adding at most $|(C_1^K C_2^K)^F \setminus (C_1^\triangleleft C_2^\triangleleft)^F|$ many elements from $C_1^\triangleleft \cup C_2^\triangleleft$ to $K$ we obtain $K' \supseteq K$ such that $(C_1^{K'} C_2^{K'})^F = (C_1^\triangleleft C_2^\triangleleft)^F$. Since $\mathcal{P}(F)$ is finite, we can do this for all $C_1, C_2 \subseteq F$, and achieve (7).

The monotonicity properties ensure that (6) and (7) and hence $\mathcal{B}(K, L, F) \simeq \mathcal{B}(L, F)$ are preserved under extensions of $K$.

**Corollary 1.** *For any finite $F$ where $\emptyset$ is not $\cdot^{\triangleleft F}$-closed, there is a finite $K \subseteq Sub(L)$ such that*

$$(B(K,L,F), \vee, \wedge, \bot, \top, \leq, \circ) \simeq (B(L,F), \vee, \wedge, \top, \bot, \leq, \circ).$$

*Proof.* For $\emptyset \neq C \subseteq F$ we have $C^{\triangleleft} \subseteq Sub(L)$. To achieve $C^{\triangleleft F} = C^{KF}$ and $(C_1^K C_2^K)^F = (C_1^{\triangleleft} C_2^{\triangleleft})^F$, only elements from $C^{\triangleleft}$ and $C_i^{\triangleleft}$ are added to $K$ in the above proof, so we can do with some $K \subseteq Sub(L)$, provided $\emptyset \neq \emptyset^{\triangleleft F}$.

If $\emptyset$ is $\cdot^{\triangleleft F}$-closed, as it often happens, there may be no finite $K \subseteq Sub(L)$ with $\emptyset^{KF} = \emptyset^{\triangleleft F} = \emptyset$. For example, when $K \subseteq Sub(L) = L \neq \Sigma^*$ and $F = \{(\epsilon, \epsilon)\}$, then $\emptyset^{\triangleleft F} = (\Sigma^*)^F = \emptyset$, but $\emptyset^{KF} = K^F = F$. Also, when $K \subseteq Sub(L)$, it may be impossible that $\circ^{KF}$ is total on $\mathcal{P}^{KF}(F)$, i.e. there may be $C_1, C_2 \in \mathcal{P}^{KF}(F)$ such that $C_1 \circ^{KF} C_2 = (C_1^K C_2^K)^F = \emptyset \neq \emptyset^{KF}$.

### 3.1 Clark's Learning Algorithm

For finite sets $K \subseteq \Sigma^*$ and $F \subseteq \Sigma^* \times \Sigma^*$ with $(\epsilon, \epsilon) \in F$, Clark relativized $G(L,V)$ to a grammar $G(K,L,F) = (V, \Sigma, P, S)$ where[2]

$$\begin{aligned}
V &:= \{C \mid C \subseteq F, C = C^{KF}\},\\
S &:= \{(\epsilon,\epsilon)\}^{KF},\\
P &:= \{(C \to w) \mid C \in V, w \in \Sigma \cup \{\epsilon\}, w \in C^K\}\\
&\quad \cup \{(C \to DE) \mid C, D, E \in V,\ (D^K E^K)^{FK} \subseteq C^K\}.
\end{aligned}$$

He then shows (Lemma 2,3,4 in [2]) that $L(G(K,L,F))$ depends monotonically on $F$ and antitonically on $K \supseteq \Sigma \cup \{\epsilon\}$. Theorem 1 in [2] claims that the following algorithm identifies $L$ in the limit, i.e. that for any oracles $T$ and $\chi_L$, $\langle G_n \mid n \in \mathbb{N}\rangle$ gets constant at some $G_n$ such that $L(G_n) = L$.

Why should $\langle G_n \mid n \in \mathbb{N}\rangle$ converge to a grammar for $L$? Call a finite set $F$ of contexts *adequate for* $L$ if $L \subseteq L(G(K,L,F))$ for every finite $\Sigma \cup \{\epsilon\} \subseteq K \subseteq \Sigma^*$. By lemma 3, there is some $F$ such that $L \subseteq L(G(L,F))$. Then any $F_n \supseteq F$ is adequate for $L$, since $G(L,F) = G(\Sigma^*, L, F)$, and $L(G(K,L,F))$ is monotone in $F$ and antitone in $K \subseteq \Sigma^*$. So one would first like to show:

$$\langle F_n \mid n \in \mathbb{N}\rangle \text{ gets constant in some } F_n \text{ that is adequate for } L,$$

so that $L \subseteq L(G(K_m, L, F_n))$ for all $m$, and then show:

$$\langle K_n \mid n \in \mathbb{N}\rangle \text{ gets constant in some } K_m \text{ such that } L = L(G(K_m, L, F_n)). \tag{8}$$

To achieve (8), Lemma 5 in [2] claims:

$$\text{For any } L \text{ and } F, \text{ there is } K \subseteq \Sigma^* \text{ such that } L(G(K,L,F)) \subseteq L. \tag{9}$$

The proof sketched in [2] only works for the infinite set $K = \Sigma^*$, but the claim is needed with finite $K \subseteq Sub(L)$ in the grammar inference algorithm. However, this strengthening of (9) is wrong:

[2] We omit a size bound $f$ on a generating subset for the $C^{KF}$ in the definition of $V$, which only serves to bound $|G(K,L,F)|$ by a polynomial in $|K|$ and $|F|$.

**Table 1.** Clark's grammar inference algorithm

Let $\Sigma$ be a finite alphabet and $L \subseteq \Sigma^*$, a language with a CNF-grammar with the finite context property, be given by oracles $\chi_L : \Sigma^* \to \mathbb{B}$ and $T : \mathbb{N} \to \Sigma^*$ for recognition and enumeration of $L$.

Produce a sequence $\langle G_n \mid n \in \mathbb{N}\rangle$ of CNF-grammars, where $w_n = T(n)$:

$$E_0 := \emptyset, \qquad E_{n+1} := E_n \cup \{w_n\},$$

$$K_0 := \Sigma \cup \{\epsilon\}, \qquad K_{n+1} := \begin{cases} Sub(E_{n+1}), & \text{if } E_{n+1} \not\subseteq L(G_n), \text{ or} \\ & G(Sub(E_{n+1}), L, F_n) \not\simeq G_n \\ K_n, & \text{else} \end{cases}$$

$$F_0 := \{(\epsilon, \epsilon)\}, \qquad F_{n+1} := \begin{cases} Con(E_{n+1}), & \text{if } E_{n+1} \not\subseteq L(G_n), \\ F_n, & \text{else} \end{cases}$$

$$G_0 := G(K_0, L, F_0). \qquad G_{n+1} := G(K_{n+1}, L, F_{n+1}).$$

*Example 1.* There is a language $L \subseteq \Sigma^*$ and a finite set of contexts $F$ with

$$L(G(K, L, F)) \not\subseteq L, \quad \text{for all finite } K \text{ such that } \Sigma \cup \{\epsilon\} \subseteq K \subseteq Sub(L).$$

Let $\Sigma = \{a\}$, $F = \{(\epsilon, \epsilon)\}$, $L = \{\epsilon, a\}$. The only finite set $K$ with $Sub(L) \supseteq K \supseteq \Sigma \cup \{\epsilon\}$ is $K = L$.

*(i)* The set $V$ of closed elements of $\mathcal{P}(F)$ contains $F$, since $F \subseteq F^{KF}$ is maximal in $\mathcal{P}(F)$. Since $F^K = \{k \in K \mid \epsilon k \epsilon \in L\} = K \cap L = L$ and $a \in L$, $(F \to a)$ is a non-branching rule of $G(K, L, F)$.

*(ii)* To see that $(F \to FF)$ is a branching rule of $G(K, L, F)$, notice that since $LL \not\subseteq L$, we have $(F^K F^K)^F = (LL)^F = \emptyset$, so $(F^K F^K)^{FK} = \emptyset^K = K = L = F^K$.

Since $F$ is the start symbol of $G(K, L, F)$, it follows that $aa \in L(G(K, L, F)) \setminus L$. This grammar, extended by the rule $F \to \epsilon$, is $G_0$, so $L \subset L(G_0)$ and $K_1 = K_0$, $F_1 = F_0$, and $G_1 = G(K_1, L, F_1) = G_0$. By induction, $\langle G_n \mid n \in \mathbb{N}\rangle$ gets constant in $G_0$, a grammar that does *not* define $L$.

### 3.2 Correcting Clark's Algorithm

Example 1 shows that Clark's condition $(D^K E^K)^{FK} \subseteq C^K$ in branching rules of $G(K, L, F)$ is too permissive. For concepts $C, D, E$, being a branching rule $(C \to DE)$ of $G(L, V)$ amounts to any of the equivalent conditions $D^{\triangleleft} E^{\triangleleft} \subseteq C^{\triangleleft}$, $(D^{\triangleleft} E^{\triangleleft})^{\triangleright\triangleleft} \subseteq C^{\triangleleft}$ or $C \subseteq (D^{\triangleleft} E^{\triangleleft})^{\triangleright}$. These are no longer equivalent in the relativized situation; for concepts $C, D, E$ relativized to $K, F$ we only have

$$D^K E^K \subseteq C^K \quad \Longrightarrow \quad C \subseteq (D^K E^K)^F \quad \Longrightarrow \quad (D^K E^K)^{FK} \subseteq C^K.$$

Though $\cdot^{FK}$ is a closure operation on $\mathcal{P}(K)$, we may have $D^K E^K \not\subseteq (D^K E^K)^{FK}$ when $D^K E^K \not\subseteq K$. We modify the definition of $G(K, L, F)$ by replacing the condition $(D^K E^K)^{FK} \subseteq C^K$ by the stronger condition[3] $C \subseteq (D^K E^K)^F$. Moreover, we exclude the empty context set from the nonterminals.

**Definition 1.** *Let $K, L \subseteq \Sigma^*$ be arbitrary sets of words, and $F \subseteq \Sigma^* \times \Sigma^*$ a finite set of contexts such that $(\epsilon, \epsilon) \in F$. Then $G(K, L, F)$ is the binary grammar $(V, \Sigma, P, S)$ where*

$$\begin{aligned} V &:= \{C \mid \emptyset \neq C \subseteq F,\ C^{KF} = C\} \\ S &:= \{(\epsilon, \epsilon)\}^{KF}, \\ P &:= \{(C \to w) \mid C \in V, w \in \Sigma \cup \{\epsilon\}, C \subseteq w^F\} \\ &\quad \cup \{(C \to DE) \mid C, D, E \in V,\ C \subseteq (D^K E^K)^F\}. \end{aligned}$$

Note that for $C = C^{KF}$, the condition $C \subseteq w^F$ is equivalent to $w \in C^K$. When $K$ and $F$ are finite, we can determine $V$, $S$ and $P$ from a decision algorithm for membership in $L$. To determine $V$, we need to know $(F \odot K) \cap L$, and to determine $P$, we need to know $(F \odot KK) \cap L$.

Our conditions $C \subseteq w^F$ for non-branching rules $(C \to w)$ and $C \subseteq (D^K E^K)^F$ for branching rules $(C \to DE)$ of $G(K, L, F)$ are monotone in $F$ and antitone in $K$. Yet, as pointed out by R. Yoshinaka[4], $L(G(K, L, F))$ is neither monotone in $F$ nor antitone in $K$. The reason is that for $F_1 \subseteq F_2$, say, the set $V_1$ of nonterminals of $G(K, L, F_1)$ is not a subset of the set $V_2$ of nonterminals of $G(K, L, F_2)$, and the embedding $\tilde{\cdot} : V_1 \to V_2$ given by $\tilde{C} := C^{KF_2}$ gives $C \subseteq \tilde{C}$ and hence can lead from $C \subseteq (D^K E^K)^{F_1}$ to $\tilde{C} \not\subseteq (\tilde{D}^K \tilde{E}^K)^{F_2}$ and does not induce a grammar homomorphism.

Before presenting a correction of Clark's algorithm, let us recapitulate his idea. A finite amount of positive information $E \subseteq L$ about $L$ gives a finite set $F = \mathit{Con}(E)$ of contexts and, through $\mathcal{B}(L, F)$, a grammar $G(L, F)$ of a sublanguage of $L$. If $F$ is big enough, $G(L, F)$ defines $L$. Each $F$ and each finite $K \subseteq \Sigma^*$ provide, through the finite $\mathcal{B}(K, L, F)$, a grammar $G(K, L, F)$ whose language is monotone in $F$ and antitone in $K$. If $F$ is big enough, $L \subseteq L(G(K, L, F))$, and if $K$ is big enough, one has $\mathcal{B}(L, F) \simeq \mathcal{B}(K, L, F)$, in which case $G(K, L, F)$ is $G(L, F)$ and defines $L$. Thus, when $E \not\subseteq L(G(K, L, \mathit{Con}(E)))$, one needs to increase $E$, and otherwise one should keep $F = \mathit{Con}(E)$ fixed and increase $K$ to make it big enough.

Since the input to the inference process consists of positive information about $L$ only, we cannot use lemma 4 directly to get $\mathcal{B}(K, L, F) \simeq \mathcal{B}(L, F)$, but need a refinement with $K \subseteq \mathit{Sub}(L)$ instead of $K \subseteq \Sigma^*$. (We get no clue on which $K \not\subseteq \mathit{Sub}(L)$ would satisfy $\emptyset = \emptyset^{\triangleleft F} = \emptyset^{KF}$, so we exclude $\emptyset$ resp. $\top$ from the nonterminals.) Moreover, it will in general be impossible to define $L$ with $G(K, L, F)$ where $K = \mathit{Sub}(E)$ and $F = \mathit{Con}(E)$ for some finite $E \subseteq L$; we

[3] It excludes $(F \to FF)$ in example 1, where $F \not\subseteq (F^K F^K)^F = \emptyset$. The weaker $C \subseteq (D^K E^K)^{FKF}$ is equivalent to Clark's $(D^K E^K)^{FK} \subseteq C^K$.

[4] Personal communication, February 2013.

may need a finite $K$ with $Sub(E) \subset K \subseteq Sub(L)$. So we have to switch between increasing $F$ on the one hand and increasing $K$ while keeping $F$ fixed on the other. Finally, of course we cannot explicitly test whether $F$ is big enough so that $L = L(G(L,F))$, or whether $K$ gives $\mathcal{B}(K,L,F) \simeq \mathcal{B}(L,F)$. We need computable substitutes for such tests.

We say that $\langle \mathcal{P}^{KF}(F), \circ^{KF} \rangle$ *and* $\langle \mathcal{P}^{\triangleleft F}(F), \circ^{\triangleleft F} \rangle$ *almost agree*, in symbols: $\langle \mathcal{P}^{KF}(F), \circ^{KF} \rangle \equiv \langle \mathcal{P}^{\triangleleft F}(F), \circ^{\triangleleft F} \rangle$, if for all non-empty $C, C_1, C_2 \subseteq F$,

$$C^{KF} = C^{\triangleleft F} \quad \text{and} \quad C_1 \circ^{KF} C_2 = C_1 \circ^{\triangleleft F} C_2.$$

Like (6) and (7) in the proof of lemma 4, this property is monotone in $K$; but now we only consider non-empty subsets of $F$.

**Lemma 5.** *Let $F \subseteq \Sigma^* \times \Sigma^*$ and $K \subseteq Sub(L)$ be finite, and $(\epsilon, \epsilon) \in F$.*

(i) *There is a finite $K \subseteq \tilde{K} \subseteq Sub(L)$ with $\langle \mathcal{P}^{\tilde{K}F}(F), \circ^{\tilde{K}F} \rangle \equiv \langle \mathcal{P}^{\triangleleft F}(F), \circ^{\triangleleft F} \rangle$.*
(ii) *If $\langle \mathcal{P}^{KF}(F), \circ^{KF} \rangle \equiv \langle \mathcal{P}^{\triangleleft F}(F), \circ^{\triangleleft F} \rangle$, then $G(K,L,F) = G(L,F)$.*

*Proof.* (i) We need to satisfy the restrictions of (6) and (7) to non-empty sets $C, C_1, C_2 \subseteq F$. As the proof of lemma 4 shows, we have to extend $K$ by elements of $C^{\triangleleft}, C_1^{\triangleleft}, C_2^{\triangleleft}$, and these are subsets of $Sub(L)$ when $C, C_1, C_2$ are non-empty.

(ii) Since $\langle \mathcal{P}^{KF}(F), \circ^{KF} \rangle$ and $\langle \mathcal{P}^{\triangleleft F}(F), \circ^{\triangleleft F} \rangle$ almost agree, $G(K,L,F)$ and $G(L,F)$ have the same nonterminals, start symbols, and non-branching rules and branching rules.

Although $\langle \mathcal{P}^{\triangleleft F}(F), \supseteq, \circ^{\triangleleft F} \rangle$ is a finite structure, we generally cannot compute it, given an oracle for membership in $L$, because possibly infinite word sets $C^{\triangleleft}$ are involved. So we cannot test whether $\langle \mathcal{P}^{KF}(F), \circ^{KF} \rangle$ almost agrees with $\langle \mathcal{P}^{\triangleleft F}(F), \circ^{\triangleleft F} \rangle$. Nor can we test whether $G(K,L,F)$ equals $G(L,F)$. But we *can* test the following weaker property, which however is *not* monotone in $K$.

We say $\circ^{KF}$ *is almost total on* $\mathcal{P}^{KF}(F)$, if for all non-empty $C_1, C_2 \in \mathcal{P}^{KF}(F)$, $C_1 \circ^{KF} C_2$ belongs to $\mathcal{P}^{KF}(F) \cup \{\emptyset\}$. Using an oracle for membership in $L$ one can check whether $\circ^{KF}$ is almost total.

**Proposition 10.** *The following conditions are equivalent:*

(i) *For all non-empty $C \subseteq F$, $C^{KF} = C^{\triangleleft F}$.*
(ii) $\mathcal{P}^{KF}(F) \setminus \{\emptyset\} = \mathcal{P}^{\triangleleft F}(F) \setminus \{\emptyset\}$.

The easy proof is left to the reader.

**Proposition 11.** *Let $F \subseteq \Sigma^* \times \Sigma^*$ and $K \subseteq \Sigma^*$. If $\langle \mathcal{P}^{KF}(F), \circ^{KF} \rangle$ and $\langle \mathcal{P}^{\triangleleft F}(F), \circ^{\triangleleft F} \rangle$ almost agree, then $\circ^{KF}$ is almost total on $\mathcal{P}^{KF}(F)$.*

*Proof.* Suppose $C_1, C_2 \in \mathcal{P}^{KF}(F)$ are nonempty, and $(C_1 \circ^{KF} C_2) \neq \emptyset$. Then

$$C_1 \circ^{KF} C_2 = C_1 \circ^{\triangleleft F} C_2 = (C_1^{\triangleleft} C_2^{\triangleleft})^F = (C_1^{\triangleleft} C_2^{\triangleleft})^{F \triangleleft F} = (C_1^{\triangleleft} C_2^{\triangleleft})^{FKF} = (C_1 \circ^{KF} C_2)^{KF},$$

and so $(C_1 \circ^{KF} C_2) \in \mathcal{P}^{KF}(F)$.

**Lemma 6.** *Suppose $F \subseteq \Sigma^* \times \Sigma^*$ is finite with $(\epsilon, \epsilon) \in F$ and $L \subseteq \Sigma^*$ has a CNF-grammar $G$ with the FCP with respect to $F$. If $\Sigma \cup \{\epsilon\} \subseteq K \subseteq \Sigma^*$ is finite and $\circ^{KF}$ is almost total on $\mathcal{P}^{KF}(F)$, then $L \subseteq L(G(K, L, F))$.*

*Proof.* By assumption, for each nonterminal $A$ of $G$ there is a finite set $\emptyset \neq C_A \subseteq F$ with $L(G, A)^{\triangleright\triangleleft} = C_A^{\triangleleft}$. We show that $A \mapsto C_A^{KF}$ induces a grammar homomorphism from $G$ to $G(K, L, F)$. For each nonterminal $A$ of $G$, $\emptyset \neq C_A^{KF}$ is $\cdot^{KF}$-closed, hence a nonterminal of $G(K, L, F)$. Let $(A \to DE)$ be a rule of $G$, so $C_A^{\triangleleft} \supseteq C_D^{\triangleleft} C_E^{\triangleleft}$, using proposition 2. Then

$$\emptyset \neq C_A \subseteq C_A^{\triangleleft F} \subseteq (C_D^{\triangleleft} C_E^{\triangleleft})^F \subseteq (C_D^K C_E^K)^F = (C_D^{KFK} C_E^{KFK})^F = C_D^{KF} \circ^{KF} C_E^{KF}.$$

By monotonicity of $\cdot^{KF}$ and since $\circ^{KF}$ is almost total on $\mathcal{P}^{KF}(F)$, we get

$$C_A^{KF} \subseteq (C_D^{KF} \circ^{KF} C_E^{KF})^{KF} \subseteq C_D^{KF} \circ^{KF} C_E^{KF},$$

and so $(C_A^{KF} \to C_D^{KF} C_E^{KF})$ is a rule of $G(K, L, F)$. Likewise, let $(A \to a)$ be a rule of $G$. Then $a \in L(G, A) \cap K \subseteq C_A^{\triangleleft} \cap K = C_A^K$, so $C_A^{KF} \subseteq a^F$, and $(C_A^{KF} \to a)$ is a rule of $G(K, L, F)$. Hence, if $A \Rightarrow_G^* w$, then $C_A^{KF} \Rightarrow_{G(K,L,F)}^* w$. We may assume that $C_S = \{(\epsilon, \epsilon)\}$, as $C_S^{\triangleleft} = L = \{(\epsilon, \epsilon)\}^{\triangleleft} = \{(\epsilon, \epsilon)\}^{\triangleleft F \triangleleft}$ since $(\epsilon, \epsilon) \in F$. Then $C_S^{KF} = \{(\epsilon, \epsilon)\}^{KF}$ is the start symbol of $G(K, L, F)$, and we have $L = L(G) \subseteq L(G(K, L, F))$.

The idea for the corrected grammar inference is as follows. Start with $F = \{(\epsilon, \epsilon)\}$ and consume increasing finite subsets $E$ of $L$ until $K = Sub(E)$ makes $\circ^{KF}$ almost total. We find such a $K$ by lemma 5 and proposition 11. If then $E \not\subseteq L(G(K, L, F))$, we know by lemma 6 that $L$ does not have a grammar with the FCP with respect to $F$ (i.e. "$F$ is not adequate for $L$"). So we update $F$ to $Con(E)$ and repeat this process, until we have $K \subseteq Sub(L)$ where $\circ^{KF}$ is almost total and $E \subseteq L(G(K, L, F))$. Then we keep $F$ fixed and increase $E$ and $K$ until $\circ^{KF}$ is almost total, and check if $E \subseteq L(G(K, L, F))$. If we no more run into the case $E \not\subseteq L(G(K, L, F))$ where $F$ is increased, we exhaust the finite subsets $K$ of $Sub(L)$ and hence reach $G(K, L, F) = G(L, F)$ by lemma 5; since this is monotone in $K$, the grammar $G(L, F)$ is the limit grammar. Then on the one hand, $E \subseteq L(G(K, L, F)) = G(L, F)$ for all finite $E \subseteq L$, hence $L \subseteq L(G(L, F))$, and on the other hand $L(G(L, F)) \subseteq L$.

Since $L$ has a grammar with the FCP, after finitely many updates of $F = Con(E)$ it has a grammar with the FCP with respect to $F$. Then by lemma 6, $L \subseteq L(G(K, L, F))$ for all $K$ where $\circ^{KF}$ is almost total, and so we do not run into the case $E \not\subseteq L(G(K, L, F))$ any more.

**Theorem 2.** *If $\emptyset \neq L \subseteq \Sigma^*$ has a CNF-grammar with the finite context property, then the algorithm of table 2 identifies $L$ in the limit, i.e. for the sequence $\langle G_n \mid n \in \mathbb{N} \rangle$ of grammars produced for any membership oracle and enumeration of $L$, there is $m$ such that $L(G_m) = L$ and $G_n = G_m$ for all $n \geq m$.*

*Proof.* Let $G$ be a grammar for $L$ with the FCP. We may assume that $G$ has no unneccesary symbols. By lemma 3, $G$ has the FCP with respect to some $F \subseteq Con(E)$ for some finite $E \subseteq L$.

**Table 2.** Grammar inference algorithm

Let $\Sigma$ be a finite alphabet and $\emptyset \neq L \subseteq \Sigma^*$, a language with a CNF-grammar with the finite context property, be given by oracles $\chi_L : \Sigma^* \to \mathbb{B}$ and $T : \mathbb{N} \to \Sigma^*$ for recognition and enumeration of $L$.

Produce a sequence $\langle G_n \mid n \in \mathbb{N} \rangle$ of CNF-grammars, where $w_n = T(n)$:

$$E_0 := \emptyset, \qquad E_{n+1} := E_n \cup \{w_n\},$$

$$K_0 := \Sigma \cup \{\epsilon\}, \qquad K_{n+1} := Sub(E_{n+1}) \cup K_0,$$

$$F_0 := \{(\epsilon, \epsilon)\}, \qquad G_{n+1} := \begin{cases} G(K_{n+1}, L, F_n) & \text{if } \circ^{K_{n+1}F_n} \text{ is almost total,} \\ G_n, & \text{else,} \end{cases}$$

$$G_0 := G(K_0, L, F_0). \qquad F_{n+1} := \begin{cases} Con(E_{n+1}) & \text{if } E_{n+1} \not\subseteq L(G_{n+1}) \text{ and } \circ^{K_{n+1}F_n} \text{ is almost total,} \\ F_n, & \text{else.} \end{cases}$$

***Claim 1*** The sequence $\langle F_n \mid n \in \mathbb{N} \rangle$ gets constant at a finite $F_{\tilde{n}} \subseteq Con(L)$.

*Proof of Claim 1*: By induction, $F_n \subseteq F_{n+1} \subseteq Con(E_{n+1})$ for all $n$. Assume that $\langle F_n \mid n \in \mathbb{N} \rangle$ does not get constant. Then neither do $\langle E_n \mid n \in \mathbb{N} \rangle$ nor $\langle K_n \mid n \in \mathbb{N} \rangle$, and there are infinitely many $n$ such that $F_n \subset F_{n+1} = Con(E_{n+1})$. Let $n$ be one of those. Then $\circ^{K_{n+1}F_n}$ is almost total and hence $G_{n+1} = G(K_{n+1}, L, F_n)$. We may assume that $F \subseteq F_{n+1}$, so $G$ has the FCP with respect to $F_{n+1}$.

Since $\langle F_n \mid n \in \mathbb{N} \rangle$ does not get constant, there is a least $m \geq n+1$ such that $\circ^{K_{m+1}F_m}$ is almost total, and then

$$G_{m+1} = G(K_{m+1}, L, F_m) = G(K_{m+1}, L, F_{n+1}).$$

By lemma 6, $L \subseteq L(G(K_{m+1}, L, F_{n+1})) = L(G_{m+1})$, so $E_{m+1} \subseteq L(G_{m+1})$ and $F_{m+1} = F_m = F_{n+1}$. Let $\tilde{m}$ be the least $k \geq m+1$ such that $\circ^{K_{k+1}F_k}$ is almost total. Then $G_{\tilde{m}} = G_{m+1}$, $F_{\tilde{m}} = F_{m+1} = F_{n+1}$ and

$$G_{\tilde{m}+1} = G(K_{\tilde{m}+1}, L, F_{\tilde{m}}) = G(K_{\tilde{m}+1}, L, F_{n+1}).$$

Since $\circ^{K_{\tilde{m}+1}F_{\tilde{m}}}$ is almost total, we also have $E_{\tilde{m}+1} \subseteq L \subseteq L(G_{\tilde{m}+1})$, again by lemma 6, so $F_{\tilde{m}+1} = F_{\tilde{m}} = F_{n+1}$. By induction, $F_{\tilde{m}+1} = F_{n+1}$ for all $\tilde{m} \geq n+1$ where $\circ^{K_{\tilde{m}+1}F_{\tilde{m}}}$ is almost total, hence for all $\tilde{m} \geq n+1$. This contradicts the assumption that $\langle F_n \mid n \in \mathbb{N} \rangle$ does not get constant.

***Claim 2***: If $\langle F_n \mid n \in \mathbb{N} \rangle$ converges to $F_{\tilde{n}}$, then $\langle G_n \mid n \in \mathbb{N} \rangle$ converges to $G(L, F_{\tilde{n}})$ and $L = L(G(L, F_{\tilde{n}}))$.

*Proof of Claim 2*: Suppose $\langle F_n \mid n \in \mathbb{N} \rangle$ converges to $F_{\tilde{n}}$. By lemma 5 there is some finite $K \subseteq Sub(L)$ such that

$$\langle \mathcal{P}^{KF_{\tilde{n}}}(F_{\tilde{n}}), \circ^{KF_{\tilde{n}}} \rangle \equiv \langle \mathcal{P}^{\triangleleft F_{\tilde{n}}}(F_{\tilde{n}}), \circ^{\triangleleft F_{\tilde{n}}} \rangle. \qquad (10)$$

Since (10) is monotone in $K$ and $\langle K_n \mid n \in \mathbb{N}\rangle$ is non-decreasing and majorizes all finite subsets of $Sub(L)$, there is $m_0 \geq \tilde{n}$ such that for all $m \geq m_0$,

$$\langle \mathcal{P}^{K_{m+1}F_{\tilde{n}}}(F_{\tilde{n}}), \circ^{K_{m+1}F_{\tilde{n}}}\rangle \equiv \langle \mathcal{P}^{\lhd F_{\tilde{n}}}(F_{\tilde{n}}), \circ^{\lhd F_{\tilde{n}}}\rangle.$$

By lemma 11, $\circ^{K_{m+1}F_{\tilde{n}}} = \circ^{K_{m+1}F_m}$ is almost total for $m \geq m_0$. Therefore $G_{m+1} = G(K_{m+1}, L, F_{\tilde{n}})$ for all $m \geq m_0$, and $G(K_{m+1}, L, F_{\tilde{n}}) = G(L, F_{\tilde{n}})$ by lemma 5. So $\langle G_n \mid n \in \mathbb{N}\rangle$ gets constant in $G(L, F_{\tilde{n}})$.

We have $L(G(L, F_{\tilde{n}})) \subseteq L$ by lemma 1. Suppose $L \not\subseteq L(G(L, F_{\tilde{n}}))$, and pick $w \in L \setminus L(G(L, F_{\tilde{n}}))$. Since $\langle E_n \mid n \in \mathbb{N}\rangle$ majorizes every finite subset of $L$, for some $m \geq m_0$ we have $E \cup \{w\} \subseteq E_{m+1} \not\subseteq L(G(L, F_{\tilde{n}})) = L(G(K_{m+1}, L, F_m))$. Then $F_{\tilde{n}} = F_{m+1} = Con(E_{m+1}) \supseteq Con(E) \supseteq F$. So the given $G$ has the FCP with respect to $F_{\tilde{n}}$. Then $L \subseteq L(G(K_{m+1}, L, F_m)) = L(G(L, F_{\tilde{n}}))$ by lemma 6, in contradiction to the assumption.

The proof does not rely on whether $L(G(K, L, F))$ is monotone in $F$ or antitone in $K$. Due to space limitations, we do not demonstrate that a grammar for the Dyck language of well-bracketed strings is correctly inferred.

To obtain polynomial update complexity, we need to limit the number of concepts. A grammar has the *f-finite context property* ($f$-FCP), if for each of its nonterminals $A$ there is a set $C_A$ of contexts such that $A^{\rhd\lhd} = C_A^{\lhd}$ and $|C_A| \leq f$, where $f \geq 1$. Restrict the algorithm of table 2 to languages having a grammar with the $f$-FCP, and to build the hypotheses $G_n$, use grammars $G_f(K, L, F)$ defined like $G(K, L, F)$, but whose nonterminals are non-empty elements from $\mathcal{P}_f^{KF}(F) := \{C^{KF} \mid C \subseteq F, |C| \leq f\}$. Then the algorithm has *polynomial update time*, i.e. the number of steps to generate the hypothesis grammar $G_n$ is bounded by a polynomial in $|E_n|$ and $\max\{|w| \mid w \in E_n\}$.

First observe that the number of steps to compute $G_f(K, L, F)$ from $K$ and $F$ is bounded by a polynomial in $|K|$ and $|F|$. Clearly, $\mathcal{P}_f^{KF}(F)$ is of size $O(|F|^f)$ and its elements can be represented in a trie of bitvectors of length $|F|$. For each $C \subseteq F$ with $|C| \leq f$ we need $O(f|K||F|)$ membership queries to determine $C^{KF}$ and $|F|$ steps to insert it into the trie. Likewise, for $C_1, C_2 \in \mathcal{P}_f^{KF}(F)$ we can compute $C_1 \circ^{KF} C_2$ and check if it belongs to $\mathcal{P}_f^{KF}(F)$ in $O(|K|^2|F|)$ steps. Therefore, we can compute $(\mathcal{P}_f^{KF}(F), \circ^{KF})$ and check if $\circ^{KF}$ is almost total on $\mathcal{P}_f^{KF}(F)$ in $O(|K|^2|F|^{2f+1})$ steps. If $\circ^{KF}$ is almost total, we can read off $G_f(K, L, F)$ from $(\mathcal{P}_f^{KF}(F), \circ^{KF})$ in $O(|V|^3|F|) = O(|F|^{3f+1})$ steps.

Next, let $e = |E_{n+1}|$ and $m$ be the maximal length of words in $E_{n+1}$. Then $Sub(E_{n+1})$ and $Con(E_{n+1})$ are determined in $O(em^2)$ steps, so $K_{n+1}$ and $F_n \subseteq Con(E_n)$ are of size $O(em^2)$, and $G_{n+1} = G_f(K_{n+1}, L, F_n)$ is polynomial in $e$ and $m$. Finally, to determine $F_{n+1}$, we must check whether $E_{n+1} \not\subseteq L(G_{n+1})$, which can be done in time polynomial in $e$ and $m$ using a CYK-recognizer ([6]).

## 4 Conclusion

We have pointed out that Clark's grammar inference algorithm may converge to a grammar for a superset of the intended language. We modified Clark's grammars $G(K, L, F)$, replaced major parts of the reasoning for the inference process, and provided proofs of the correctness of the algorithm. We have thus shown that one can learn a grammar for $L$, if $L$ does have a CNF grammar with the finite context property, and can do so in the framework of relativized syntactic concepts of $L$. Some experts seemed to think this was impossible[5], as [2] and [8] rely heavily on the (anti)monotonicity of $L(G(K, L, F))$ in $(K)F$.

Yoshinaka's[8] "dual" approach uses grammars $H(K, L, F)$ that differ from our $G(K, L, F)$ by admitting arbitrary $C \subseteq F$ as nonterminals in order to make $L(H(K, L, F))$ monotone in $F$ and antitone in $K$. This simplifies the reasoning, but leads to a limit grammar with many "copies" of the same rule.

If we consider syntactic concepts of $L$ as the only linguistically relevant notions to be used in describing $L$, we would like to do syntactic analysis in terms of "concept arithmetic", i.e. using $\mathcal{B}(L)$ rather than $\mathcal{P}(\Sigma^*)$. It remains to be developed what this amounts to, in particular when $L$ is not context-free.

**Acknowledgement.** I wish to thank A.Clark for email discussions and W. Buszkowski and R. Yoshinaka for pointing out mistakes in a draft version and for a hint to Jipsen e.a.[4].

## References

1. Clark, A.: A learnable representation for syntax using residuated lattices. In: de Groote, P., Egg, M., Kallmeyer, L. (eds.) FG 2009. LNCS, vol. 5591, pp. 183–198. Springer, Heidelberg (2011)
2. Clark, A.: Learning context free grammars with the syntactic concept lattice. In: Sempere, J.M., García, P. (eds.) ICGI 2010. LNCS (LNAI), vol. 6339, pp. 38–51. Springer, Heidelberg (2010)
3. Harris, Z.S.: From morpheme to utterance. Language 22(3), 161–183 (1946)
4. Jipsen, P., Tsinakis, C.: A survey of residuated lattices. In: Martinez, J. (ed.) Ordered Algebraic Structures, pp. 19–56. Kluwer (2002)
5. Leiß, H.: Learning CFGs with the finite context property. A note on A. Clark's algorithm. Universität München, CIS, Manuscript (July 2012)
6. Hopcroft, J.E., Ullman, J.D.: Introduction to Automata Theory, Languages, and Computation. Addison-Wesley (1979)
7. Wurm, C.: Completeness of full Lambek calculus for syntactic concept lattices. In: Morrill, G., Nederhof, M.-J. (eds.) Formal Grammar 2012 and 2013. LNCS, vol. 8036, pp. 126–141. Springer, Heidelberg (2013)
8. Yoshinaka, R.: Towards dual approaches for learning context-free grammars based on syntactic concept lattices. In: Mauri, G., Leporati, A. (eds.) DLT 2011. LNCS, vol. 6795, pp. 429–440. Springer, Heidelberg (2011)

[5] Ryo Yoshinaka: "I am afraid relativized lattices do not have quite a right property to base on for learning.", personal communication, February 2013

# A Dynamic Categorial Grammar*

Scott Martin[1] and Carl Pollard[2]

[1] Natural Language Understanding and Artificial Intelligence Laboratory
Nuance Communications, 1198 East Arques Avenue
Sunnyvale, California 94085, USA
scott.martin@nuance.com
[2] Department of Linguistics
The Ohio State University, 1712 Neil Avenue
Columbus, Ohio 43210, USA
pollard@ling.ohio-state.edu

**Abstract.** We present a compositional, dynamic categorial grammar for discourse analysis that captures the core insights of dynamic semantics: indefinites do not quantify but introduce discourse referents; definites are anaphoric to previously-mentioned discourse referents; discourse referents have their 'lifespan' limited by certain operators. The categorial grammar formalism we propose is strongly lexicalist and derives linguistic signs with a syntactic division of labor separating surface form from the underlying combinatorics. We argue that this formalism compares favorably with earlier efforts on several counts. It does not require any complicated or idiosyncratic machinery such as specialized assignments, states, or continuations, and encodes the requirement that a certain discourse referent be present in the discourse context using dependent types, rather than e.g. partial functions. The dynamic semantics itself is a straightforward extension of an underlying static semantics that is fully (hyper)intensional, avoiding many unsavory problems associated with standard possible worlds approaches.

**Keywords:** categorial grammar, dynamic semantics, compositionality, dependent type theory, hyperintensionality.

## 1 Introduction

In dynamic semantics, the interpretation of sentences both depends upon and acts upon the utterance context. When the classic dynamic semantic frameworks of discourse representation theory (DRT, [15]) and file change semantics (FCS, [13]) were first introduced, they attracted a lot of attention because they provided analyses of phenomena (such as donkey anaphora, cross-sentential anaphora, presupposition satisfaction, and the novelty condition on indefinites) that the then-predominant semantic framework of Montague semantics (MS, [24]) could

* Thanks are due to audiences at Ohio State University who critiqued early drafts of this work, especially to the joint linguistics/philosophy seminar in the spring of 2011 and the pragmatics reading group.

G. Morrill et al. (Eds.): Formal Grammar 2014, LNCS 8612, pp. 138–154, 2014.

not account for. However, in formal terms, classic dynamic semantics compared unfavorably with MS, which provided an explicit (albeit awkward) compositional interface to syntax and whose semantic theory was expressed in a higher-order language essentially equivalent to the classical simple theory of types ([3,6,14]). Neither DRT nor FCS was compositional; the theoretical content of FCS was expressed (not always as precisely as might have been hoped for) only in the metalanguage, and DRT struck many as excessively syntactic and procedural.

Early attempts at logical formalization of dynamic semantics such as dynamic predicate logic (DMG, [8]) and dynamic Montague grammar (DPL, [9]) addressed the compositionality issue, but suffered from idiosyncratic, inelegant handling of variables (failure of alphabetic variance, destructive assignments, scope extension of existential quantifiers, etc.). Muskens [25] seems to have been the first to demonstrate that the fundamental insights of dynamic semantics could be captured in a compositional way without going beyond the expressive limits of classical type theory. But his approach incorporated some nonstandard—and unnecessary—features, such as an additional basic type for *states* and an explicit encoding of DRT accessibility conditions. More recent type-theoretic embodiments of dynamic semantics [1,4,11], including our earlier efforts [19,20,22,23] are free of these particular defects, but room for improvement remains.

In this paper, we propose a new framework for compositional type-theoretic dynamic semantics which improves on previous proposals in the following respects: (1) It comes equipped with a straightforward interface to a linear-logic-based categorial grammar (LCG) along the general lines of [10,26,27], etc. (2) Although grammars and the derivations they generate make no reference to possible worlds (or extensions of meanings at them), the underlying semantic theory is fully (hyper)intensional from the get-go, so that it can be straightforwardly extended to handle propositional attitudes, evidentiality [17], supplements [21], interrogatives [31], etc. (3) Since it permits, but does not require, the modeling of (static) propositions as sets of possible worlds, one can choose between the familiarity of (intensional) MS and a weaker, hyperintensional, static underpinning [28,29,30] that avoids MS's notorious foundational problems (e.g. the granularity problem). (4) It straightforwardly models contexts as functions from tuples of entities ('discourse referents', abbreviated 'DRs' ) to propositions ('common ground', abbreviated 'CG'), and updates as functions from contexts to contexts, as in FCS, obviating the need for states [25] or continuations [11]. (5) Following the original insights of both DRT and FCS, the semantics of indefinites ('dynamic existential quantification') is not defined in terms of static existential quantification, so there is no need for any notion of scope extension. (6) Updates are explicitly bifurcated into the 'carryover' from the input context and the new content proffered by the current utterance, thereby providing a hook for an envisioned extension covering utterance acceptance and rejection. (7) There is no requirement that common grounds remain consistent; instead, and more realistically, *perceived* inconsistency could be grounds for rejecting proffered content.

## 2 Syntactic Framework

In place of Montague's categorial grammar, we use a linear-logic-based form of categorial grammar (hereafter, LCG) broadly similar to de Groote's ACG [10] or Muskens's λ-grammar [26]. An LCG generates tripartite **signs**, such as:

$$\vdash \text{pedro} ; \text{NP} ; \mathsf{p} : \text{e}$$
$$\vdash \text{donkey} ; \text{N} ; \mathsf{donkey} : \text{p}_1$$

The three components of a sign are called, respectively, the **phenogrammatical** component (**pheno** for short), the **tectogrammatical** component (**tecto** for short), and the **semantic** component (**semantics** for short). (The semantic types will be explained in due course.) In the tecto, there is only one connective ⊸ (linear implication) instead of directional slashes / and \.

The pheno component of a sign need not be a string (basic type s), but can be a (possibly higher order) function over strings that tells how a functor is to be ordered with respect to its arguments:

$$\vdash \lambda_s.s \cdot \text{brays} ; \text{NP} \multimap \text{S} ; \mathsf{bray} : \text{p}_1$$
$$\vdash \lambda_{st}.s \cdot \text{beats} \cdot t ; \text{NP} \multimap \text{NP} \multimap \text{S} ; \mathsf{beat} : \text{p}_2$$
$$\vdash \lambda_{sf}.f\ (\text{every} \cdot s) ; \text{N} \multimap (\text{NP} \multimap \text{S}) \multimap \text{S} ; \mathsf{every} : \text{dt}$$

Here $s, t$ : s and $f$ : s → s. The higher-order pheno theory axiomatizes strings to form a monoid with identity **e** : s (null string) and associative operation · : s → s → s (concatenation, written infix). All other pheno constants employed, representing word phonologies, are of type s, including **e** (null string).

Besides lexical entries, which serve as nonlogical axioms, there is a schema of logical axioms

$$p : P ; A ; z : B \vdash p : P ; A ; z : B \ ,$$

instances of which correspond to traces in mainstream generative grammar (MGG), or to Montague's 'syntactic variables'. That is: a trace is a hypothetical sign, with variables for pheno and semantics. Here the type metavariables $P$, $A$, and $B$ range, respectively, over **pheno types** (s and implicative types over s), **tecto types** (basic tecto types such as N, NP and S, and linear implicative types over them), and **sense types** (to be defined in due course).

For example, an NP trace looks like this:

$$t : \text{s} ; \text{NP} ; x : \text{e} \vdash t : \text{s} ; \text{NP} ; x : \text{e}$$

As in the implicative fragment of intuitionistic linear propositional logic, the only tectogrammatical rules are **modus ponens**:

$$\frac{\Gamma \vdash f : A \rightarrow D ; B \multimap E ; g : C \rightarrow F \qquad \Delta \vdash a : A ; B ; c : C}{\Gamma, \Delta \vdash f\ a : D ; E ; g\ c : F} \ ,$$

and **hypothetical proof**:

$$\frac{\Gamma, p : P ; A ; z : B \vdash a : C ; D ; b : E}{\Gamma \vdash \lambda_p.a : P \rightarrow C ; A \multimap D ; \lambda_z.b : B \rightarrow E} \ ,$$

which are, roughly, LCG's counterparts of MGG's Merge and (the trace-binding aspect of) Move. These correspond, respectively, to application and abstraction in the pheno and semantic components.

Finally, the LCG counterparts of Montague's analysis trees are (pheno- and semantics-labeled) sequent-style linear-logic proof trees whose leaves are axioms (i.e. either lexical entries or traces). The following three proofs use only modus ponens:

$$\frac{\vdash \lambda_s.s \cdot \text{brayed} ; \text{NP} \multimap \text{S} ; \mathsf{bray} \qquad \vdash \text{chiqita} ; \text{NP} ; \mathsf{c}}{\vdash \text{chiqita} \cdot \text{brayed} ; \text{S} ; \mathsf{bray\ c}}$$

$$\frac{\dfrac{\vdash \lambda_{st}.s \cdot \text{beats} \cdot t ; \text{NP} \multimap \text{NP} \multimap \text{S} ; \mathsf{beat} \qquad \vdash \text{pedro} ; \text{NP} ; \mathsf{p}}{\vdash \lambda_t.\text{pedro} \cdot \text{beats} \cdot t ; \text{NP} \multimap \text{S} ; \mathsf{beat\ p}} \qquad \vdash \text{chiqita} ; \text{NP} ; \mathsf{c}}{\vdash \text{pedro} \cdot \text{beats} \cdot \text{chiquita} ; \text{S} ; \mathsf{beat\ p\ c}}$$

$$\frac{\dfrac{\vdash \lambda_{sf}.f\ (\text{every} \cdot s) ; \text{Det} ; \mathsf{every} \qquad \vdash \text{donkey} ; \text{N} ; \mathsf{donkey}}{\vdash \lambda_f.f\ \text{every} \cdot \text{donkey} ; \text{QP} ; \mathsf{every\ donkey}} \qquad \vdash \lambda_s.s \cdot \text{brays} ; \text{NP} \multimap \text{S} ; \mathsf{bray}}{\vdash \text{every} \cdot \text{donkey} \cdot \text{brays} ; \text{S} ; \mathsf{every\ donkey\ bray}}$$

In the last of these, QP abbreviates (NP $\multimap$ S) $\multimap$ S, and Det abbreviates N $\multimap$ QP. And the following proof uses a trace and an instance of hypothetical proof to 'lower' *every donkey* into the object position. Here we abbreviate subtrees by numerical labels:

(1)
$$\frac{\vdash \lambda_{sf}.f\ (\text{every} \cdot s) ; \text{Det} ; \mathsf{every} \qquad \vdash \text{donkey} ; \text{N} ; \mathsf{donkey}}{\vdash \lambda_f.f\ \text{every} \cdot \text{donkey} ; \text{QP} ; \mathsf{every\ donkey}}$$

(2)
$$\frac{\dfrac{\dfrac{\vdash \lambda_{st}.s \cdot \text{beats} \cdot t ; \text{NP} \multimap \text{NP} \multimap \text{S} ; \mathsf{beat} \qquad \vdash \text{pedro} ; \text{NP} ; \mathsf{p}}{\vdash \lambda_t.\text{pedro} \cdot \text{beats} \cdot t ; \text{NP} \multimap \text{S} ; \mathsf{beat\ p}} \qquad s ; \text{NP} ; x \vdash s ; \text{NP} ; x}{s ; \text{NP} ; x \vdash \text{pedro} \cdot \text{beats} \cdot s ; \text{S} ; \mathsf{beat\ p}\ x}}{\vdash \lambda_s.\text{pedro} \cdot \text{beats} \cdot s ; \text{NP} \multimap \text{S} ; \lambda_x.\mathsf{beat\ p}\ x}$$

$$\frac{(1) \qquad\qquad (2)}{\vdash \text{pedro} \cdot \text{beats} \cdot \text{every} \cdot \text{donkey} ; \text{S} ; \mathsf{every\ donkey}\ (\lambda_x.\mathsf{beat\ p}\ x)}$$

This technology, due to Oehrle [27], plays the same role in LCG that quantifier lowering does in Montague grammar.

## 3 The Underlying Logic and Notational Conventions

Our semantic theory is couched in a classical higher order logic (HOL) with entities (e), propositions (p, cf. [32]), and natural numbers (n) as basic types, in addition to truth values (t) courtesy of the logic itself. Unlike MS, where propositions are sets of worlds and so the boolean structure on them is parasitic on that of the truth values, we axiomatize (in the following section) that propositions (in the sense of static declarative sentence meanings) form a preboolean algebra relative to (propositional) entailment. Besides the usual cartesian type constructors $\rightarrow$ (exponential), $\times$ (product), and T (unit type), we also make use of dependent products ($\Pi_{n:\mathrm{n}}.A$) and sums ($\Sigma_{n:\mathrm{n}}.A$) that depend on a natural number.

We adopt the following notational conventions. For any type $A$, the $n$-fold cartesian product is written $A^n$; hence $A^0 = \mathrm{T}$, $A^1 = A$, and $A^{n+1} = A^n \times A$ ($n > 0$). Implication associates to the right, so $A \to B \to C$ abbreviates $A \to (B \to C)$, and similarly for the linear implication $\multimap$ used in the tectogrammar. Outermost parentheses of terms are often deleted. Application associates to the left, so $f\ b\ a$ abbreviates $(f\ b)\ a$. For terms whose type is a cartesian power $A^n$ ($n > 0$), we often write the vector notation $\mathbf{x}$ in place of $(x_0, \ldots, x_{n-1})$. The first occurrence of a 'vector' variable $\mathbf{x}$ is often mnemonically superscripted with its number of components, thus $\mathbf{x}^n$. Abstraction on product types is written either $\lambda_{\mathbf{x}}.a$ or $\lambda_{x_0,\ldots,x_{n-1}}.a$, while abstraction on multiple variables *without* commas abbreviates *successive* abstraction, e.g. $\lambda_{xy}.a$ abbreviates $\lambda_x.\lambda_y.a$. If $x : A$, $N : B$, and $M : (A \to B) \to C$, we abbreviate $M\ \lambda_x.N$ to $M_x.N$.

# 4 The Underlying Static Semantic Theory

## 4.1 Static Semantic Types

The underlying static semantic theory is **agnostic hyperintensional semantics (AHS, [28,30])**, a possible worlds semantics with fine-grained meanings. 'Agnostic' here means that the theory is indifferent as to whether propositions are sets of worlds (as in MS) or worlds are (in one-to-one correspondence with) maximal consistent sets of propositions (as in the 'first-wave' possible-worlds theories of Wittgenstein and C.I. Lewis), though the latter is by far our personal preference. (In fact, AHS is *logically weaker* than Montague semantics (MS): adding one axiom turns it into MS.)

We assume the basic types e (entities), p ((static) propositions), and w (worlds). (The type w is never mentioned in the grammar or in analyses of expressions; it only comes up when using the semantic theory to reason about extensions.) It's also convenient to define (for $n \geq 0$) the types of $n$-ary *static properties* by:

$$\mathrm{p}_0 =_{\mathrm{def}} \mathrm{p}$$

$$\mathrm{p}_{n+1} =_{\mathrm{def}} \mathrm{e} \to \mathrm{p}_n$$

For example, the types for *static generalized quantifiers (over entities)* and *static generalized determiners (over entities)* are, respectively, of types:

$$\mathrm{q} =_{\mathrm{def}} \mathrm{p}_1 \to \mathrm{p}$$

$$\mathrm{dt} =_{\mathrm{def}} \mathrm{p}_1 \to \mathrm{p}_1 \to \mathrm{p}$$

## 4.2 Static Semantic Constants

Side-by-side with the usual truth-value connectives and quantifiers of the underlying HOL (true, false, $\wedge$, $\vee$, $\to$, $\neg$, $\exists$, and $\forall$), there are the propositional connectives truth, falsity, and, or, implies, and not, and the propositional quantifiers $\mathsf{exists}_A$, and $\mathsf{forall}_A$. The latter are polymorphic over sense types (see following subsection) $A$. Some determiner meanings can be defined in terms of these (omitting the type subscripts):

$$\mathsf{every} =_{\mathrm{def}} \lambda_{PQ}.\mathsf{forall}_x.(P\ x)\ \mathsf{implies}\ (Q\ x)$$
$$\mathsf{a} =_{\mathrm{def}} \lambda_{PQ}.\mathsf{exists}_x.(P\ x)\ \mathsf{and}\ (Q\ x)$$

These connectives and quantifiers are subject to meaning postulates that relate them to their truth-value counterparts in the expected way (see [28] or [30] for details). As a consequence, AHS is *finer-grained* than MS: senses which agree in extension at every world need not be identical. For example, not all inconsistent CGs are interchangeable, even though they all entail every proposition. What matters from the dynamic perspective is whether a discourse participants (DPs) can *detect* that a CG would be rendered inconsistent by conjoining to it the proffered content of the current utterance (which would constitute grounds for declining to admit it to the CG).

### 4.3 Static Meanings and Their Extensions

Types to which static meanings can belong are called **sense types.** These are T, e, p, and function types formed from these. Each sense type $A$ has a corresponding **extension** type Ext($A$) defined as follows:

$$\begin{aligned}
&\mathrm{Ext}(\mathrm{T}) = \mathrm{T}\\
&\mathrm{Ext}(\mathrm{e}) = \mathrm{e}\\
&\mathrm{Ext}(\mathrm{p}) = \mathrm{t} \qquad \text{(following Frege [5])}\\
&\mathrm{Ext}(A \to B) = A \to \mathrm{Ext}(B)
\end{aligned}$$

For each sense type $A$, $A \to \mathrm{p}$ is called the type of $A$-**properties**, and $A \to \mathrm{t}$ is called the type of $A$-**sets**. If $w$ is a world and $a : A$ a sense, then the **extension** of $a$ at $w$ is written $a\ @\ w$. (So @ is really a family of constants $@_A : A \to \mathrm{w} \to \mathrm{Ext}(A)$, written infix.) Following, roughly, Kripke [18], we assume every entity is its own extension at every world:

$$\vdash \forall_{x:\mathrm{e}}.\forall_{w:\mathrm{w}}.x\ @\ w = x$$

For any proposition $p$, $p\ @\ w$ is called, following [5], the **truth value** of $p$ at $w$. The extensions of senses with a functional type $A \to B$ are given by

$$\vdash \forall_{f:A\to B}.\forall_{w:\mathrm{w}}.f\ @\ w = \lambda_x.(f\ x)\ @\ w\ .$$

In particular,

$$\vdash \forall_{P:\mathrm{p}_1}.\forall_{w:\mathrm{w}}.P\ @\ w = \lambda_x.(P\ x)\ @\ w\ .$$

For example, if **w** is a world, then the extension at **w** of the donkey property is the set of all entities $a$ such that the proposition that $a$ is a donkey is true at **w**.

# 5 Hyperintensional Dynamic Semantics

## 5.1 Contexts

To model contexts, we begin with our counterparts to FCS's *assignments*, namely *tuples* of entities. The linear positions in these tuples play the 'addressing' role played by DRs (in DRT) or file cards (in FCS). What about contexts? On the Stalnaker/Lewis conception, this is supposed to be the set of propositions in the CG, or else a single proposition which is the conjunction of those. We modify that view to treat the CG not as a proposition (type p), but rather as *a function from tuples of entities to propositions* (type $e^n \to p$), where $n$ is the number of 'live' DRs.[1]. The philosophy behind this typing is that the DPs don't really have a proposition in common, since in general the identity of the DRs that the discourse is about (typically, introduced into the discourse by uses of indefinite NPs) are not known. Rather, what they have in common is only a function from $n$-tuples of entities to propositions, which would give rise to a proposition in the obvious way if only the identities of the entities were known. To put it another way: the DRs are 'identified' only in terms of what the CG says about them. To give a highly simplified example, suppose (counterfactually!) that there are actual 'out-of-the-blue' utterances where the input CG is empty.[2] Then the output context from an out-of-the-blue utterance of *a farmer beat a donkey* will be

$$\lambda_{x,y}.(\mathsf{farmer}\ x)\ \mathsf{and}\ (\mathsf{donkey}\ y)\ \mathsf{and}\ (\mathsf{beat}\ x\ y)\ ,$$

or, using the equivalent vector notation,

$$\lambda_{\mathbf{x}^2}.(\mathsf{farmer}\ x_0)\ \mathsf{and}\ (\mathsf{donkey}\ x_1)\ \mathsf{and}\ (\mathsf{beat}\ x_0\ x_1)\ .$$

Note that this is just an uncurried binary static property. In particular there is no existential quantification. This reflects the fundamental insight of many versions of dynamic semantics that indefinites (and also definites) are *not* quantificational in nature. (However, as we will see, they have the same dynamic semantic type as 'truly quantificational' NPs.) Intuitively, the context corresponds not to an existential proposition, but to mutual acceptance that *whichever farmer it is that we're talking about* beats *whichever donkey it is that we're talking about.* (We could turn this uncurried property into a proposition by applying the static existential quantifier $\mathsf{exists}_{e \times e}$ to it, and then define the context to be 'true' in any world where that proposition is true.)

Based on these considerations, we now *define* the type of $n$-ary **contexts** $c_n$ to be simply p if $n = 0$ and $e^n \to p$ if $n > 0$; and the type c of **contexts** is defined to be the dependent sum of all these:

$$c =_{\text{def}} \Sigma_{n:\text{n}}.c_n$$

[1] However, we continue to use the type p for CGs where $n = 0$, in preference to the mathematically more elegant but notationally awkward $T \to p$.

[2] Technically, this is modeled by truth, the designated top element in the preboolean algebra of static propositions.

Also, we define the **arity** of an $n$-ary context $c$, written $|c|$, to be $n$. (So, since technically, since a member of an n-indexed sum is an ordered pair of a natural number $n$ and a member of the $n$-th cofactor, the arity of a context is just its first component.) Intuitively, $|c|$ is the number of DRs that $c$ is about. For example, $|c| = 2$ if $c$ is the context $\lambda_{x,y}.(\mathsf{farmer}\ x)\ \mathsf{and}\ (\mathsf{donkey}\ y)\ \mathsf{and}\ (\mathsf{beat}\ x\ y)$ discussed above. As we'll see, the DRs are potential targets of subsequent anaphora.

In particular, a nullary context (same as a static proposition) is, intuitively, a context where the DPs have nothing in common to talk about. A special case of a nullary context is the **null context**

$$\mathrm{T} =_{\mathrm{def}} \mathsf{truth}\ ,$$

also known as **out of the blue**, where $\mathsf{truth} : \mathrm{p}$ is some obvious necessary truth. This models the content where the DPs have no DRs to talk about given in advance, and not even anything they have agreed to take for granted. Realistically, discourse is never completely out of the blue; even the driver and the hitchhiker can agree that the driver picked the hitchhiker up, and have some DRs to talk about (the weather, them Bucks, the car, the boring scenery, etc.).

When we consider anaphora, it will be important to have a handle on how many DRs the context knows about, because we will analyze anaphoric expressions (e.g. definite pronouns) as essentially $n$-ways *ambiguous*, where $n$ is the arity of the utterance context. For any natural number $n$, an anaphoric reference to the $n$-th DR will only interpretable in a context whose arity is greater than $n$. The type of such entities, called $\mathrm{c}_{>n}$, is defined as the dependent sum

$$\mathrm{c}_{>n} =_{\mathrm{def}} \mathrm{c}_{\geq n+1}\ ,$$

where

$$\mathrm{c}_{\geq m} =_{\mathrm{def}} \Sigma_{n:\mathrm{n}}.\mathrm{c}_{m+n}\ .$$

## 5.2 Toward Dynamic Senses

To dynamicize LCG, we have to replace the static senses with dynamic ones. For example, as we'll see, the dynamic counterpart of type e is the type n of natural numbers (thought of as DRs). And the dynamic counterpart of type p will be the type u of **updates**, which correspond roughly to FCS's context change potentials (CCPs). To a first approximation, updates are functions from contexts to contexts (type $\mathrm{c} \rightarrow \mathrm{c}$), but there is a catch: if an update makes anaphoric reference to something in the context, then it is only defined on contexts that contain a suitable antecedent to which the anaphoric reference can be resolved. We'll use dependent typing to give u a more subtle definition than $\mathrm{c} \rightarrow \mathrm{c}$ that solves this problem (and which does not require partial functions, which our type theory does not countenance).

Again, to a first approximation, dynamic counterparts of other sense types are obtained in the expected way from these basic dynamic types. For example, the dynamic counterpart of $e \to p$ (properties of entities) is (to a first approximation) the type $n \to u$ of *dynamic properties*; and the dynamic counterpart of $(e \to p) \to p$ (GQs) is (to a first approximation) the type $(n \to u) \to u$ of *dynamic generalized quantifiers (DGQs)*.

### 5.3 Contents vs. Updates

In discourse, when an utterance is accepted, the update carries over the common ground of the input context, while conjoining to it new content contributed by the utterance. To capture this fundamental intuition, we distinguish between two different things of the same type: updates, which correspond to FCS's CCPs, and **contents**, the meaning contribution of the new utterance (here, we consider only assertions). For any content $k$ which is accepted by the DPs, the induced update is obtained by applying to it a certain function called $\mathsf{cc}$ (mnemonic for 'context change'). As we'll see, $\mathsf{cc}$ applies first to a content $k$ and then to an input context $c$ to produce the new context $\mathsf{cc}\ k\ c$ for the next utterance. Thus the update $\mathsf{cc}\ k$ is itself a function which converts an input context into a new context into which the newly accepted content $k$ has been incorporated.

So, naively, it *looks* as though the type u for both contents and updates should be $c \to c$, and the type for $\mathsf{cc}$ should be $(c \to c) \to (c \to c)$. But there is a subtlety: different contents (and the updates they induce upon acceptance) have different *degrees*: the number of new DRs that are introduced. As we'll see,

$$|\mathsf{cc}\ k| = |\mathsf{cc}\ k\ c| = |c| + |k|\ ,$$

so the arity of the output context is the arity of the input context plus the degree of the utterance content. (Like the arity of a context, the degree of a content $k$ is written $|k|$.)

Now, for any natural number $n$, the type of $n$-degree updates is that of a function that maps a context $c$ to a new context whose degree is $|c| + n$. Accordingly, we define the type of $n$-**degree updates** to be the dependent product

$$\mathrm{u}_n =_{\mathrm{def}} \Pi_{c:\mathrm{c}}.\mathrm{c}_{|c|+n}\ .$$

Then the type u of **updates** is the dependent sum of the $\mathrm{u}_n$ as $n$ ranges over all natural numbers:

$$\mathrm{u} =_{\mathrm{def}} \Sigma_{n:\mathrm{n}}.\mathrm{u}_n = \Sigma_{n:\mathrm{n}}.\Pi_{c:\mathrm{c}}.\mathrm{c}_{|c|+n}$$

The types of contents are defined similarly, as $\mathrm{k}_n =_{\mathrm{def}} \mathrm{u}_n$ and $\mathrm{k} =_{\mathrm{def}} \mathrm{u}$. With these type definitions in place, we can now define the context change function as follows:

$$\mathsf{cc} =_{\mathrm{def}} \lambda_{k:\mathrm{k}}.\lambda_{c:\mathrm{c}}.\lambda_{\mathbf{x}^{|c|},\mathbf{y}^{|k|}}.(c\ \mathbf{x})\ \mathsf{and}\ (k\ c\ \mathbf{x},\mathbf{y})$$

That is: $\mathsf{cc}$ takes as arguments a content $k$ with degree $n = |k|$ and a context $c$ with arity $m = |c|$ and returns a new context $\lambda_{\mathbf{x}^m,\mathbf{y}^n}.(c\ \mathbf{x})\ \mathsf{and}\ (k\ c\ \mathbf{x},\mathbf{y})$. This new context is a function that maps any $m + n$ entities to a (static) proposition which is itself the conjunction of two propositions: (1) $(c\ \mathbf{x})$, the 'carryover', which is what the input context had already established about the first $m$ DRs; and (2) $(k\ c\ \mathbf{x}, \mathbf{y})$, the new contribution, which is what the current utterance's content says about *all* the DRs (the $m$ original ones plus the $n$ new ones that it introduces) *in* that context. For example, the content of *it's raining* is the 0-degree content

$$\text{RAIN} =_{\text{def}} \lambda_{c:\text{c}}.\lambda_{\mathbf{x}|c|}.\mathsf{rain}\ .$$

So the associated update is

$$\vdash \mathsf{cc}\ \text{RAIN} = \lambda_{c:\text{c}}.\lambda_{\mathbf{x}|c|}.(c\ \mathbf{x})\ \mathsf{and}\ (\text{RAIN}\ c\ \mathbf{x}) = \lambda_{c:\text{c}}.\lambda_{\mathbf{x}|c|}.(c\ \mathbf{x})\ \mathsf{and}\ \mathsf{rain}\ .$$

## 5.4 From Linear Categorial Grammar to Dynamic Categorial Grammar

Almost all the work involved in dynamicizing an LCG consists of replacing word meanings with their dynamic counterparts, since the logical rules and axioms (traces) of LCG carry over to DyCG unchanged. We will turn to that task presently. However, we also need one new, nonlogical, grammar rule, **continue**, whose purpose is to continue a discourse (tecto type D) by the addition of the next accepted utterance:

$$\frac{\vdash s\,;\text{D}\,;u \qquad \vdash t\,;\text{S}\,;k}{\vdash s\cdot t\,;\text{D}\,;\lambda_{c:\text{c}}.\mathsf{cc}\ k\ (u\ c)}$$

Note that the sequent contexts are empty: binding of traces, and therefore '*wh*-movement' and 'quantifying in' are impossible across root clause boundaries.

And finally, we need an axiom for the **null discourse**, to ground the recursive construction of discourses:

$$\vdash \mathbf{e}\,;\text{D}\,;\lambda_{c:\text{c}}.c$$

Note that using the null discourse as the first premise in the continue rule yields the derived rule

$$\frac{\vdash s\,;\text{S}\,;k}{\vdash s\,;\text{D}\,;\mathsf{cc}\ k}\ ,$$

which says that any sentence can be 'promoted' to a single-sentence discourse.

## 5.5 The Dynamic Connectives

*The content negation* NOT A fundamental insight of dynamic semantics is that an *indefinite* inside the scope of negation *cannot* antecede definite anaphora in the *subsequent* discourse, but a *definite* in the scope of negation *can* be anteceded by an indefinite in *prior* discourse:

(3) a. Pedro has a donkey$_i$. It$_i$ $\left\{\begin{array}{c}\text{is}\\ \text{isn't}\end{array}\right\}$ friendly.

b. Pedro doesn't have a donkey$_i$. # It$_i$ $\left\{\begin{array}{c}\text{is}\\ \text{isn't}\end{array}\right\}$ friendly.

Such facts are accounted for by the following definition for the **content negation** NOT : $\mathrm{k} \to \mathrm{k}_0$:

$$\text{NOT} =_{\text{def}} \lambda_{k:\mathrm{k}}.\lambda_{c:\mathrm{c}}.\lambda_{\mathbf{x}^{|c|}}.\mathsf{not}\ \mathsf{exists}_{\mathbf{y}^{|k|}}.k\ c\ \mathbf{x}, \mathbf{y}$$

This has the effect that DRs introduced within the scope of content negation are existentially bound within the scope of (static) propositional negation, and therefore inaccessible as antecedents for subsequent definite anaphora.

The effect of *double* content negation on any content $k$ is to existentially bind any new DRs that it introduces:

$$\vdash \text{NOT}\ (\text{NOT}\ k) \equiv \lambda_{c:\mathrm{c}}.\lambda_{\mathbf{x}^{|c|}}.\mathsf{exists}_{\mathbf{y}^{|k|}}.k\ c\ \mathbf{x}, \mathbf{y}$$

From this it follows that a content of degree 0 (i.e. one which introduces no new DRs) is equivalent to its own double negation. In particular, since the content negation of any content is itself of degree 0, content negation is equivalent to *triple* content negation.

*The content conjunction* AND Another fundamental insight of dynamic semantics is that sentential conjunction is not commutative:

(4) a. A farmer walked in and he sat down. (*Who is he?*)

b. He sat down and a farmer walked in. (*Who is he?*)

We capture this insight with the following definition of **content conjunction**, of type $\Pi_{h:\mathrm{k}}.\Pi_{k:\mathrm{k}}.\Pi_{c:\mathrm{c}}.\mathrm{c}_{|c|+|h|+|k|}$:

$$\text{AND} =_{\text{def}} \lambda_{h:\mathrm{k}}.\lambda_{k:\mathrm{k}}.\lambda_{c:\mathrm{c}}.\lambda_{\mathbf{x}^{|c|},\mathbf{y}^{|h|},\mathbf{z}^{|k|}}.(h\ c\ \mathbf{x}, \mathbf{y})\ \mathsf{and}\ (k\ (\mathsf{cc}\ h\ c)\ \mathbf{x}, \mathbf{y}, \mathbf{z})$$

Crucially, the input context (cc $h$ $c$) for the *second* conjunct is created by applying the update induced by the *first* conjunct to *its* input context $c$. It is not hard to show that the update induced by conjoined declarative utterances is the same as the function composition of the updates induced by the conjuncts:

$$\vdash \forall_{hk}.(\mathsf{cc}\ (h\ \text{AND}\ k)) = \lambda_{c:\mathrm{c}}.\mathsf{cc}\ k\ (\mathsf{cc}\ h\ c)$$

*The content disjunction* OR As in other versions of dynamic semantics, we define **content disjunction** by DeMorgan duality:

$$\text{OR} =_{\text{def}} \lambda_{h:\mathrm{k}}.\lambda_{k:\mathrm{k}}.\text{NOT}\ ((\text{NOT}\ h)\ \text{AND}\ (\text{NOT}\ k))$$

This predicts that indefinites within either disjunct can't antecede subsequent definite anaphora:

(5) a. Either a donkey brayed or someone is making barnyard noises. # It's friendly.

b. Either someone is making barnyard noises, or a donkey brayed. # It's friendly.

c. # Either a donkey brayed, or it's friendly.

*The content implication* IMPLIES Our definition of **content implication** is modeled on a valid but nonstandard equivalence for static propositional implication:

$$\mathsf{implies} =_{\text{def}} \lambda_{pq}.(\mathsf{not}\ p)\ \mathsf{or}\ (p\ \mathsf{and}\ q)$$
$$\text{IMPLIES} =_{\text{def}} \lambda_{hk}.(\text{NOT}\ h)\ \text{OR}\ (h\ \text{AND}\ k)$$

This predicts that indefinites in either the antecedent or the consequent of the conditional cannot antecede definite anaphora in a subsequent sentence, but an indefinite in the antecedent *can* antecede definite anaphora in the consequent:

(6) a. If a donkey brayed, it's hungry. # We better feed it.

b. If Pedro is a farmer, he has a donkey. # He better feed it.

An additional virtue of this definition of content implication is that it gives rise to so-called *weak* readings of conditional sentences:

(7) a. If you have a donkey, I'll buy it.

b. If you have a donkey, I'll buy a donkey you have. (*weak*)

c. If you have a donkey, I'll buy every donkey you have. (*strong*).

There are known pragmatic strengthening strategies for inferring strong understandings from weak meanings, but if a strong semantics is used (say, based on a different static equivalence $\vdash (p\ \mathsf{implies}\ q) \equiv (\mathsf{not}\ (p\ \mathsf{and}\ (\mathsf{not}\ q)))$, then it is hard to explain where weak readings come from.

## 5.6 Dynamic Generalized Quantifiers (DGQs)

*Three kinds of noun phrase* We turn next to the dynamic meanings of indefinite referring expressions (e.g. *a donkey*), definite referring expressions (e.g. *it, the donkey*), and 'truly quantificational' NPs (e.g. *every donkey, no donkey*). Although these will all be of the same type q (DGQs), they differ radically in their discourse behavior. A use of an indefinite in an utterance introduces a new DR into its output context, while a definite 'picks up' or 'continues' an already existing DR:

(8) A donkey$_i$ brayed. It$_i$ was hungry.

But a use of a truly quantificational NP renders any DRs introduced within either its restriction or its scope inaccessible to the subsequent discourse:

(9) a. $\left\{\begin{matrix}\text{Every}\\\text{No}\end{matrix}\right\}$ farmer that owned a donkey$_i$ was unhappy. # It$_i$ was lazy.

b. $\left\{\begin{matrix}\text{Every}\\\text{No}\end{matrix}\right\}$ farmer owned a donkey$_i$. # It$_i$ was lazy.

However, an indefinite introduced in the restriction of a truly quantificational NP is accessible from its scope:

(10) $\left\{\begin{matrix}\text{Every}\\\text{No}\end{matrix}\right\}$ farmer that owns a donkey$_i$ beats it$_i$.

*Dynamic properties* In static semantics, a GQ is a property of properties of entities (type $\mathrm{p}_1 \to \mathrm{p}$). Since, in dynamic semantics, n and u are the respective counterparts of static sense types e and p, it seems reasonable to assume that the type for the dynamic counterparts of properties of, and GQs over, entities would be, respectively, $\mathrm{n} \to \mathrm{k}$ and $(\mathrm{n} \to \mathrm{k}) \to \mathrm{k}$. However, some delicacy is called for, because we have to ensure that in applying a dynamic property to a DR, the resulting content is one which is defined on contexts which 'know about' the DR in question. Additionally, we have to take into consideration that a given dynamic property (e.g. *farmer that owns a donkey*) may itself introduce new DRs. Accordingly, we define (for each $i$) the type $\mathrm{d}_{1,i}$ of unary **dynamic properties** of degree $i$ as

$$\mathrm{d}_{1,i} =_{\mathrm{def}} \Pi_{n:\mathrm{n}}.\Pi_{c:\mathrm{c}_{>n}}.\mathrm{c}_{|c|+i}\ ,$$

and the type of unary dynamic properties as the dependent sum

$$\mathrm{d}_1 =_{\mathrm{def}} \Sigma_{i:\mathrm{n}}.\Pi_{n:\mathrm{n}}.\Pi_{c:\mathrm{c}_{>n}}.\mathrm{c}_{|c|+i}\ .$$

In due course, we will define dynamic counterparts $\mathrm{d}_n$ for all the static types $\mathrm{p}_n$.

*Dynamic counterparts of static properties* In dynamic semantics, the static-property senses of common nouns, predicative adjectives, and intransitive verbs have to be replaced by their dynamic counterparts, which are unary dynamic properties (of degree 0, since they introduce no DRs). This change is effected by the unary **dynamicization** function $\mathrm{dyn}_1 : \mathrm{p}_1 \to \mathrm{d}_{1,0}$ defined as follows:

$$\mathrm{dyn}_1 =_{\mathrm{def}} \lambda_{P:\mathrm{p}_1}.\lambda_{n:\mathrm{n}}.\lambda_{c:\mathrm{c}_{>n}}.\lambda_{\mathbf{x}^{|c|}}.P\ x_n$$

For example, the dynamic meaning of the common noun *donkey* is:

$$\mathrm{DONKEY} =_{\mathrm{def}} (\mathrm{dyn}_1\ \mathsf{donkey}) = \lambda_{n:\mathrm{n}}.\lambda_{c:\mathrm{c}_{>n}}.\lambda_{\mathbf{x}^{|c|}}.\mathsf{donkey}\ x_n$$

This maps a DR $n$ and a context $c$ which knows about $n$ to a content which asserts that (whichever entity corresponds to) $n$ is a donkey. More generally, for each $n : \mathrm{n}$, there is a type $\mathrm{d}_n$ of $n$-ary **dynamic relations**, the dynamic counterparts of the static types $\mathrm{p}_n$. As in the unary case, each of these is a dependent sum

$$\mathrm{d}_n = \Sigma_{i:\mathrm{n}}.\mathrm{d}_{n,i}\ ,$$

where $i$ ranges over the degree (number of newly introduced DRs) of the relation. For example, for $n = 2$:

$$\mathrm{d}_{2,i} =_{\mathrm{def}} \Pi_{m:\mathrm{n}}.\Pi_{n:\mathrm{n}}.\Pi_{c:\mathrm{c}_{>(\mathbf{max}\ m\ n)}}.\mathrm{c}_{|c|+i}$$

A special case of this is:

$$\mathrm{d}_{2,0} =_{\mathrm{def}} \Pi_{m:\mathrm{n}}.\Pi_{n:\mathrm{n}}.\Pi_{c:\mathrm{c}_{>(\mathbf{max}\ m\ n)}}.\mathrm{c}_{|c|}$$

Also as in the unary case, for each $n : \mathrm{n}$, there is a function $\mathrm{dyn}_n : \mathrm{p}_n \to \mathrm{d}_n$ that maps each static relation to its dynamic counterpart. For $n < 3$, these are:

$$\begin{aligned}
\mathrm{dyn}_0 &=_{\mathrm{def}} \lambda_{p:\mathrm{p}}.\lambda_{c:\mathrm{c}}.\lambda_{\mathbf{x}^{|c|}}.p \\
\mathrm{dyn}_1 &=_{\mathrm{def}} \lambda_{P:\mathrm{p}_1}.\lambda_{n:\mathrm{n}}.\lambda_{c:\mathrm{c}_{>n}}.\lambda_{\mathbf{x}^{|c|}}.P\ x_n \\
\mathrm{dyn}_2 &=_{\mathrm{def}} \lambda_{R:\mathrm{p}_2}.\lambda_{m:\mathrm{n}}.\lambda_{n:\mathrm{n}}.\lambda_{c:\mathrm{c}_{>(\mathbf{max}\ m\ n)}}.\lambda_{\mathbf{x}^{|c|}}.R\ x_m\ x_n
\end{aligned}$$

For example:

$$\begin{aligned}
\text{RAIN} &= \lambda_{c:\mathrm{c}}.\lambda_{\mathbf{x}^{|c|}}.\mathsf{rain} \\
\text{DONKEY} &= \lambda_{n:\mathrm{n}}.\lambda_{c:\mathrm{c}_{>n}}.\lambda_{\mathbf{x}^{|c|}}.\mathsf{donkey}\ x_n \\
\text{OWN} &= \lambda_{m:\mathrm{n}}.\lambda_{n:\mathrm{n}}.\lambda_{c:\mathrm{c}_{>(\mathbf{max}\ m\ n)}}.\lambda_{\mathbf{x}^{|c|}}.\mathsf{own}\ x_m\ x_n
\end{aligned}$$

*Indefinites* To give a dynamic meaning for indefinites, we start with the **context extension** function $(\cdot)^+$ of type $\Pi_{c:\mathrm{c}}.\mathrm{c}_{|c|+1}$, defined as follows:

$$(\cdot)^+ =_{\mathrm{def}} \lambda_{c:\mathrm{c}}.\lambda_{\mathbf{x}^{|c|},y}.c\ \mathbf{x}$$

This just adds a new DR to any context. In terms of this, we now define the dynamic 'existential' quantifier EXISTS to be the following function of type $\Pi_{n:\mathrm{n}}.\Pi_{D:\mathrm{d}_{1,n}}.\mathrm{k}_{n+1}$ (i.e. it maps a dynamic property to a content of degree one greater than that of the dynamic property):

$$\text{EXISTS} =_{\mathrm{def}} \lambda_{D:\mathrm{d}_1}.\lambda_{c:\mathrm{c}}.D\ |c|\ c^+$$

Then the dynamic meaning of the indefinite determiner is defined by analogy with its static counterpart:

$$\begin{aligned}
\mathsf{a} &=_{\mathrm{def}} \lambda_{PQ}.\mathsf{exists}_x.(P\ x)\ \mathsf{and}\ (Q\ x) \\
\text{A} &=_{\mathrm{def}} \lambda_{DE}.\text{EXISTS}_n.(D\ n)\ \text{AND}\ (E\ n)
\end{aligned}$$

For example, the (degree 1) content of *a donkey brays* is

$$\vdash \text{A DONKEY BRAY} = \lambda_{c:\mathrm{c}}.\lambda_{\mathbf{x}^{|c|},y}.(\mathsf{donkey}\ y)\ \mathsf{and}\ (\mathsf{bray}\ y)\ .$$

The fact that the variable $y$ corresponding to the DR for the donkey is only λ-bound (not exists-bound) has as a consequence that this DR will be accessible to subsequent definite anaphora.

*Definite anaphora* To keep within space bounds, we provide here a simplified, Montague-like, treatment of anaphora in terms of lexical ambiguity as to which DR is the antecedent; [20] and [21] describe a mechanism (improving on the `sel` function of [11]) for selecting the (presupposed) DR that satisfies the definite expression's descriptive content. For example, the dynamic meaning of the $n$-th definite pronoun *it* is

$$\begin{aligned}
\text{IT}^n &=_{\mathrm{def}} \lambda_{D:\mathrm{d}_1}.\lambda_{c:\mathrm{c}_{>n}}.D\ n\ c \\
&= \lambda_{D:\mathrm{d}_1}.\lambda_{c:\mathrm{c}_{>n}}.\lambda_{\mathbf{x}^{|c|}}.D\ n\ c\ \mathbf{x}\ .
\end{aligned}$$

Unlike an indefinite, which introduces a new DR, the definite simply resumes an old one.

*A 'truly quantificational' DGQ* Again by analogy with the static case, we define the dynamic universal quantifier FORALL as follows:

$$\mathsf{forall} =_{\text{def}} \lambda_P.\mathsf{not}\ (\mathsf{exists}_x.(\mathsf{not}\ (P\ x)))$$
$$\text{FORALL} =_{\text{def}} \lambda_D.\text{NOT}\ (\text{EXISTS}_n.(\text{NOT}\ (D\ n)))$$

Then, again analogizing to the static case, we define the dynamic universal determiner EVERY:

$$\mathsf{every} =_{\text{def}} \lambda_{PQ}.\mathsf{forall}_x.(P\ x)\ \mathsf{implies}\ (Q\ x)$$
$$\text{EVERY} =_{\text{def}} \lambda_{DE}.\text{FORALL}_n.(D\ n)\ \text{IMPLIES}\ (E\ n)$$

This in turn can be shown to be equivalent to

$$\lambda_{DE}.\text{NOT}\ (\text{EXISTS}_n.((\text{NOT}\ (\text{NOT}(D\ n)))\ \text{AND}\ (\text{NOT}\ ((D\ n)\ \text{AND}\ (E\ n)))))$$

For example, a simple universal sentence like *Every donkey brays* ends up with the content EVERY DONKEY BRAY, which can be shown to be equivalent to

$$\lambda_{\mathsf{c:c}}.\lambda_{\mathbf{x}|\mathsf{c}|}.\mathsf{not}\ \mathsf{exists}_y.(\mathsf{donkey}\ y)\ \mathsf{and}\ (\mathsf{not}\ (\mathsf{bray}\ y))\ .$$

As desired, the fact that the variable $y$ corresponding to the DR for the donkey is exists-bound (within the scope of negation) has as a consequence that this DR is *in*accessible to subsequent definite anaphora.

*Donkey anaphora* We conclude with a DyCG derivation for the classic universal donkey sentence *Every farmer that owns a donkey beats it.* Because we based our semantics for EVERY on the 'weak' dynamic implication IMPLIES, our analysis produces the weak reading (that every farmer that owns a donkey beats a donkey that s/he owns), which again is a desirable state of affairs (cf. [2,16]). The lexical entries employed are:

$$\vdash \lambda_{sf}.f\ (\text{every} \cdot s)\ ;\ \text{N} \multimap (\text{NP} \multimap \text{S}) \multimap \text{S}\ ;\ \text{EVERY}$$
$$\vdash \text{farmer}\ ;\ \text{N}\ ;\ \text{FARMER}$$
$$\vdash \lambda_{sf}.s \cdot \text{that} \cdot (f\ \mathbf{e})\ ;\ \text{N} \multimap (\text{NP} \multimap \text{S}) \multimap \text{N}\ ;\ \text{THAT}$$
$$\vdash \lambda_{st}.s \cdot \text{owns} \cdot t\ ;\ \text{NP} \multimap \text{NP} \multimap \text{S}\ ;\ \text{OWN}$$
$$\vdash \lambda_{sf}.f\ (\text{a} \cdot s)\ ;\ \text{N} \multimap (\text{NP} \multimap \text{S}) \multimap \text{S}\ ;\ \text{A}$$
$$\vdash \text{donkey}\ ;\ \text{N}\ ;\ \text{DONKEY}$$
$$\vdash \lambda_{st}.s \cdot \text{beats} \cdot t\ ;\ \text{NP} \multimap \text{NP} \multimap \text{S}\ ;\ \text{BEAT}$$
$$\vdash \lambda_f.f\ \text{it}\ ;\ (\text{NP} \multimap \text{S}) \multimap \text{S}\ ;\ \text{IT}^n$$

And the endsequent of the derivation is

$$\vdash \text{every} \cdot \text{farmer} \cdot \text{that} \cdot \text{owns} \cdot \text{a} \cdot \text{donkey} \cdot \text{beats} \cdot \text{it}\ ;\ \text{S}\ ;$$
$$\text{EVERY}\ (\text{FARMER THAT}\ \lambda_m.(\text{A DONKEY})_n.(\text{OWN}\ m\ n))\ \lambda_m.\text{IT}^i.(\text{BEAT}\ m)\ .$$

In a given context, this will produce the (weak) donkey-anaphora reading provided $i$ is selected to be whichever natural number corresponds to the DR introduced by *a donkey*. As mentioned above, [20] gives a more realistic treatment for the definite pronoun *it* as selecting the most salient, informationally unique discourse referent with the property of being nonhuman.

## 6 Conclusion

This dynamic semantics is not only fully compositional and expressed in pure (dependent) type theory, it also captures all of the central insights of the dynamic tradition, with indefinites introducing discourse referents, definites selecting their antecedents from the input context, and 'accessibility constraints' captured by existentially binding variables in the scope of negation and operators defined in terms of it. No idiosyncratic machinery is required, and we overcome the problem in which discourse contexts must have a certain arity by a mild use of dependent types, rather than extending the type theory with partial functions. As we have shown, this theory can be seen as a straightforward extension of the underlying static semantics, achieved by adding a type of discourse contexts and replacing entities with natural number indices into the list of discourse referents the incoming context 'knows about.' Since the underlying static semantics is hyperintensional, we avoid certain foundational problems with possible worlds approaches while allowing the grammar to focus on the dynamic senses of expressions, rather than bothering with worlds and extensions at them.

## References

1. Beaver, D.I.: Presupposition and Assertion in Dynamic Semantics. CSLI Publications (2001)
2. Chierchia, G.: The Dynamics of Meaning: Anaphora, Presupposition, and the Theory of Grammar. University of Chicago Press (1995)
3. Church, A.: A formulation of the simple theory of types. Journal of Symbolic Logic 5(2), 56–68 (1940)
4. van Eijck, J., Unger, C.: Computational Semantics with Functional Programming. Cambridge University Press (2010)
5. Frege, G.: Über Sinn und Bedeutung. Zeitschrift für Philosophie und philosophische Kritik 100, 25–50 (1892), English translation titled On Sense and Reference in [7], pp. 56–78
6. Gallin, D.: Intensional and Higher Order Modal Logic, Mathematics Studies, vol. 19. North-Holland, Amsterdam (1975)
7. Geach, P., Black, M.: Translations from the Philosophical Writings of Gottlob Frege. Blackwell, Oxford (1952)
8. Groenendijk, J., Stokhof, M.: Dynamic Montague grammar. In: Kálmán, L., Pólos, L. (eds.) Papers from the Second Symposium on Logic and Language. Akadémiai Kiadó (1990)
9. Groenendijk, J., Stokhof, M.: Dynamic predicate logic. Linguistics and Philosophy 14(1), 39–100 (1991)

10. de Groote, P.: Towards abstract categorial grammars. In: Association for Computational Linguistics, Proceedings of the 39th Annual Meeting and 10th Conference of the European Chapter (2001)
11. de Groote, P.: Towards a Montagovian account of dynamics. In: Proceedings of the 16th Conference on Semantics and Linguistic Theory (2006)
12. de Groote, P., Nederhof, M.-J. (eds.): Formal Grammar 2010/2011. LNCS, vol. 7395. Springer, Heidelberg (2012)
13. Heim, I.: The Semantics of Definite and Indefinite Noun Phrases. Ph.D. thesis, University of Massachusetts, Amherst (1982)
14. Henkin, L.: Completeness in the theory of types. Journal of Symbolic Logic 15(2), 81–91 (1950)
15. Kamp, H.: A theory of truth and semantic representation. In: Groenendijk, J., Janssen, T., Stokhof, M. (eds.) Formal Methods in the Study of Language. Mathematical Center, Amsterdam (1981)
16. Kanazawa, M.: Weak vs. strong readings of donkey sentences and monotonicity inference in a dynamic setting. Linguistics and Philosophy 17(2), 109–158 (1994)
17. Kierstead, G., Martin, S.: A multistratal account of the projective Tagalog evidential 'daw'. In: Proceedings of the 22nd Conference on Semantics and Linguistic Theory (2012)
18. Kripke, S.: Naming and Necessity. Harvard University Press (1980)
19. Martin, S.: Weak familiarity and anaphoric accessibility in dynamic semantics. In: de Groote, Nederhof (eds.) [12], pp. 287–306
20. Martin, S.: The Dynamics of Sense and Implicature. Ph.D. thesis, Ohio State University (2013)
21. Martin, S.: Supplemental update (In preparation), unpublished manuscript
22. Martin, S., Pollard, C.: A higher-order theory of presupposition. Studia Logica 100(4), 729–754 (2012)
23. Martin, S., Pollard, C.: Hyperintensional dynamic semantics: Analyzing definiteness with enriched contexts. In: de Groote, Nederhof (eds.) [12], pp. 114–129
24. Montague, R.: The proper treatment of quantification in ordinary English. In: Thomason, R. (ed.) Formal Philosophy: Selected Papers of Richard Montague, pp. 247–270. Yale University Press, New Haven (1974)
25. Muskens, R.: Combining Montague semantics and discourse representation theory. Linguistics and Philosophy 19(2), 143–186 (1996)
26. Muskens, R.: Separating syntax and combinatorics in categorial grammar. Research on Language and Computation 5(3), 267–285 (2007)
27. Oehrle, R.T.: Term-labeled categorial type systems. Linguistics and Philosophy 17(6), 633–678 (1994)
28. Plummer, A., Pollard, C.: Agnostic possible worlds semantics. In: Béchet, D., Dikovsky, A. (eds.) LACL 2012. LNCS, vol. 7351, pp. 201–212. Springer, Heidelberg (2012)
29. Pollard, C.: Hyperintensions. Journal of Logic and Computation 18(2), 257–282 (2008)
30. Pollard, C.: Agnostic hyperintensional semantics. Synthese (in press)
31. Pollard, C., Yasavul, M.: Anaphoric clefts: the myth of exhaustivity (In preparation), to appear in Proceedings of CLS 2014 (2014)
32. Thomason, R.: A model theory for propositional attitudes. Linguistics and Philosophy 4(1), 47–70 (1980)

# On Minimalism of Analysis by Reduction by Restarting Automata*

Martin Plátek[1], Dana Pardubská[2], and Markéta Lopatková[1]

[1] Charles University in Prague, Faculty of Mathematics and Physics
Malostranské nám. 25, 118 00 Prague, Czech Republic
martin.platek@mff.cuni.cz, lopatkova@ufal.mff.cuni.cz
[2] Comenius University in Bratislava,
Faculty of Mathematics, Physics and Informatics,
Mlynská dolina, 842 48 Bratislava, Slovak Republic
pardubska@dcs.fmph.uniba.sk

**Abstract.** The paper provides linguistic observations as a motivation for a formal study of an analysis by reduction. It concentrates on a study of the whole mechanism through a class of restarting automata with meta-instructions using pebbles, with delete and shift operations (DS-automata). Four types of (in)finite sets defined by these automata are considered as linguistically relevant: basic languages on word forms marked with grammatical categories, proper languages on unmarked word forms, categorial languages on grammatical categories, and sets of reductions (reduction languages). The equivalence of proper languages is considered for a weak equivalence of DS-automata, and the equivalence of reduction languages for a strong equivalence of DS-automata.

The complexity of a language is naturally measured by the number of *pebbles*, the number of *deletions*, and the number of *word order shifts* used in a single reduction step. We have obtained unbounded hierarchies (scales) for all four types of classes of finite languages considered here, as well as for Chomsky's classes of infinite languages. The scales make it possible to estimate relevant complexity issues of analysis by reduction for natural languages.

## 1 Introduction

The method of analysis by reduction (AR) plays an important role in a lexicalized syntax of natural languages. It consists in a stepwise simplification of a sentence, which profits from the integration of the sentence syntactic structure and the corresponding grammatical categories.

To model AR, various types of restarting automata can be found in literature [6, 7], which allow one to study dependencies in a natural language. Unfortunately, these types of automata are not able to adequately cope with a word

* The paper reports on the research supported by the grants of GAČR No. P202/10/1333 and partially P103/10/0783. The third author is supported by the Slovak Grant Agency for Science (VEGA) under the contract No. 1/0979/12.

G. Morrill et al. (Eds.): Formal Grammar 2014, LNCS 8612, pp. 155–170, 2014.

order freedom frequently present in Czech sentences. In this paper we present a formal model of the analysis by reduction that is, in addition to a *deletion*, enriched with a *word order shift*, an operation reflecting a word order freedom of natural languages [4]. Section 2 provides a linguistic motivation and informal description of the process.

The core sections 3 and 4 provide a formal study of the whole mechanism through a refined class of restarting automata (DS-automata), and their descriptional complexity based on the number of pebbles, on the number of deletions, and on the number of word order shifts used within a single meta-instruction. Using these measures, we are able to argue that natural languages (e.g. Czech) can be described using rather simple reductions. Our paper refines the notion of window size [5, 7] by the number of pebbles.

Four types of (in)finite sets defined by DS-automata are the most relevant: basic languages on word forms marked with their linguistic categories, sets of reductions on basic languages forming reduction languages, proper languages on unmarked word forms, and categorial languages on pure categories. Inspired by Chomsky [2], we consider the equivalence of proper languages as the *weak equivalence* (close to the weak equivalence by formal automata and grammars), and the equivalence of reduction languages as the linguistically finest *strong equivalence* between DS-automata.

Formal parts of this article are based on the descriptional complexity of selected classes of finite languages and traditional Chomsky classes of infinite languages. Note that the concentration on finite languages comes from the domain of interest; we can to a certain extent claim that the core vocabulary of a natural language may be considered finite and that in a normal everyday use of a language the comprehensible/understandable sentence may be considered finite as well. We introduce the map- and mrp- properties of restarting automata, resp. meta-instructions – these properties in some sense characterize the minimality of reductions. If a correct sentence of a natural language undergoes the process of analysis by reduction, we require that it remains correct in each step of the reduction. The obtained results give the theoretical background for an incremental transfer from finite (linguistic) observations (as in [4]) to adequate, fully lexicalized, formal descriptions (models) of natural languages based on sentence reductions that are applicable on infinite languages.

## 2 Analysis by Reduction

*Analysis by reduction* (AR) helps to identify syntactic structure and the corresponding grammatical categories of the analyzed language. AR is based upon a stepwise simplification of an analyzed sentence, see [4]. It defines possible sequences of reductions in the sentence – each step of AR consists in *deleting* at least one word of the input sentence and thus its shortening. Here we allow that a deletion of a word is accompanied by a *shift* of some word(s) to another word order position(s) in the sentence.

Let us stress the basic constraints imposed on each reduction step of AR:

(i) individual words (word forms), their morphological characteristics and/or their syntactic categories must be preserved in the course of AR;
(ii) a grammatically correct sentence must remain correct after its simplification;
(iii) shortening of any reduction would violate the correctness principle (ii);
(iv) a sentence which contains a correct sentence (or its permutation) as a subsequence, must be further reduced;
(v) an application of the shift operation is limited only to cases when a shift is enforced by the correctness principle (ii); i.e., a simple deletion would result in an incorrect word order.

Note that the possible order of reductions reflects dependency relations between individual sentence members, i.e., relations between governing and dependent nodes, as it is described in [7].

Let us illustrate the basic principles of AR on the following example. The sentence undergoing AR is represented as a string of word forms (words and punctuation) enriched with their disambiguated lexical, morphological and syntactic categories.[1]

*Example 1.*
(1) [*Petr*,Sb] [*se*,AuxT] [*bojí*,Pred] [*o*,AuxP] [*otce*,Obj] [.,AuxK]
'Peter – REFL – worries – about – father – .'
'Peter worries about his father.'

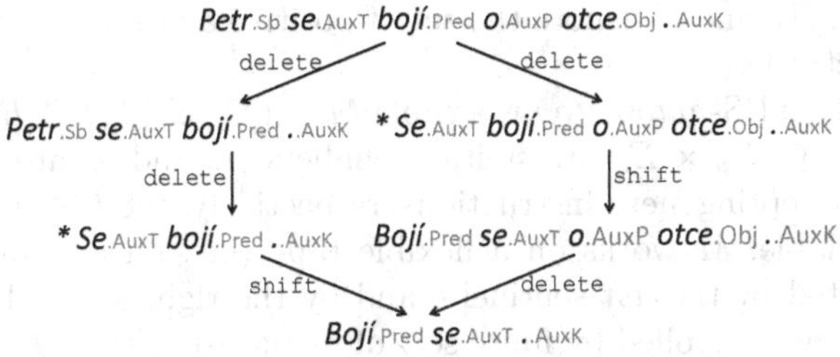


**Fig. 1.** The schema of AR for sentence (1)

Our example sentence can be simplified in two ways:
(i) either by deletion of the prepositional group *o otce* 'about father' (according to the correctness constraint on the simplified sentence, the pair of word forms must be deleted in a single step, see the left branch of the scheme);

[1] For the simplicity, only lexical categories (i.e., original word forms and punctuation like full stop in the example) as they appear in the sentence and syntactic categories (like predicate (Pred), subject (Sb), object (Obj), auxiliary words (AuxT, AuxP, AuxK)) are displayed in the examples; see [7] for more detailed description.

(ii) or by deleting the subject *Petr* (the right part of the scheme); however, the simple simplification would result in the incorrect word order variant starting with the clitic *se* (such position of a clitic is forbidden in Czech); thus the shift operation is enforced $\rightarrow_{shift}$ *Bojí se o otce.* '(he) worries about his father.'. The reduction proceeds in a similar way in both branches of AR until the minimal correct simplified sentence *Bojí se.* '(He) worries.' is obtained. This sentence cannot be further correctly reduced.

## 3 Restarting Automata with Delete and Shift Operation

In order to model the analysis by reduction with shifts, we introduce a restarting automaton that uses a limited number of pebbles, and that performs several deletions and shifts within one meta-instruction – a DS-automaton. The DS-automaton is a refinement of the so called sRL-automaton in [7]; here, the automaton is enriched with the shift operation, and with categorial and basic alphabets.

DS-automata are suitable for modeling AR – these automata make it possible to check the whole input sentence and mark selected words with pebbles prior to any changes. It resembles a linguist who can read the whole sentence first, and then reduce the sentence in a correct way. To enable simulation of various orders of reductions, we choose a nondeterministic model of the automaton. We distinguish three alphabets (or vocabularies): a proper alphabet $\Sigma_p$ that is used to model individual word forms, an alphabet $\Sigma_c$ of categories and a basic alphabet $\Gamma$. Since only symbols from the basic alphabet can appear on a tape of a DS-automaton, $\Gamma$ is also called a tape alphabet. (In what follows, $\lambda$ denotes an empty word, $\mathbb{N}_+$ and $\mathbb{N}$ denote the set of positive and the set of nonnegative integers, respectively.)

More formally, a DS-*automaton* is a tuple $M = (\Sigma_p, \Sigma_c, \Gamma, ¢, \$, R, A, k)$, where $\Sigma_p$, $\Sigma_c$, and $\Gamma \subseteq \Sigma_p \times \Sigma_c$ are finite alphabets, $R$ and $A$ are finite sets of restarting and accepting meta-instructions, respectively, and $k \in \mathbb{N}$ is the number of pebbles available. $M$ works on a flexible tape (i.e., on a string of symbols from $\Gamma$) delimited by the left sentinel ¢ and by the right sentinel \$ ($¢, \$ \notin \Gamma$). Its computation is controlled by finite sets of meta-instructions $R$ and $A$, and it makes use of $k$ pebbles $p_1, \cdots, p_k$.

A projection from $\Gamma^*$ to $\Sigma_p^*$ and $\Sigma_c^*$, respectively, is convenient – we define two homomorphisms, a *proper homomorphism* $h_p : \Gamma \rightarrow \Sigma_p$ and a *categorial homomorphism* $h_c : \Gamma \rightarrow \Sigma_c$ in the obvious way: $h_p([a, b]) = a$, and $h_c([a, b]) = b$ for each $[a, b] \in \Gamma$. For technical reasons, we define $h_p(¢) = h_c(¢) = ¢$.

Each computation of a DS-automaton consists of several phases called *cycles*, and a last halting phase called a *tail*. In each cycle, the automaton performs three passes through the tape with symbols from $\Gamma$. During the first pass, it marks certain symbols of a processed sentence with pebbles according to some meta-instruction $I$; then during the second pass, it performs the shift operations as described by the chosen meta-instruction $I$; and during the third pass, it performs the delete operations as described by the meta-instruction $I$. The operations are applied only on symbols marked by pebbles. Moreover, we allow the left sentinel to be pebbled as well since we sometimes need to shift some symbols just behind it. However, although pebbled, ¢ is treated differently from other tape symbols; it can neither be deleted nor can it be shifted.

In each accepting tail, the automaton according to a meta-instruction $I_{acc}$ from $A$ halts and accepts the analyzed sentence.

In accordance with the linguistic motivation, the meta-instructions check only the categorial part of tape symbols from $\Gamma$.[2]

**Restarting Meta-instructions.** Each cycle of the DS-automaton $M$ is controlled by a single restarting meta-instruction $I \in R$ of the form

$$I = (E_0, a_1, E_1, \dots a_s, E_s; O_{sh}; O_d; \; \mathsf{Restart}) \tag{1}$$

where:
- each $E_i$, $0 \le i \le s$ is a regular language over $\Sigma_c \cup \{\lambda\}$, $E_i$ is called the $i$-th context of $I$;
- $a_i \in \Sigma_c \cup \{¢\}$ (for $1 \le i \le s \le k$) indicates that each $a_i$ is marked with the pebble $p_i$;
- $O_{sh} = o_1, \cdots, o_{p_{sh}}, o_j \in \{sh[i,l] \mid 1 \le i, j \le s, \; i \ne l\}$, is a sequence of shifting operations performed in the second phase: if $o_j = sh[i,l]$ then it shifts the tape symbol marked with $p_i$ to the position behind the symbol with $p_l$;
- $O_d = d_1, \cdots, d_{p_d}, d_j \in \{\mathrm{dl}[i] \mid 1 \le i \le s\}$, is a sequence of delete operations performed in the third phase: if $d_j = \mathrm{dl}[i]$ then it deletes the tape symbol marked by $p_i$;
- ¢ is neither deleted nor shifted within $I$.

We require an 'exclusivity' of the shift operation: each symbol $a_i$ can be shifted only once (as a maximum); moreover, if $a_i$ is shifted then it cannot be deleted within the same meta-instruction. Formally, if $O_{sh}$ contains the shift operation $\mathrm{sh}[i,l]$ for some $i$ then no other $\mathrm{sh}[i,r]$ can be in $O_{sh}$; moreover, $O_d$ cannot contain the $dl[i]$ operation.

Each computation of $M$ on the input $w \in \Gamma^*$ starts with the *tape inscription* $¢w\$$. After a nondeterministic choice of a cycle $C$ realizing the guessed restarting meta-instruction $I$, $M$ nondeterministically marks tape symbols $b_1, \dots, b_s$ by pebbles in accordance with $I$: it finds a factorization $w = v_0 b_1 v_1 b_2 \dots v_{s-1} b_s v_s$, $0 \le i \le s$, $1 \le j \le s$ such that $h_c(v_i) \in E_i$, $h_c(b_j) = a_j$ in the first pass. Then $M$ applies the implied sequence of shifts $O_{sh}$ during the second pass, and the implied sequence of deletions during the third pass. If the factorization is not found within the first pass, the automaton gets stuck (and thus it rejects $w$). Notice that due to regularity of individual $E_i$'s the instruction $I$ can be (nondeterministically) identified within one pass over $¢w\$$.

[2] When considering only categorial symbols as a context we avoid both the problem of data sparsity and the problem of a very large alphabet $\Sigma_p$ (i.e., lexicon with hundred of thousands word forms for a natural language).

At the end of each cycle, the Restart operation removes all pebbles from the new tape inscription $w'$ and places the head on the left sentinel. We write $w \vdash^I w'$.

Remember that none of the sentinels can be deleted/shifted and that $M$ is required to execute at least one delete operation during each (restarting) cycle.

If no further cycle is performed, each accepting computation necessarily finishes in a tail performed according to one of the accepting meta-instructions.

**Accepting Meta-instructions.** Tails of accepting computations are described by a set of accepting meta-instructions $A$, each of the form:

$$I_{\text{acc}} = (a_1 \dots a_s, \mathsf{Accept}), \tag{2}$$

where $a_i$ are symbols from $\Sigma_c$.

The tail performed by the meta-instruction $I_{\text{acc}}$ starts with the inscription on the tape ¢$z$\$; if $h_c(z) = a_1 \cdots a_s$ then $M$ accepts $z$ (we write $z \vdash^{I_{acc}} \mathsf{Accept}$), and the whole computation as well. Otherwise, the computation halts with rejection.

We denote by $u \vdash_M v$ the reduction of $u$ into $v$ performed during one cycle of $M$ (that begins with the tape inscription ¢$u$\$ and ends with the tape inscription ¢$v$\$) and by $\vdash_M^*$ the reflexive and transitive closure of $\vdash_M$. We say that $u_1, u_2, \dots, u_n, \mathsf{Accept}$ is an *accepting computation* of $M$ if $u \vdash_M u_1, u_1 \vdash_M u_2, \cdots, u_{n-1} \vdash_M u_n, u_n \vdash_M \mathsf{Accept}$.

A string $w \in \Gamma^*$ is *accepted* by $M$, if $w \vdash_M^* u$ with $(u, \mathsf{Accept}) \in A$. By $L(M)$ we denote the language of words accepted by $M$; we say that $M$ recognizes (accepts) the *basic (tape) language* $L(M)$. We say that $L(M,p) = \{h_p(w) \in \Sigma_p^* \mid w \in L(M)\}$ is the *proper language* of $M$. Analogously, $L(M,c) = \{h_c(w) \in \Sigma_p^* \mid w \in L(M)\}$ is called the *categorial language* of $M$.

Since the number of reductions performed within an accepting computation is of (linguistic) interest, we denote by $L_n(M)$ the language of all sentences accepted by at most $n$ cycles of $M$; $L_0(M)$ is the set of sentences accepted directly by accepting meta-instructions.

Further, we define the *reduction language of* $M$ as $RED(M) = \{u \to v \mid u \vdash_M v, u, v \in L(M)\}$, and $RED_n(M) = \{u \to v \in RED(M) \mid v \in L_{n-1}(M)\}$. Note that $L_n(M)$ and $RED_n(M)$ are finite for any $n \in \mathbb{N}$.

The notations $L_n(M,p)$ and $L_n(M,c)$ denote the proper and categorial variants of $L_n(M)$, respectively.

**Backward Correctness Preserving Property** (bcpp) Realize that each meta-instruction $I$ of the DS-automaton $M$ is *backward correctness preserving*:

$$\big(v \in L(M) \text{ and } u \vdash^I v\big) \Rightarrow (u \in L(M))$$

We will see that this bcpp property plays a crucial role in the study of analysis by reduction.

We naturally suppose that any restarting meta-instruction of $M$ can be associated with some reduction from $RED(M)$. Two conditions formulated below, the map- and mrp- properties, reflect the linguists' preference of reductions being as simple as possible.

**Minimal Accepting Property** (map-): If $w \in L_{\mathrm{O}}(M)$ then neither proper subsequence of $w$ nor any permutation of such subsequence belongs to $L_{\mathrm{O}}(M)$.

**Minimal Reduction Property** (mrp-): Let $u \vdash_M v$ for a cycle $C$ realizing the restarting meta-instruction controlled by the sequence of operations $O = o_1, o_2, \cdots, o_p$. Let $\tilde{v}$ be obtained from $u$ by a (new) restarting meta-instruction controlled by a proper subsequence $\tilde{O}$ of $O$. Then $\tilde{v} \notin L(M)$. The property implies that a meta-instruction is in some sense minimal as none of the deletions/shifts performed by $u \rightarrow v$ can be left out in order to obtain a reducible or acceptable sentence.

If $M$ fulfils the map- and mrp- conditions then $RED(M)$ is said to create a *normalized reduction language of* $M$, and that $M$ is *normalized*. We will show in Section 4 that every DS-automaton can be transformed to a normalized one that recognizes the same proper language.

*Example 2.* The notions of restarting and accepting meta-instructions are illustrated on the analysis of our example sentence (1) from Section 2. The respective DS-automaton $M_{ex}$ is described by two restarting meta-instructions $I_{r_1}$ and $I_{r_2}$, and one accepting meta-instruction $I_{acc}$. It formalizes both branches of AR of the sentence [*Petr*,Sb] [*se*,AuxT] [*bojí*,Pred] [*o*,AuxP] [*otce*,Obj] [.,AuxK] 'Peter worries about his father.', (see the scheme in Section 2).

$I_{r_1} = (E_0^1, a_1^1, E_1^1, a_2^1, E_2^1, a_3^1, E_3^1; o_1^1, o_2^1; \mathsf{Restart})$, where $o_1^1 = \mathrm{dl}[2]$, $o_2^1 = \mathrm{dl}[3]$, $E_0^1 = \{\mathsf{Sb\ AuxT}, \lambda\}$, $a_1^1 = \mathsf{Pred}$, $E_1^1 = \{\lambda, \mathsf{AuxT}\}$, $a_2^1 = \mathsf{AuxP}$, $E_2^1 = \{\lambda\}$, $a_3^1 = \mathsf{Obj}$, $E_3^1 = \{\mathsf{AuxK}\}$;

$I_{r_2} = (E_0^2, a_1^2, E_1^2, a_2^2, E_2^2, a_3^2, E_3^2; o_1^2; o_2^2; \mathsf{Restart})$, where $o_1^2 = \mathrm{sh}[2,3]$; $o_2^2 = \mathrm{dl}[1]$, $E_0^2 = \{\lambda\}$, $a_1^2 = \mathsf{Sb}$, $E_1^2 = \{\lambda\}$, $a_2^2 = \mathsf{AuxT}$, $E_2^2 = \{\lambda\}$, $a_3^2 = \mathsf{Pred}$, $E_3^2 = \{\mathsf{AuxK}, \mathsf{AuxP\ Obj\ AuxK}\}$;

$I_{acc} = (\mathsf{Pred\ AuxT\ AuxK}, \mathsf{Accept})$.

The computation corresponding to the left branch consists in two cycles. Within the first cycle driven by the meta-instruction $I_{r_1}$, $M_{ex}$ puts pebbles $p_1$, $p_2$, and $p_3$ on the symbols containing the words [*bojí*,Pred] (with the left (tape) context [*Petr*,Sb][*se*,AuxT] and empty right context), on [*o*,AuxP], and on [*otce*,Obj], respectively; the operations dl[2] and dl[3] delete the words [*o*,AuxP] (marked with the pebble $p_2$) and [*otce*,Obj] (marked with the pebble $p_3$), respectively. Then the automaton restarts and removes the pebbles from the processed sentence. Similarly in the second cycle realizing restarting meta-instruction $I_{r_2}$, $M_{ex}$ marks the respective words and then in the second pass the operation sh[2, 3] shifts the word [*se*,AuxT] with the pebble $p_2$ to the right of the word [*bojí*,Pred] (with the pebble $p_3$); in the third pass, the operation dl[1] deletes the word [*Petr*,Sb] (marked with $p_1$). Finally, accepting instruction $I_{acc}$ just accepts the remaining words. Similarly for the right branch (starting with $I_{r_2}$ and followed by $I_{r_1}$ and $I_{acc}$ instructions).

For the sake of simplicity, let $\alpha_0, \alpha_1$, $\alpha_2$, and $\beta_1$ be defined as follows:

$\alpha_0$= [*bojí*,Pred] [*se*,AuxT] [.,AuxK];
$\alpha_1$= [*Petr*,Sb] [*se*,AuxT] [*bojí*,Pred] [.,AuxK];

$\beta_1$= [*bojí*,Pred] [*se*,AuxT] [*o*,AuxP] [*otce*,Obj] [.,AuxK], and
$\alpha_2$= [*Petr*,Sb] [*se*,AuxT] [*bojí*,Pred] [*o*,AuxP] [*otce*,Obj] [.,AuxK].

Applying the relevant definitions we get:

$L_0(M_{ex}) = \{\alpha_0\}$ as $\alpha_0 \vdash^{I_{acc}}$ Accept;
$L_1(M_{ex}) = \{\alpha_0, \alpha_1, \beta_1\}$, as $\alpha_1 \vdash^{I_{r2}} \alpha_0$, and $\beta_1 \vdash^{I_{r1}} \alpha_0$;
$L_2(M_{ex}) = \{\alpha_0, \alpha_1, \beta_1, \alpha_2\}$ since $\alpha_2$ is the only word not in $L_1(M_{ex})$ for which $\alpha_2 \vdash_{M_{ex}} \alpha, \alpha \in L_1(M)$ (as $\alpha_2 \vdash^{I_{r1}} \alpha_1$, or $\alpha_2 \vdash^{I_{r2}} \beta_1$);
$L(M_{ex}) = L_2(M_{ex})$ as no $\alpha \in L_2(M_{ex})$ and $\beta \notin L_2(M_{ex})$ exist for which $\beta \vdash_{M_{ex}} \alpha$ holds.

The proper homomorphism of $M_{ex}$ removes linguistic categories Sb, AuxT, Pred, AuxP, Obj, and AuxK; so the sentence [*Petr*,Sb] [*se*,AuxT] [*bojí*,Pred] [*o*,AuxP] [*otce*,Obj] [.,AuxK] (from the basic language of $M_{ex}$) is mapped onto the sentence *Petr se bojí o otce.* 'Peter worries about his father.' (from the proper language of $M_{ex}$). Similarly, the string Sb AuxT Pred AuxP Obj AuxK is a sentence of the categorial language of $M_{ex}$.

## 4 Results

With respect to the linguistic motivation, we focus on the number of deletions and/or shifts in individual restarting meta-instructions and we use particular abbreviations for automata/languages with a restriction on these complexity measures. In particular, prefix DS- is used to identify the delete-shift automata without any restrictions, and D- is used for automata with deletions only. Further, the prefix (k)- is used to indicate that at most $k$ pebbles are available in one meta-instruction. As a special case, (0)- means that the automaton contains only accepting meta-instructions and thus it accepts in tail computations only. We use the syllable d(i)- for automata with at most $i$ deletions in one meta-instruction and s(j)- for automata with at most $j$ shifts in a single meta-instruction. The requirements for normalized reduction languages are denoted by map-, and mrp-.

For each type X of restarting automata, we use $\mathcal{L}$(X), $\mathcal{LP}$(X), $\mathcal{LC}$(X) to denote the class of all basic, proper and categorial languages, recognizable by automata of this type. Analogously, $\mathcal{RED}$(X) denotes the class of all reduction languages of these automata. Further, $\mathcal{L}_n$(X), $\mathcal{LP}_n$(X), $\mathcal{LC}_n$(X) denote the classes of basic, proper, and categorial languages defined by at most $n$ reductions of X-automata; $\mathcal{RED}_n$(X) denotes analogical notion for reduction languages. Proper inclusions are denoted by $\subset$.

FIN, REG, (D)CFL, and CSL are used for classes of finite, regular, (deterministic) context-free, and context-sensitive languages, respectively, and FINR, REGR, (D)CFR, and CSR for the classes of reduction languages defined by DS-automata for which their basic languages are from FIN, REG, (D)CFL, and CSL, respectively.

First part of our results is devoted to the map- and mrp- properties, their influence on the delete and shift complexities and on the computational power. Then we show delete, shift and pebble hierarchies.

Let $M_{ex}$ be the automaton (implicitly) given in Example 2. Let us recall that $L(M_{ex}) = L_2(M_{ex})$. Analyzing its meta-instructions we see that automaton $M_{ex}$ is in fact map-mrp-s(1)-d(2)-DS-automaton implying $L(M_{ex}) = L_2(M_{ex}) \in \mathsf{FIN}$. Relaxing the conditions map- and mrp-, the finite language $L(M_{ex})$ can obviously be recognized by accepting tails only. The situation changes when the automaton recognizing $L(M_{ex})$ is required to fulfil the condition map-. Then, not only that it cannot accept this finite language by accepting tails only; it requires (at least) two pebbles and two deletions in one restarting meta-instruction. Thus, it can be used for separation.

**Proposition 1.** *Let $M_{ex}$ be the automaton implicitly given in Example 2. Then*

1. $L(M_{ex}) \in \mathcal{L}_2(\mathsf{map\text{-}d(2)\text{-}DS}) \setminus \mathcal{L}(\mathsf{map\text{-}d(1)\text{-}DS})$,
2. $RED(M_{ex}) \in \mathcal{RED}_2(\mathsf{map\text{-}d(2)\text{-}DS}) \setminus \mathcal{RED}(\mathsf{(1)\text{-}DS})$,
3. $L_1(M_{ex}) \in \mathcal{L}_1(\mathsf{map\text{-}s(1)\text{-}d(1)\text{-}DS}) \setminus \mathcal{L}(\mathsf{map\text{-}(1)\text{-}DS})$,
4. $L(M_{ex}) \in \mathcal{L}(\mathsf{(0)\text{-}D})$.

**Proof.** Take the sentence [*Petr*,Sb] [*se*,AuxT] [*bojí*,Pred][.,AuxK] from $L_1(M_{ex})$. The single reduction of the sentence follows the map-principle, as *bojí se* is a permutation of a correct subsequence *se bojí*. The reduction consists of one delete and one shift operations; they are unambiguously given by the backward correctness preserving property (bcpp). Two operations on two different positions imply automaton with at least two pebbles, thus $L_1(M_{ex}) \notin \mathcal{L}(\mathsf{map\text{-}(1)\text{-}DS})$, and $RED(M_{ex}) \notin \mathcal{RED}(\mathsf{(1)\text{-}DS})$. Similarly for assertions *3*, and *4* is obvious. □

Consider shifts and deletions being the only operations allowed on (a set of) words. Based on these operations, we can naturally define the partial order $\succ_L$ on the set $L$ of words. We say that $u$ *syntactically precedes* $v$ in $L$ and write $u \succ_L v$ iff:

1. $u, v \in L$, $|u| > |v|$;
2. $v$ can be obtained from $u$ by a sequence $O$ of deletions and shifts applied on $u$; $u \overset{O}{\to} v$;
3. the application of any proper subsequence $O'$ of $O$ on $u$ would end up with a word outside $L$.

By $\succ_L^+$ we denote the transitive, nonreflexive closure of $\succ_L$. Obviously, for a DS-automaton $M$, $u \to v \in RED(M)$ implies $u \succ_{L(M)}^+ v$ and $h_p(u) \succ_{L(M,p)}^+ h_p(v)$.

We define the set $L_{\succ_L}^{min} = \{v \in L \mid \neg\exists u \in L : v \succ_L u\}$ as the set of minimal words in $L$. Then, for the map-DS-automaton $M$:

$$L_0(M) = L_{\succ_{L(M)}}^{min}.$$

For $w \in L$ we denote by $\sigma(w)$ any sequence $\sigma(w) = w_0, w_1, \ldots, w_n$ such that $w = w_0$, $w_{i-1} \succ_L w_i$, $1 \le i \le n$ and $w_n \in L_{\succ_L}^{min}$. We call $\sigma(w)$ the $\succ_L$-sequence of $w$. Realize that $\succ_L$-sequence of $w$ needs not be uniquely given by $L, w$.

Note that every pair $u, v$ with $u \succ_L v$ implicitly defines one or more sequences $O$ of deletions and shifts that transforms $u$ into $v$. For technical reasons, we will only work with sequences of minimal length and will, for every pair $u \succ_L v$, denote one of them as $O(u,v)$. Since the number of deletions and shifts in $O(u,v)$ is determined unambiguously by the length of it, we denote by $S(u \succ_L v)$ the number of shifts, and by $D(u \succ_L v)$ the number of deletions of $O(u,v)$. Let us stress that the minimal number of shifts and deletions needed to transform $u$ into $v$ is a kind of edit-distance between words $u, v$. Not surprisingly, it will be shown that these numbers are related to the numbers of deletions and shifts used in meta-instructions of the corresponding DS-automaton. For that, we introduce several delete and shift complexities.

By $D_\omega(L) = \max\{D(u \succ_L v); u, v \in L\}$ we denote the *delete upper bound* of $L$ (or of $\succ_L$). Analogously, $S_\omega(L) = \max\{S(u \succ_L v); u, v \in L\}$ denotes the *shift upper bound of* $L$.

For any word $w$ and its $\succ_L$-sequence $\sigma(w) = w_0, w_1, \ldots, w_n$ we define $D(\sigma(w)) = \max_i\{D(w_{i-1} \succ_L w_i)\}$ and the *delete lower bound* of $w$ with respect to $L$ as $D_\ell(L, w) = \min_{\sigma(w)}\{D(\sigma(w))\}$. For shifts, $S(\sigma(w))$ and $S_\ell(L, w)$ are defined analogously.

For $L_1 \subseteq L_2$ the delete and shift lower bounds of $L_1$ with respect to $L_2$ are defined in the following way: $D_\ell(L_1, L_2) = max_{w \in L_1}\{D_\ell(L_2, w)\}$ and $S_\ell(L_1, L_2) = max_{w \in L_1}\{S_\ell(L_2, w)\}$.

We call the DS-automaton *reduced* if each of its meta-instruction uses exactly as many pebbles as it is needed to realize the involved sequence of operations.

A close relation between the complexity of meta-instructions of map- and/or mrp-DS-automata and the above defined upper and lower bounds is formulated in the following Theorem 1.

**Theorem 1.** *Let $M$ be a reduced* map-mrp-s(i)-d(j)-DS-*automaton, and* $LM \in \{L(M), L(M, c)\}$. *Then the following holds:*

1. $u \to v \in RED(M)$ *implies* $u \succ_{L(M)} v$;
2. $S_\ell(LM, LM) \leq i \leq S_\omega(LM)$, *and* $D_\ell(LM, LM) \leq j \leq D_\omega(LM)$;
3. ($L \subseteq LM$ *and* $r \leq D_\ell(L, LM) - 1$) *implies* $LM \notin \mathcal{L}(\text{map-d(r)-DS})$;
4. ($L \subseteq LM$ *and* $r \leq S_\ell(L, LM) - 1$) *implies* $LM \notin \mathcal{L}(\text{map-s(r)-DS})$.

Notice that without the mrp- condition the assertion 1. of Theorem 1 would not hold.

**Proof.** Assertion 1. directly follows from the definition of the mrp- condition; together with the map- property it implies the lower bounds in assertion 2. The upper bounds in 2. follow from the definition of $D_\omega$ and $S_\omega$. To get 3. and 4., realize that $M$ is reduced map-DS-automaton and that $u \vdash_M v$ implies $u \succ_M^+ v$. □

As shown in Theorem 2, map-D-automata are powerful enough for categorial recognition of deterministic CF-languages and proper recognition of all CF-languages. As a corollary of Theorem 3 we even get that every CF-language is a proper language of some map-mrp-DS-automaton.

**Theorem 2.** DCFL $\subset \mathcal{LC}(\text{map-D})$, CFL $\subset \mathcal{LP}(\text{map-D})$.

**Proof.** To make use of two useful results from [6], we need to introduce an extended class of D-automata. By $D_{reg}$, we denote a class of D-automata whose accepting meta-instructions are of the form $I_a = (L_r, \mathsf{Accept})$, where $L_r$ is a regular language. $I_a$ accepts each $v \in L_r$.

*1. Proof of* $\mathsf{DCFL} \subset \mathcal{LC}(\mathsf{map\text{-}D})$. The relevant assertion from Proposition 3.8. in [6] can be reformulated as $\mathsf{DCFL} \subset \mathcal{LC}(\mathsf{D}_{reg})$. To prove our proposition it is therefore sufficient to transform the given $\mathsf{D}_{reg}$-automaton to a map-D-automaton. Thus, let $M$ be a $\mathsf{D}_{reg}$-automaton such that $L(M, c) \in \mathsf{DCFL}$. Based on the size of $|L_0(M, c)|$ we distinguish two cases:

(a) If $L_0(M, c)$ is a finite language then it is easy to check whether it fulfils the condition map-. If not, then there obviously exists map-D-automaton $A_0$ recognizing exactly the language $L_0(M, c)$. The automaton $A_0$ simulates the leftmost branch of the syntactic precedence $\succ_{L_0(M,c)}$. Then, it suffices to substitute original accepting meta-instructions of $M$ by all of $A_0$'s restarting and accepting meta-instructions.

(b) If $L_0(M, c)$ is not a finite language then $L_0(M, c)$ is still $\in \mathsf{REG}$ and there is a deterministic finite automaton $A$ recognizing $L_0(M, c)$. From $A$ we can construct deterministic D-automaton $A_0$ whose restarting meta-instructions always delete by the simulating of a rightmost cycle of $A$ and accepting meta-instructions accepts only cycle-free words. Here, by a cycle in a word $w$ we mean such subword $u$ of $w$ that – based on pumping lemma – could be iterated (within the computation on $w$, the automaton starts reading $u$ in the same state as it leaves it). Replacing accepting meta-instructions of $M$ by all meta-instructions of $A_0$, we get a D-automaton that according to (a) can be modified to an equivalent map-D-automaton recognizing $L(M, c)$.

*2. Proof of* $\mathsf{CFL} \subset \mathcal{LP}(\mathsf{map\text{-}D})$. It directly follows from Proposition 3.4. in [6] that $\mathsf{CFL} \subset \mathcal{LP}(\mathsf{D}_{reg})$ which – based upon the above given construction for DCFL – implies $\mathsf{CFL} \subset \mathcal{LP}(\mathsf{map\text{-}D})$ . □

Adopting the above given constructions we show in the next theorem that as for the proper languages, the linguistic requirement of normalization preserves the power of DS-automata.

**Theorem 3.** *Let* $X \in \{\mathsf{DS}, \mathsf{D}\}$. *Then* $\mathcal{LP}(\mathsf{X}) = \mathcal{LP}(\mathsf{map\text{-}mrp\text{-}X})$.

**Proof.** Let us prove the theorem by explaining how a DS-automaton for a proper language can be transformed onto one with the map- and mrp- properties. We will show the construction for D-automata first and secondly we will explain how the construction can be adopted for automata with shifts. Note that based on the proof of Theorem 2 we can suppose that the original D-automaton already posseses the map- property.

Let us start with the proof of $\mathcal{LP}(\mathsf{D}) = \mathcal{LP}(\mathsf{map\text{-}mrp\text{-}D})$. It is obvious that $\mathcal{LP}(\mathsf{map\text{-}mrp\text{-}D}) \subseteq \mathcal{LP}(\mathsf{D})$. To show the opposite inequality we start with a map-D-automaton $M = (\Sigma_p, \Sigma_c, \Gamma, ¢, \$, R, A, k)$ and we construct a map-mrp-D-automaton $M_D = (\Sigma_p, \Sigma_c^D, \Gamma_D, ¢, \$, R_D, A_D, k_D)$ with an enriched set of categories $\Sigma_c^D$ and a new basic alphabet $\Gamma_D$.

Based on any individual accepting computation $\mathcal{C}_\mathcal{A}$ of $M$ on $w$, where $h_c(w) = w_1 \dots w_n$, each symbol $w_j$ can be unambiguously associated with either restarting or accepting meta-instruction $I(j)$ and with an index $p_j$ identifying either the pebble that is within realization of the restarting meta-instruction $I(j)$ put on it or that identifies the position of $w_j$ within the accepting meta-instruction $I(j)$; if $w_j$ is deleted within $\mathcal{C}_\mathcal{A}$ then $I(j)$ is that restarting meta-instruction that has deleted it, otherwise $I(j)$ is the accepting meta-instruction applied in $\mathcal{C}_\mathcal{A}$. To be able to insert this information into $w$ we enrich the categorial alphabet $\Sigma_c$ of $M$: $\Sigma_c^D = \{[x,i,r] \ : \ x \in \Sigma_c, i \in \{1, \dots |R \cup A|\}, r \leq \ell$, where $\ell = \max\{k, \max\{s; (a_1, \dots a_s;\ \mathsf{Accept}) \in A\}\}$.

The order of meta-instructions in $R \cup A$: If $I = (E_0, a_1, E_1, \dots a_s, E_s; O_{sh}; O_p;$ $\mathsf{Restart}) \in R$ is $i$-th in the ordering then put $I(i) = (E_0, [a_1, i, 1], E_1, \dots [a_s, i, s],$ $E_s; O_{sh}; O_p\ \mathsf{Restart})$ into $R_D$. Analogously, if $I_{acc} = (a_1, \dots a_s;\ \mathsf{Accept})$ is $j$-th then put the accepting meta-instruction $I_{acc}(j) = ([a_1, j, 1], \dots [a_s, j, s];\ \mathsf{Accept})$ to $A_D$.

Realize that the proper alphabet $\Sigma_p$ has not been changed, thus $L(M,p) = L(M_2,p)$:

- if $u \vdash_{M_D} v$ then there are $x, y$ such that $x \vdash_M y$, and $h_p(x) = h_p(u), h_p(y) = h_p(v)$;
- $\forall x, y \in L(M)$, $x \vdash_M y$ there are $x_2, y_2 \in L(M_D)$ such that $x_2 \vdash_{M_D} y_2$, and $h_p(x_2) = h_p(x), h_p(y_2) = h_p(y)$

Due to the index of the instruction and the position in that instruction associated with individual symbols as described above, the obtained D-automaton $M_2$ fulfills both the mrp- and map- conditions.

To finish the proof of the theorem $\mathcal{LP}(\mathsf{DS}) = \mathcal{LP}(\mathsf{map\text{-}mrp\text{-}DS})$, we explain how information about shifts can be handled using the idea of the above given construction. Let $M = (\Sigma_p, \Sigma_c, \Gamma, ¢, \$, R, A, k)$ be the map-DS-automaton. We will construct map-mrp-DS-automaton $M_{DS} = (\Sigma_p, \Sigma_c^{DS}, \Gamma^{DS}, ¢, \$, R_{DS}, A_{DS}, k_{DS})$ with an enriched set of categories $\Sigma_c^{DS} = \{[x,i,r] \ : \ x \in \Sigma_c, i \in \{1, \dots |R \cup A|\}, r \leq \ell$, where $\ell = \max\{k, \max\{s; (a_1, \dots a_s;\ \mathsf{Accept}) \in A\}\}$.

Within a restarting meta-instruction $I$ some pebbles are put on symbols that are to be deleted and some of them on those moved by $I$. As described earlier, the index $r$ in the triple $[x,i,r]$ is associated either with pebble put on the original symbol $x$ within the meta-instruction that deletes the symbol or with the position of $x$ in the accepting meta-instruction. Thus, to realize restarting meta-instruction $I$ that involves shifts, we will simply ignore the index in those triples that are moved by $I$.

Fix any ordering of meta-instructions in $R \cup A$. If $I_{acc} = (a_1, \dots a_s;\ \mathsf{Accept})$ is $j$-th in that ordering then put accepting meta-instruction $I_{acc}(j) = ([a_1, j, 1],$ $\dots [a_s, j, s];\ \mathsf{Accept})$ to $A_{DS}$.

Let $I = (E_0, a_1, E_1, \dots a_s, E_s; O_{sh}; O_d;\ \mathsf{Restart}) \in R$ be a restarting meta-instruction of $M$ and $O_d = \{dl[j_1], \dots, dl[j_d]\}$ is the set of all delete operations involved in $I$. If $I$ is $j$-th in the ordering then add the set of restarting meta-instructions $I(j) = \{(E_0, [a_1, i_1, 1], E_1, \dots [a_s, i_s, s], E_s; O_{sh}; O_d;\ \mathsf{Restart}) \mid i_j = j$ for $j \in \{j_1, \dots, j_d\}\}$ into $R_{DS}$.

Realize that the set of restarting meta-instructions of $M_{DS}$ is larger than that of $M$ if at least one of $M$'s restarting meta-instructions involves shift. However, fixing the accepting computation $\mathcal{C}_{\mathcal{A}}$ of $M$ on $w$, the transformation of $w = w_1 \dots w_n \in \Gamma^*$ to $W = W_1 \dots W_n = [w_1, i_1, r_1] \dots [w_n, i_n, r_n] \in \Gamma_{DS}^*$ such that $h_p(w) \in L(M,p) \Longleftrightarrow h_p(W) \in L(M_{DS}, p)$ can be done by a deterministic algorithm:

1. initialize $W = w$;
2. let $I_1, I_2, \dots, I_t$ be a sequence of meta-instructions corresponding to $\mathcal{C}_{\mathcal{A}}$, $I_1, \dots, I_{t-1} \in R, I_t \in A$;
3. let $I_t = (a_1 \dots a_s;\ \mathsf{Accept})$ be $j_t$-th in the ordering of $R \cup A$, where $a_1 = w_{i_1}, \dots, a_s = w_{i_s}$; set $W_{i_m}$ to $[w_{i_m}, j_t, m]$;
4. for $q = t-1$ *downto* 1 let $I_q = (E_0, a_1, E_1, \dots a_s, E_s; O_{sh}; O_d;\ \mathsf{Restart})$ be $j_q$-th in the ordering of $R \cup A$, where $a_1 = w_{i_1}, \dots, a_s = w_{i_s}$ and $p_1, \dots, p_d$ are the pebbles put on those symbols deleted by $I_q$; set $W_{i_{p_i}}$ to $[w_{i_{p_i}}, q_t, p_i]$.

Thanks to this deterministic process the indices $i, r$ in triples $[x, i, r]$ guarantee both the map- and mrp- property of the constructed DS-automaton $M_{DS}$. □

**Corollary 1.** $\mathsf{CFL} \subset \mathcal{LP}(\text{map-mrp-D})$

It is easy to see that $\mathcal{LC}(\mathsf{X})$ is a subset of $\mathcal{LP}(\mathsf{X})$ for $X \in \{D, DS\}$ and that every computation of restarting automaton can directly be simulated in linear space implying $\mathcal{LC}(\mathsf{X}) \subseteq \mathcal{LP}(\mathsf{X}) \subseteq \mathsf{CSL}$. These inequalities are shown as valid with the help of separation languages $L_e$ and $L_{a2b}$ (see Proposition 2). The lower bound parts of both Propositions are mainly based on counting argument combined with consequences of pumping lemma for regular languages.

**Proposition 2.** $L_e = \{ a^{2^n} \mid n \in \mathbb{N} \} \in \mathsf{CSL} \setminus \mathcal{LP}(\mathsf{DS})$,
$L_{a2b} = \{ a^n b^n \mid n \geq 1 \} \cup \{ a^n b^m \mid m > 2n > 0 \} \in \mathcal{LP}((3)\text{-D}) \setminus \mathcal{LC}(\mathsf{DS})$

**Corollary 2.** *Let* $X \in \{\mathsf{DS}, \mathsf{D}\}$. *Then* $\mathcal{LC}(\mathsf{X}) \subset \mathcal{LP}(\mathsf{X}) \subset \mathsf{CSL}$.

In order to show the delete and shift hierarchies, we define two classes of sample languages. Let $j \in \mathbb{N}_+$, $i \in \mathbb{N}$, $\Sigma = \{P, b, s\}$, $\Delta_j = \{c, a_1, a_2, \dots, a_j\}$, $\Lambda = \{\lambda\}$:

$$LS(j,i) = \{ Ps^i\{b^j\}^+, \{b^j\}^+ s^i, s^i \}, \qquad Le(j) = \{ a_1^n a_2^n \cdots a_j^n \mid n > 0 \}.$$

The construction of relevant DS-automata and delete and shift complexities of these languages are given in Lemma 1. It is easy to see that all above defined classes of languages belong to CSL, languages $LS(j,i)$ are infinite regular, $Le(2) \in \mathsf{CFL}$, and $Le(j) \in \mathsf{CSL} \setminus \mathsf{CFL}$ for $j \geq 3$.

**Lemma 1.** *Let* $\mathsf{X} = \text{map-mrp}$, $\mathcal{LX} \in \{\mathcal{LC}, \mathcal{LP}\}$, $j \in \mathbb{N}_+$, $i \in \mathbb{N}$. *Then:*

*(a)* $LS(j,i) \in \mathcal{LX}(\text{d(j)-s(i)-X-DS})$ *(b)* $LS(j,i) \in \mathcal{LX}((\max\{j, i+2\})\text{-X-DS})$
*(c)* $Le(j) \in \mathcal{LX}(\text{d(j)-X-D})$ *(d)* $Le(j) \in \mathcal{LX}((j)\text{-X-D})$

**Proof.** The proof is done by an informal construction of DS-automata.
(a) We describe a d(j)-s(i)-map-mrp-DS-automaton $MS(j,i)$ such that $LS(j,i) = L(MS(j,i),c) = L(MS(j,i),p)$, and it uses max{j, i+2} pebbles. $MS(j,i)$ works with the basic alphabet {[P,P], [b,b], [s,s]}, and categorial and proper alphabets equal to {P, b, s}. The automaton $MS(j,i)$ simulates the leftmost syntactic precedence of any word of $LS(j,i)$); we have three possibilities for one cycle:

- the word $[P,P][s,s]^i\{[b,b]^j\}^n$ is changed to $\{[b,b]^j\}^n[s,s]^i$; for this, $i+2$ pebbles are used to mark the symbol $[P,P]$, all symbols $[s,s]^i$ and the last symbol $[b,b]$ first; then $[s,s]^i$ are shifted after $[b,b]$'s and $[P,P]$ is deleted;
- the prefix $[b,b]^j$ is marked with pebbles and deleted;
- the word $[s,s]^i$ is accepted in a tail computation.

(b) Here we describe a d(j)-map-mrp-D-automaton $Me(j)$ such that $Le(j) = L(Me(j),c) = L(Me(j),p)$, and it uses $j$ pebbles. The automaton $Me(j)$ it simulates always the leftmost syntactic precedence for any word from $Le(j)$; in one cycle the automaton marks by pebbles and deletes one copy of $[a_1,a_1],[a_2,a_2],\dots,[a_j,a_j]$ in a word longer then $j$ and accepts the word $[a_1,a_1][a_2,a_2]\dots[a_j,a_j]$ in a tail computation.

It is not hard to see that the described automata fulfill the map- and mrp-conditions. □

Now we are ready for our separation results; the very robust pebble hierarchy is dealt with in the following part, shift and delete hierarchies are given afterwards.

**Theorem 4.** *Let* $i > 0$, $X \in \{\mathsf{DS}, \mathsf{D}\}$, $Y \in \{\lambda, \mathsf{map}, \mathsf{map\text{-}mrp}\}$, $\mathcal{LX} \in \{\mathcal{L}, \mathcal{LC}, \mathcal{LP}, \mathcal{RED}\}$. *Then* $\mathcal{LX}$(Y-(i)-X) $\subset$ $\mathcal{LX}$(Y-(i+1)-X).

**Proof.** To prove the proposition, we consider the sequence of above defined languages $Le(j) = \{\, a_1^n a_2^n \cdots a_j^n \mid n > 0\}$, $j > 1$.

For the upper bound consider the automaton $Me(j)$ from the proof of Lemma 1b. Its categorial and proper languages are both equal to $Le(j)$ and the automaton uses exactly $j$ deletions and pebbles in a cycle.

The lower bound parts are based on Theorem 1. For each DS-automaton $M$ recognizing $Le(j)$, the set $L_0(M)$ is finite, thus $RED(M)$ is nonempty; any reduction from $RED(M)$ is forced by some syntactic precedences from $\succ_{Le(j)}$. Realize that if $u \succ_{Le(j)} v$ then $|u| = |v| + j$; $v$ can be obtained from $u$ by $j$ deletions for which $M$ uses at least $j$ pebbles. This way we get the desired hierarchies for basic, categorial and reduction languages.

For proper languages hierarchy realize that for any DS-automaton $M$ with proper homomorphism $h_p$, each relation $u \succ_{L(M)} v$ implies the existence of at least one relation of the form $h_p(u) \succ^+_{L(M,p)} h_p(v)$; from this, $D_\ell(L(M),L(M)) \geq D_\ell(L(M,p),L(M,p))$ follows. This means that the pebble complexity of every DS-automaton with the proper language $Le(j)$ is at least $D_\ell(Le(j),Le(j))$ and thus at least $j$. □

The subsequent results show the existence of infinite pebble hierarchies even on classes of finite sub-languages of DS-automata. Note that the hierarchies for reduction languages are more robust than for the other types of languages; they hold also without the map- and mrp- conditions. The language $Le(j)$ represents a core of the corresponding proofs.

**Theorem 5.** *Let* $n > 0$, $i > 0$, $X \in \{\mathsf{D}, \mathsf{DS}\}$, $Y \in \{\mathsf{map}, \mathsf{map\text{-}mrp}\}$, $\mathcal{LX} \in \{\mathcal{L}, \mathcal{LC}, \mathcal{LP}, \mathcal{RED}\}$.
*Then* $\mathcal{LX}_n(\mathsf{Y\text{-}(i)\text{-}X}) \subset \mathcal{LX}_n(\mathsf{Y\text{-}(i{+}1)\text{-}X})$, *and* $\mathcal{RED}_n(\mathsf{(i)\text{-}X}) \subset \mathcal{RED}_n(\mathsf{(i{+}1)\text{-}X})$.

**Theorem 6.** *For* $i > 2, j \geq 0$, $Y \in \{\mathsf{FIN}, \mathsf{REG} \setminus \mathsf{FIN}, \mathsf{CFL} \setminus \mathsf{REG}, \mathsf{CSL} \setminus \mathsf{CFL}\}$, $X \in \{\mathsf{map}, \mathsf{map\text{-}mrp}\}$, $\mathcal{LX} \in \{\mathcal{L}, \mathcal{LC}, \mathcal{LP}\}$ *we have the following proper inclusions:*
(a) $Y \cap \mathcal{LX}(\mathsf{d(i)\text{-}s(j)\text{-}X\text{-}DS}) \subset Y \cap \mathcal{LX}(\mathsf{d(i{+}1)\text{-}s(j)\text{-}X\text{-}DS})$,
(b) $Y \cap \mathcal{LX}(\mathsf{d(i)\text{-}s(j)\text{-}X\text{-}DS}) \subset Y \cap \mathcal{LX}(\mathsf{d(i)\text{-}s(j{+}1)\text{-}X\text{-}DS})$.

**Proof.** To separate $\mathcal{LX}(\mathsf{map\text{-}d(i)\text{-}s(j)\text{-}DS})$ from $\mathcal{LX}(\mathsf{map\text{-}d(i{+}1)\text{-}s(j)\text{-}DS})$ and $\mathcal{LX}(\mathsf{map\text{-}d(i)\text{-}s(j{+}1)\text{-}DS})$ we use the languages $LS(i,j)$, and their syntactic precedences. The number of shift operations is forced by the map- property, otherwise the language $LS(i,j)$ could also be recognized with a D-automaton without shifts simply by deleting the sufix $[b,b]^i$ from $[P,P][s,s]^j\{[b,b]^i\}^n$, for $n > 1$, instead. The number of deletions in one restarting instruction is also determined by the map- property that forces the instruction to delete $[b,b]^i$ from the proper prefix of $\{[b,b]^i\}^n[s,s]^j$. Since $LS(i,j)$ is infinite regular, the proof for $Y = \mathsf{REG} \setminus \mathsf{FIN}$ follows.

For remaining classes we use the languages obtained by a small modification of $LS(i,j)$: $L_{\mathsf{FIN}}(i,j) = \{Ps^jb^i, b^is^j, s^j\}$, $L_{\mathsf{CFL}\setminus\mathsf{REG}}(i,j) = LS(i,j) \cup Le(2)$, and $L_{\mathsf{CSL}\setminus\mathsf{CFL}}(i,j) = LS(i,j) \cup Le(3)$. □

The automata from the previous theorem can be used to prove the similar hierarchical results for reduction languages even with the absence of the map- and mrp- conditions. Moreover, hierarchies similar to those in Theorems 5 and 6 hold also for finite languages defined by $n$ reductions.

## Conclusion and Perspectives

We have presented a class of restarting automata (DS-automata), which formalize lexicalization in a similar way as categorial grammars (see e.g. [1,2]). This class of automata – similarly as categorial grammars – allows us to introduce (in a natural way) basic languages (on word forms marked with categories), proper languages (on unmarked word forms), and categorial languages (on grammatical categories). Further, they allow to introduce reduction languages – this concept is quite natural for DS-automata and the analysis by reduction.

We have introduced also the minimalist map- and mrp- properties, which were used for the normalization of DS-automata; these properties ensure similar and transparent hierarchies for classes of finite and infinite languages. Note that relaxation of these minimalist properties often leads to different results for classes of finite and infinite languages. The normalized DS-automata formalize the notion of analysis by reduction.

Reduction languages allow for explicit description of the integration of individual (disambiguated) word forms into a sentence structure. While proper languages play the role of input languages for weak equivalence of DS-automata (and other types of automata or grammars), reduction languages serve for linguistically more relevant strong equivalence of DS-automata.

Based on [4], we estimate that roughly seven deletions in one reduction step suffice to analyze adequately any sentence (not containing coordination in its structure) from the Prague Dependency Treebank [3]. As for the shift complexity, we have only been able to find reductions of Czech sentences with at most one shift in a single reduction step. From this point of view, normalized reductions in natural languages are quite simple. The information stored in morphological lexicons of individual natural languages is in fact modeled by the information contained in the basic (tape) alphabet of DS-automata. On the other hand, meta-instructions model syntactic potential of individual words (i.e., information stored in valency lexicons and a grammar component of a natural language description).

We have already used the analysis by reduction for explaining the basics of dependency syntax of Czech (see e.g. [7]). However, it can be used for explanation of basic issues of lexicalized syntax based on (even discontinuous) constituents as well – in such a case, individual (restarting) meta-instructions of a normalized description correspond to individual types of constituents. The proposed type of strong equivalence and three proposed types of very robust complexity measures can serve for both types of syntactic methods.

Finally, we strongly believe that for linguistic applications, (relatively simple) star-free languages are sufficient as contexts in meta-instructions since the main information contained in a context is that the context cannot contain some special subwords (or some simple symbols as, eg., punctuation symbols).

## References


1. Ajdukiewicz, K.: Die syntaktische Konnexität. Studia Philosophica I, 1–27 (1935)
2. Chomsky, N.: Formal Properties of Grammar. In: Handbook of Mathematical Psychology, vol. II, pp. 323–418. John Wiley and Sons, Inc. (1963)
3. Hajič, J., Panevová, J., Hajičová, E., Sgall, P., Pajas, P., Štěpánek, J., Havelka, J., Mikulová, M., Žabokrtský, Z., Ševčíková-Razímová, M.: Prague Dependency Treebank 2.0. Linguistic Data Consortium, Philadelphia (2006)
4. Kuboň, V., Lopatková, M., Plátek, M.: On Formalization of Word Order Properties. In: Gelbukh, A. (ed.) CICLing 2012, Part I. LNCS, vol. 7181, pp. 130–141. Springer, Heidelberg (2012)
5. Mráz, F.: Lookahead Hierarchies of Restarting Automata. Journal of Automata, Languages and Combinatorics 6(4), 493–506 (2001)
6. Mráz, F., Otto, F., Plátek, M.: The degree of Word-Expansion of Lexicalized RRWW-automata: A New Measure for the Degree of Nondeterminism of (Context-Free) Languages. Theoretical Computer Science 410(37), 3530–3538 (2009)
7. Plátek, M., Mráz, F., Lopatková, M. (In)Dependencies in Functional Generative Description by Restarting Automata. In: Bordihn, H., et al. (eds.) Proceedings of NCMA 2010. books@ocg.at, vol. 263, pp. 155–170. Österreichische Computer Gesellschaft, Wien (2010)

# The Conjoinability Relation in Discontinuous Lambek Calculus

Alexey Sorokin

Moscow State University, Faculty of Mathematics and Mechanics
Moscow Institute of Physics and Technology,
Faculty of Innovations and High Technologies, Russia

**Abstract.** In 2013 Sorokin proved that the criterion of type conjoinability in 1-discontinuous Lambek calculus is the equality of interpretations in the free abelian group generated by primitive types. We extend the method to obtain the analogous result in full discontinuous Lambek calculus. It holds that the criterion is exactly the same as in 1-discontinuous Lambek calculus.

## 1 Introduction

Lambek calculus was introduced by Joachim Lambek in 1958 in order to model the syntactic structure of natural languages. In 1994 M. Pentus proved that Lambek grammars generate exactly context-free languages ([8]). Since context-free languages are well-known to be too weak for adequate representation of natural languages, G. Morrill introduced a generalization of Lambek calculus, the so-called discontinuous Lambek calculus ([5], [6]), which extends the standard Lambek calculus with discontinuous connectives. The calculus obtained has enough power to simulate various discontinuous linguistic phenomena, as shown in [6].

Let $A$ and $B$ be types of a particular categorial calculus. A type $C$ is called a join for $A$ and $B$ (in this calculus) if both the sequents $A \to C$ and $B \to C$ are derivable. In this case the types $A$ and $B$ are called conjoinable. The conjoinability problem is very interesting from the algebraic point of view. For example, two types are conjoinable in Lambek calculus iff they have the same interpretation in the free group generated by the primitive types (this criterion was proved in [7]). If we replace the free group by the free abelian group, then we obtain the criterion of conjoinability in commutative Lambek calculus. It is worth noting that the criterion of conjoinability in Lambek-Grishin calculus also uses the interpretation in a free abelian group ([4]), though this calculus lacks commutativity (and even associativity). The same conjoinability criterion holds for 1-discontinuous Lambek calculus ([9]), although this calculus is not commutative either. We generalise this result to obtain the criterion for full discontinuous Lambek calculus.

It is worth noting that the character of conjoinability of the categorial calculus has deep connection with generative power of this calculus. The calculi where

G. Morrill et al. (Eds.): Formal Grammar 2014, LNCS 8612, pp. 171–184, 2014.

conjoinability criteria use the free group interpretation, such as Lambek calculus or pregroup calculus, usually generate the family of context-free languages ([8], [1], [2]). Even the method of translating the categorial grammar to equivalent context-free one is essentially the same for both calculi and uses the properties of binary reductions in noncommutative free groups (see Lemma 8 in [8] and Lemma 3.7 in [1]). By contrast, there is no appropriate binary reduction in free abelian groups so the method of Pentus cannot be used for the calculi with abelian characterization of conjoinability. Therefore we lack the equivalence theorems for such calculi as commutative Lambek calculus or discontinuous Lambek calculus.

The nature of this correspondence is uncovered if we note that the target type of the grammar is in fact an "infinite join" for the sequences of types from the lexicon. Hence the study of the conjonability relation for a particular calculus might be helpful to characterize the generative power of this calculus. Therefore we consider the conjoinability relation in discontinuous Lambek calculus in order to come closer to the characterization of the family of languages it recognizes.

## 2 Discontinuous Lambek Calculus

Let Pr be a countable ranked set of primitive types and $rk\colon \mathrm{Pr} \to \mathbb{N}$ be a rank function. Let $I, J \notin \mathrm{Pr}$ be two distinguished constants, we refer to the elements of the set $\mathrm{Base} = \mathrm{Pr} \cup \{I, J\}$ as basic types. Then the set Tp of discontinuous types is the smallest ranked set satisfying the following conditions ($s$ is a sort function which extends the rank function to the set of types):

1. $\mathrm{Pr} \subset \mathrm{Tp}$, $\forall A \in \mathrm{Pr}\ s(A) = rk(A)$,
2. $I \in \mathrm{Tp}$, $s(I) = 0$,
3. $J \in \mathrm{Tp}$, $s(J) = 1$,
4. $\forall A, B \in \mathrm{Tp}\ ((s(A) \geq s(B)) \Rightarrow (A/B), (B\backslash A) \in \mathrm{Tp},\ s(A/B) = s(B\backslash A) = s(A) - s(B))$,
5. $\forall A, B \in \mathrm{Tp}\ (A \cdot B) \in \mathrm{Tp},\ s(A \cdot B) = s(A) + s(B)$,
6. $\forall A, B \in \mathrm{Tp}((s(A) \geq s(B)-1) \Rightarrow (B \downarrow A) \in \mathrm{Tp}, s(B \downarrow A) = s(A)-s(B)+1)$,
7. $\forall A, B \in \mathrm{Tp}\ ((s(A) \geq s(B)) \Rightarrow (A \uparrow B \in \mathrm{Tp}),\ s(A \uparrow B) = s(A) - s(B) + 1)$,
8. $\forall A, B \in \mathrm{Tp}\ ((s(A) \geq 1) \Rightarrow A \odot B \in \mathrm{Tp},\ s(A \odot B) = s(A) + s(B) - 1)$.

Let $\Sigma$ be some alphabet, say, the alphabet of syntactic categories (for example, $s$ is a sentence category, $np$ stands for noun phrases etc.). Then types are interpreted as formal languages, which are subsets of the set $\Sigma^*$. The connective $\cdot$ is interpreted as concatenation and $\backslash$ and $/$ are interpreted as its left and right residuals: the type $B/C$ means "something that becomes $B$ after appending $C$ to the right", and $C\backslash B$ means "something that becomes $B$ after appending $C$ to the left". $I$ is the constant for the language containing only the empty word.

Let 1 be the separator, $1 \notin \Sigma$, let us consider the words over the extended alphabet $\Sigma_1 = \Sigma \cup \{1\}$. Then the result of applying $\odot_j$ to the words $w_1, w_2 \in \Sigma^*$ equals the word obtained after replacing the $j$-th separator in $w_1$ by $w_2$, if $w_1$ contains less than $j$ separators, the result is undefined. The operations $\uparrow_j$ and $\downarrow_j$ are the analogues of $/$ and $\backslash$, respectively: $B \uparrow_j C$ means "something that becomes $B$ after replacing its $j$-th separator by $C$" and $C \downarrow_j B$ is interpreted as "something that yields $B$ after replacing $j$-th separator in $C$ by itself". Note that the sort of a type equals the number of separators in the words contained in its interpretation.

This informal interpretation is reflected in the axiomatics of discontinuous Lambek calculus. We give its Hilbert-style interpretation HDL, introduced in [10], since it is more convenient for our purposes. The calculus HDL is equivalent to the sequential Gentzen-style calculus DL studied in [5]. The sequents of the calculus HDL have the form $A \to B$ where $A, B \in \mathrm{Tp}$ and $s(A) = s(B)$. We denote the derivability of the sequent $A \to B$ by $\mathrm{HDL} \vdash A \to B$. The axiomatics of HDL include the axioms $A \to A$, for $A \in \mathrm{Tp}$, the rules for the connectives:

$$\frac{A \to C/B}{A \cdot B \to C} \qquad \frac{A \cdot B \to C}{A \to C/B}$$
$$\frac{B \to A\backslash C}{A \cdot B \to C} \qquad \frac{A \cdot B \to C}{B \to A\backslash C}$$
$$\frac{A \to C \uparrow_j B}{A \odot_j B \to C} \qquad \frac{A \odot_j B \to C}{A \to C \uparrow_j B}$$
$$\frac{B \to A \downarrow_j C}{A \odot_j B \to C} \qquad \frac{A \odot_j B \to C}{B \to A \downarrow_j C},$$

the identity axioms for constants:

$$A \cdot I \leftrightarrow A \leftrightarrow I \cdot A,$$
$$J \odot_1 A \leftrightarrow A \leftrightarrow A \odot_j J, \text{ if } j \le s(A),$$

the mixed associativity axioms:

$$(A \cdot B) \cdot C \leftrightarrow A \cdot (B \cdot C),$$
$$(A \odot_i B) \odot_j C \leftrightarrow (A \odot_j C) \odot_{i+s(B)-1} B, \text{ if } j < i,$$
$$(A \odot_i B) \odot_j C \leftrightarrow A \odot_j (B \odot_{j-i+1} C), \text{ if } i \le j < i + s(B),$$
$$(A \odot_i B) \odot_j C \leftrightarrow (A \odot_{j+1-s(B)} C) \odot_i B, \text{ if } i + s(B) \le j,$$

the axioms of interactions between "continuous" and "discontinuous" connectives:

$$A \cdot B \leftrightarrow (A \cdot J) \odot_{s(A)+1} B \leftrightarrow (J \cdot B) \odot_1 B$$

and the transitivity rule:

$$\frac{A \to B \qquad B \to C}{A \to C}.$$

*Example 1.* Let $s(A) = s(B) = 1$, then the sequent $(A \uparrow_1 B) \downarrow_1 A \to A/(B\backslash A)$ is derivable in HDL.

*Proof.*

$$\dfrac{\dfrac{\dfrac{\dfrac{\dfrac{B\backslash A \to B\backslash A}{B \cdot (B\backslash A) \to A}}{(J \cdot (B\backslash A)) \odot_1 B \to A}}{(J \cdot (B\backslash A)) \to A \uparrow_1 B} \quad \dfrac{(A \uparrow_1 B) \downarrow_1 A \to (A \uparrow_1 B) \downarrow_1 A}{A \uparrow_1 B \to A \uparrow_1 ((A \uparrow_1 B) \downarrow_1 A)}}{\dfrac{(J \cdot (B\backslash A)) \to A \uparrow_1 ((A \uparrow_1 B) \downarrow_1 A)}{(J \cdot (B\backslash A)) \odot_1 ((A \uparrow_1 B) \downarrow_1 A) \to A}}}{\dfrac{((A \uparrow_1 B) \downarrow_1 A) \cdot (B\backslash A) \to A}{(A \uparrow_1 B) \downarrow_1 A \to A/(B\backslash A)}}$$

The following lemma demonstrates the monotonicity properties of discontinuous Lambek calculus. Its proof immediately follows from the transitivity of the derivability relation.

**Lemma 1.** *Let* $A_1, B_1, A_2, B_2 \in \mathrm{Tp}$ *be such that all the types in the rules are correctly defined. Then the following rules are admissible in the calculus* HDL*:*

$$\dfrac{A_1 \to A_2 \quad B_1 \to B_2}{A_1 \cdot B_1 \to A_2 \cdot B_2} \qquad \dfrac{A_1 \to A_2 \quad B_1 \to B_2}{A_1 \odot_j B_1 \to A_2 \odot_j B_2}$$

$$\dfrac{A_1 \to A_2 \quad B_1 \to B_2}{A_1/B_2 \to A_2/B_1} \qquad \dfrac{A_1 \to A_2 \quad B_1 \to B_2}{B_2\backslash A_1 \to B_1\backslash A_2}$$

$$\dfrac{A_1 \to A_2 \quad B_1 \to B_2}{A_1 \uparrow_j B_2 \to A_2 \uparrow_j B_1} \qquad \dfrac{A_1 \to A_2 \quad B_1 \to B_2}{B_2 \downarrow_j A_1 \to B_1 \downarrow_j A_2.}$$

Let $\mathrm{Tp}_k$ be the set of all types that do not contain subtypes of sort greater than $k$ (including the type itself). We denote by $\mathrm{HDL}_k$ the calculus obtained if only the types from $\mathrm{Tp}_k$ are allowed. Such calculi naturally correspond to their sequential counterparts $\mathrm{DL}_k$ of the sequential fragment DL: $\mathrm{DL}_k$ is a fragment of the calculus $\mathrm{DL}_k$ permitting only the subtypes and subconfigurations of sort $k$ or less. Any sequent provable in DL also has a proof in some of the calculi $\mathrm{DL}_k$, we take as $k$ the maximal sort of types and configurations involved in the proof.

Additionally, DL possesses cut-elimination ([5]), so a usual backward proof search is possible. It is not difficult to mention that if a $\mathrm{DL}_k$-sequent is derivable in the full calculus DL, then the premises in the last rule of its derivation are again $\mathrm{DL}_k$-sequents. So any $\mathrm{DL}_k$-sequent, derivable in the full calculus DL, is derivable in the calculus $\mathrm{DL}_k$ itself. Since the translation from HDL to DL given in [10] maps $\mathrm{HDL}_k$-sequents to $\mathrm{DL}_k$-sequents and vice versa, we conclude that if $\mathrm{HDL} \vdash A \to B$ then it also holds that $\mathrm{HDL}_l \vdash A \to B$ where $l$ is the smallest number such that the types $A$ and $B$ both belong to $\mathrm{Tp}_l$.

## 3 Conjoinability in Discontinuous Lambek Calculus

In this section we define the conjoinability relation for discontinuous Lambek calculus. This relation was first defined in [3] for the basic Lambek calculus L.

In [7] M. Pentus proved that two types of the calculus L are conjoinable iff they have the same interpretation in the free group generated by primitive types. The main goal of our work is to obtain the conjoinability criterion for the calculus HDL and its fragments $\mathrm{HDL}_k$.

**Definition 1.** *The types $A$ and $B$ are said to be conjoinable if there exists a type $C$ such that the sequents $A \to C$ and $B \to C$ are both derivable. The type $C$ is called the join type.*

*Example 2.* Let $s(A) = s(B) = 1$, then the types $(A \uparrow_1 B) \downarrow_1 A$ and $B$ are conjoinable in $\mathrm{HDL}_k$ for any $k \geq 1$ with the join $(A/(B\backslash A))$. The sequent $(A \uparrow_1 B) \downarrow_1 B \to A/(B\backslash A)$ was proved in Example 1, the other sequent's derivation is

$$\dfrac{\dfrac{B\backslash A \to B\backslash A}{B \cdot (B\backslash A) \to A}}{B \to A/(B\backslash A)}$$

We denote the conjoinability relation by $\sim$. If we need to explicitly specify that the conjoinability in the calculus $\mathrm{HDL}_k$ is considered, we use the notation $\sim_k$. The following lemma uses the arguments from [3].

**Lemma 2.** *The conjoinability relation is a congruence on types.*

*Proof.* Let us at first prove that it is an equivalence relation. The reflexivity and symmetry are obvious. Consider the transitivity, let the statements $A_1 \sim B$ and $B \sim A_2$ hold. Let $C_1$ and $C_2$ be the join types for these pairs. Then it is easy to verify that the type $(B/C_1)\backslash B/(C_2\backslash B)$ is the join type for both the types $A_1$ and $A_2$. The congruence property follows from the monotonicity properties of discontinuous Lambek calculus. The lemma is proved.

The main goal of the present work is to prove the conjoinability criterion in any of the calculi $\mathrm{HDL}_k$. As a consequence we obtain an analogous criterion for the embracing calculi HDL. Let us fix some arbitrary positive $k$ which will be the same in the rest of the proof. Let Pr be the set of all primitive types and $\mathrm{Pr}_k$ denote the set of primitive types whose sort does not exceed $k$. Let $\alpha$ be an additional element such that $\alpha \notin \mathrm{Pr}$. We denote by $\mathcal{F}$ the free abelian group generated by the set $\mathrm{Pr} \cup \{\alpha\}$ and $\mathcal{F}_k$ stands for the free abelian group generated by the set $\mathrm{Pr}_k \cup \{\alpha\}$. Let 1 be the neutral element of the group $\mathcal{F}$ defined; it can be thought as the neutral element of any of the groups $\mathcal{F}_k$ as well.

For any type $A \in \mathrm{Pr}$ we define its interpretation $[\![A]\!] \in \mathcal{F}$. The interpretation mapping is defined recursively in the following way:

1. $[\![p]\!] = p$, for $p \in \mathrm{Pr}$,
2. $[\![I]\!] = 1$,
3. $[\![J]\!] = \alpha$,
4. $[\![B\backslash C]\!] = [\![C/B]\!] = [\![C]\!][\![B]\!]^{-1}$,
5. $[\![B \cdot C]\!] = [\![B]\!][\![C]\!]$,

6. $[\![C \uparrow_i B]\!] = [\![B \downarrow_j C]\!] = [\![C]\!]\alpha[\![B]\!]^{-1}$, for any indexes $i, j$ such that these types are correctly defined.
7. $[\![B \odot_i C]\!] = [\![B]\!]\alpha^{-1}[\![C]\!]$ for any index $i$ such that the type is correctly defined.

For any type $A \in \mathrm{Tp}$ and any basic type $p$ we introduce two counters $|A|_p^+$ and $|A|_p^-$ for the positive and negative occurrences of $p$ in $A$. The counters are defined recursively in the following way:

1. $|p|_p^+ = 1, |p|_p^- = 0$, if $p \in \mathrm{Base}$,
2. $|q|_p^+ = |q|_p^- = 0$, if $p, q \in \mathrm{Base}, p \neq q$,
3. $|A \cdot B|_p^+ = |A \odot_j B|_p^+ = |A|_p^+ + |B|_p^+$, $|A \cdot B|_p^- = |A \odot_j B|_p^- = |A|_p^- + |B|_p^-$ for any $j$ such that the considered type is correct,
4. $|A/B|_p^+ = |B\backslash A|_p^+ = |A|_p^+ + |B|_p^-$, $|A/B|_p^- = |B\backslash A|_p^- = |A|_p^- + |B|_p^+$,
5. $|A \uparrow_i B|_p^+ = |B \downarrow_j A|_p^+ = |A|_p^+ + |B|_p^-$, $|A \uparrow_i B|_p^- = |B \downarrow_j A|_p^- = |A|_p^- + |B|_p^+$ for any $i, j$ such that the considered type are correct.

*Example 3.* Let $p, q, r \in \mathrm{Pr}_2$, $s(p) = 0, s(q) = 1, s(r) = 2$, $A = r \downarrow_2 (((p \cdot r)/q) \uparrow_1 (r/(q \cdot p)))$, $B = (r \downarrow_1 (J \cdot p)) \odot_2 (p/(J\backslash q))$. Then $|A|_p^+ = 2$, $|A|_q^+ = 1$, $|A|_r^+ = 1$, $|A|_p^- = 0$, $|A|_q^- = 1$, $|A|_r^- = 2$, $[\![A]\!] = p^2r^{-1}\alpha^2$; $|B|_p^+ = 2$, $|B|_q^- = |B|_r^- = 1$, $|B|_J^+ = 2$, $|B|_J^- = 0$, $[\![B]\!] = p^2q^{-1}r^{-1}\alpha^2$.

The notion of positive and negative occurrences and the corresponding counts are extended to the connectives of the discontinuous Lambek calculus in the natural way. The formal definition is below:

1. $|p|_*^+ = |p|_*^- = 0$, if $p \in \mathrm{Base}, * \in \{\cdot, /, \backslash, \odot_j, \uparrow_j, \downarrow_j\}$,
2. $|A * B|_*^+ = |A|_*^+ + |B|_*^+ + 1, |A * B|_\star^+ = |A|_\star^+ + |B|_\star^+$, if $* \in \{\cdot, \odot_i\}, \star \in \{\cdot, /, \backslash, \odot_j, \uparrow_j, \downarrow_j\}, \star \neq *$,
3. $|A * B|_\star^- = |A|_\star^- + |B|_\star^-$, if $* \in \{\cdot, \odot_i\}, \star \in \{\cdot, /, \backslash, \odot_j, \uparrow_j, \downarrow_j\}$,
4. $|A * B|_*^+ = |A|_*^+ + |B|_*^- + 1, |A * B|_\star^+ = |A|_\star^+ + |B|_\star^-$, if $* \in \{/, \uparrow_i\}, \star \in \{\cdot, /, \backslash, \odot_j, \uparrow_j, \downarrow_j\}, \star \neq *$,
5. $|A * B|_\star^- = |A|_\star^- + |B|_\star^+$, if $* \in \{/, \uparrow_i\}, \star \in \{\cdot, /, \backslash, \odot_j, \uparrow_j, \downarrow_j\}$,
6. $|B * A|_*^+ = |A|_*^+ + |B|_*^- + 1, |B * A|_\star^+ = |A|_\star^+ + |B|_\star^-$, if $* \in \{\backslash, \downarrow_i\}, \star \in \{\cdot, /, \backslash, \odot_j, \uparrow_j, \downarrow_j\}, \star \neq *$,
7. $|B * A|_\star^- = |A|_\star^- + |B|_\star^+$, if $* \in \{\backslash, \downarrow_i\}, \star \in \{\cdot, /, \backslash, \odot_j, \uparrow_j, \downarrow_j\}$.

The counters introduced allow us to express the interpretation of any type $A$ in the closed form. The following lemma is proved by induction on type structure.

**Lemma 3.** *For any type $A \in \mathrm{Tp}$ its interpretation has the form*

$$[\![A]\!] = \alpha^{([\![A]\!]_J + [\![A]\!]_\uparrow + [\![A]\!]_\downarrow - [\![A]\!]_\odot)} \circ \prod_{p \in \mathrm{Pr}_1} p^{[\![A]\!]_p}$$

.

**Lemma 4.** *The condition $[\![A]\!] = [\![B]\!]$ is necessary for the sequent $A \to B$ to be derivable in* HDL.

*Proof.* Induction on derivation length.

Since $\mathrm{HDL}_k$ is the fragment of HDL it follows that equality of interpretations is also necessary for derivability in all the calculi $\mathrm{HDL}_k$. The following corollary is apparent by the definition of conjoinability:

**Corollary 1.** *The condition* $[\![A]\!] \sim [\![B]\!]$ *is necessary for the statements* $A \sim B$ *and* $A \sim_k B$ *for any types* $A, B \in \mathrm{Tp}_k$

Thus, the equality of interpretations is a necessary condition of conjoinability. The remaining part of the paper is devoted to the proof of its sufficiency.

## 4 Proof of the Criterion

In this section we establish that the equality of interpretations in free abelian group is the necessary condition for the types to be conjoinable and, hence, the criterion of conjoinability. We prove the claim for the calculus $\mathrm{HDL}_k$ for an arbitrary natural $k \geq 1$. Some of the technical steps are borrowed from the proof of the corresponding conjoinability criterion in the calculus $\mathrm{HDL}_1$, given in [9]. However, we need to be especially careful with the types encountered in proofs to stay always inside the set $\mathrm{Tp}_k$ of correct types. To the rest of the paper we fix some arbitrary $k \geq 1$ and work with the calculus $\mathrm{HDL}_k$ and the free abelian group $\mathcal{F}_k$ defined in the previous section. We omit the subscript when writing the conjoinability sign $\sim_k$.

In the lemma below we list the sequents derivable in $\mathrm{HDL}_k$ which we need in the following. All the proofs immediately follow from the axioms or are the axioms themselves (as in the last two parts).

**Lemma 5.** *The following statements are derivable in the calculus* $\mathrm{HDL}_k$. *The derivability of the sequent also means that the right part of it is in* $\mathrm{Tp}_k$ *whenever the left part is.*

1. $(A/B) \cdot B \to A$; $B \cdot (B\backslash A) \to A$,
2. $A \to (A \cdot B)/B$; $A \to B\backslash(B \cdot A)$,
3. $A \cdot (B/C) \to (A \cdot B)/C$; $(C\backslash B) \cdot A \to C\backslash(B \cdot A)$,
4. $B \odot_j (B \downarrow_j A) \to A$ *for any* $j \leq s(B)$;
   $(A \uparrow_j B) \odot_j B \to A$ *for any* $j \leq s(A) - s(B) + 1$,
5. $A \cdot J \to (A \cdot B) \uparrow_{s(A)+1} B$; $J \cdot A \to (B \cdot A) \uparrow_1 B$,
6. $A \cdot B \to (A \cdot J) \odot_{s(A)+1} B$; $A \cdot B \to (J \cdot B) \odot_1 A$,
7. $A \cdot I \to A$; $I \cdot A \to A$.

The next lemma gives the examples of the "neutral" types with respect to concatenation.

**Lemma 6.** *For any two types* $A, B \in \mathrm{Tp}_k$ *the statements* $A \cdot (B/B) \sim A \cdot (B\backslash B) \sim (B/B) \cdot A \sim (B\backslash B) \cdot A \sim A$ *are valid.*

*Proof.* The conjoinability of the types $A$ and $A\cdot(B/B)$ follows from the derivability of the sequents $A \to (A\cdot(B/B))/(B/B)$ and $A\cdot(B/B) \to (A\cdot(B/B))/(B/B)$.

The former sequent follows from the second statement of Lemma 5, the proof of the latter is given below. Since $s(B/B) = 0$, the conditions $A \in \mathrm{Tp}_k$ and $B \in \mathrm{Tp}_k$ imply that $A \cdot (B/B) \in \mathrm{Tp}_k$. The conjoinability of the types $A$ and $(B\backslash B) \cdot A$ is proved analogously. Since the sequents $A \to (A \cdot (B\backslash B))/(B\backslash B)$ and $A \cdot (B\backslash B) \to (A \cdot (B\backslash B))/(B\backslash B)$ also derivable, the types $A$ and $A \cdot (B\backslash B)$ are conjoinable as well. The lemma is proved.

$$\dfrac{\dfrac{A \to A \qquad \dfrac{\dfrac{\dfrac{(B/B)\cdot(B/B) \to (B/B)\cdot(B/B)}{(B/B)\cdot(B/B)\cdot B \to (B/B)\cdot B} \qquad \dfrac{B/B \to B/B}{(B/B)\cdot B \to B}}{(B/B)\cdot(B/B)\cdot B \to B}}{(B/B)\cdot(B/B) \to B/B}}{A \cdot (B/B)\cdot(B/B) \to A\cdot(B/B)}}{A\cdot(B/B) \to (A\cdot(B/B))/(B/B)}$$

For an arbitrary type $A \in \mathrm{Tp}$ we denote by $A^i$ the type $\underbrace{A \cdot \ldots \cdot A}_{i \text{ times}}$. We also set $A^0 = I$ for any type $A$.

The following lemma shows the "abelian" nature of concatenation in discontinuous Lambek calculus. It also claims that the left and right division connectives are "indistinguishable" with respect to conjoinability.

**Lemma 7**

1. *For any type $A \in \mathrm{Tp}_k$ such that $s(A) = 0$ it holds that $A \cdot J \sim J \cdot A$.*
2. *For any type $A \in \mathrm{Tp}_k$ such that $s(A) < k$ it holds that $A \cdot J \sim J \cdot A$.*
3. *For any type $A \in \mathrm{Tp}_k$ such that $s(A) = 0$ and arbitrary type $B \in \mathrm{Tp}_k$ it holds that $A \cdot B \sim B \cdot A$.*
4. *For any types $A, B \in \mathrm{Tp}_k$ such that $A \cdot B \in \mathrm{Tp}_k$ it holds that $A \cdot B \sim B \cdot A$.*
5. *For any types $A, B \in \mathrm{Tp}_k$ such that $A/B \in \mathrm{Tp}_k$ it holds that $A/B \sim B\backslash A$.*
6. *For any types $A, B, C \in \mathrm{Tp}_k$ such that $(A/B)/C \in \mathrm{Tp}_k$ it holds that $(A/B)/C \sim (A/C)/B$.*

*Proof*

1. Since $s(A) = 0$ the statement 5 of Lemma 5 implies that $A \cdot J \sim (A \cdot A) \uparrow_1 A \sim J \cdot A$, which is what was required. Note that $s((A \cdot A) \uparrow_1 A) = 1$ hence it belongs to $\mathrm{Tp}_k$.
2. Applying statement 1 of Lemma 6 $s(A)$ times, we obtain that $(A/J^{s(A)}) \cdot J^{s(A)} \sim A$, it follows that $A \cdot J \sim (A/J^{s(A)}) \cdot J^{s(A)} \cdot J = (A/J^{s(A)}) \cdot J \cdot J^{s(A)}$. Since $s(A/J^{s(A)}) = 0$ then $(A/J^{s(A)}) \cdot J \cdot J^{s(A)} \sim J \cdot (A/J^{s(A)}) \cdot J^{s(A)} \sim J \cdot A$. Obviously, $(A/J^{s(A)}) \cdot J \cdot J^{s(A)} \in \mathrm{Tp}_k$, so we obtain the desired statement combining these two chains of conjoinabilities.
3. Since $s(A) = 0$, then the statements $A \cdot B \sim (A \cdot J) \odot_1 B$ and $(J \cdot A) \odot_1 B \sim B \cdot A$ follow from the statement 6 of Lemma 5. In the current lemma we have also proved that $A \cdot J \sim J \cdot A$ which yields the desired statement by Lemma 2.

4. We have already proved the statement for the case $s(A) = 0$ so we assume that $s(A) > 0$. Then we write $A \cdot B \sim (A/J^{s(A)}) \cdot J^{s(A)} \cdot B$ and obtain by statements 2 and 3 of the current lemma that $(A/J^{s(A)}) \cdot J^{s(A)} \cdot B \sim (A/J^{s(A)}) \cdot B \cdot J^{s(A)} \sim B \cdot (A/J^{s(A)}) \cdot J^{s(A)} \sim B \cdot A$, which is what was required.
5. The desired statement follows from the chain of conjoinabilities $A/B \sim (B \cdot (B\backslash A))/B \sim ((B\backslash A) \cdot B)/B \sim (B\backslash A) \cdot (B/B) \sim B\backslash A$. The first part follows from the first statement of Lemma 5 and Lemma 2, the third statement follows from statement 3 of Lemma 5, the second from statement 4 of the current lemma, and the fourth part — from Lemma 6.
6. The condition $(A/B)/C \in \mathrm{Tp}_k$ implies that $s(C) \leq s(A/B)$, then $s(B) + s(C) \leq s(A) \leq k$, so we derive that $C \cdot B \in \mathrm{Tp}_k$. The statement $(A/B)/C \sim A/(C \cdot B)$ follows from the derivability of the sequent $(A/B)/C \to A/(C \cdot B)$. Analogously we prove that $(A/C)/B \sim A/(B \cdot C)$. From the second statement of the current lemma and Lemma 2 it follows that $A/(C \cdot B) \sim A/(B \cdot C)$. Combining these three statements we obtain the desired result.

The next lemma shows that the subscripts of the connectives $\odot_j, \downarrow_j, \uparrow_j$ are meaningless for conjoinability.

**Lemma 8**

1. *For any types $A, B \in \mathrm{Tp}_k$ and any indexes $i, j$, such that the corresponding types are defined and belong to $\mathrm{Tp}_k$ it holds that $A \odot_i B \sim A \odot_j B$.*
2. *For any types $A, B \in \mathrm{Tp}_k$ and any indexes $i, j$, such that the corresponding types are defined and belong to $\mathrm{Tp}_k$ it holds that $A \uparrow_i B \sim A \uparrow_j B$.*
3. *For any types $A, B \in \mathrm{Tp}_k$ and any indexes $i, j$, such that the corresponding types are defined and belong to $\mathrm{Tp}_k$ it holds that $B \downarrow_i A \sim B \downarrow_j A$.*

*Proof*

1. Without loss of generality we assume that $i < j \leq s(A)$. Then $A \odot_i B \sim (J^j \cdot (J^j\backslash A)) \odot_i B \sim (J^j \odot_i B) \cdot (J^j\backslash A) \sim J^{i-1} \cdot B \cdot J^{j-i} \cdot (J^j\backslash A) \sim J^{j-1} \cdot B \cdot (J^j\backslash A) \sim (J^{j-1} \cdot J \cdot (J^j\backslash A)) \odot_j B \sim A \odot_j B$, which was required. The first and the last conjoinabilities follow from statement 1, the second, the third and the fifth — from statement 6 of Lemma 5, the fourth similarity follows from Lemma 7.
2. From the first statement of the current lemma and Lemma 5 we deduce that $A \uparrow_i B \sim ((A \uparrow_j B) \odot_j B) \uparrow_i B \sim ((A \uparrow_j B) \odot_i B) \uparrow_i B \sim A \uparrow_j B$, which is what was required.
3. By the same arguments as in the previous case we obtain that $B \downarrow_i A \sim B \downarrow_i (B \odot_j (B \downarrow_j A)) \sim B \downarrow_i (B \odot_i (B \downarrow_j A)) \sim B \downarrow_j A$, which is what was required. The lemma is proved.

The type $A \in \mathrm{Tp}_k$ is continuous if it does not contain any discontinuous connectives $\odot_j, \downarrow_j, \uparrow_j$. The next lemma shows that for any type there is a conjoinable continuous type, so we may restrict our attention to continuous types only.

**Lemma 9**

1. *For any types $A, B \in \mathrm{Tp}_k$ such that $A \uparrow_j B \in \mathrm{Tp}_k$ it holds that $A \uparrow_j B \sim (A/B) \cdot J$.*
2. *For any types $A, B \in \mathrm{Tp}_k$ such that $B \downarrow_j A \in \mathrm{Tp}_k$ it holds that $B \downarrow_j A \sim A/(B/J)$.*
3. *For any types $A, B \in \mathrm{Tp}_k$ such that $A \odot_j B \in \mathrm{Tp}_k$ it holds that $A \odot_j B \sim (A/J) \cdot B$.*

*Proof*

1. The condition $A \uparrow_j B \in \mathrm{Tp}_k$ implies that $s(B) \le s(A)$ and $s(A) - s(B) + 1 \le k$, consequently $(A/B) \cdot J \in \mathrm{Tp}_k$. Since all correct types of the form $A \uparrow_j B$ are conjoinable, it suffices to prove the statement for one value of $j$, let us take $j = s(A) - s(B) + 1 = s(A/B) + 1$. Then the statement follows from the derivability of the sequent $(A/B) \cdot J \to A \uparrow_{(s(A/B)+1)} B$:

$$\cfrac{\cfrac{\cfrac{A/B \to A/B}{(A/B) \cdot B \to A}}{((A/B) \cdot J) \odot_{s(A/B)+1} B \to A}}{(A/B) \cdot J \to A \uparrow_{(s(A/B)+1)} B}$$

2. Since $B \downarrow_j A \in \mathrm{Tp}_k$, then $s(B) \ge 1$ and $s(A) \ge s(B) - 1$, which implies that $A/(B/J) \in \mathrm{Tp}_k$. As in the previous part of the lemma, it suffices to prove the lemma for one particular $j$, let us prove it for $j = 1$. Then we obtain that $A/(B/J) \sim A/(J\backslash B)$ by part 5 of Lemma 7 and Lemma 2, also $B \sim J \cdot (J\backslash B)$ by the first part of Lemma 5. Let us prove that the sequent $A/(J\backslash B) \to (J \cdot (J\backslash B)) \downarrow_1 A$ is derivable:

$$\cfrac{\cfrac{\cfrac{A/(J\backslash B) \to A/(J\backslash B)}{(A/(J\backslash B)) \cdot (J\backslash B) \to A}}{(J \cdot (J\backslash B)) \odot_1 (A/(J\backslash B)) \to A}}{A/(J\backslash B) \to (J \cdot (J\backslash B)) \downarrow_1 A}$$

   Since the sequent $(J \cdot (J\backslash B)) \downarrow_1 A \to B \downarrow_1 A$ is also derivable, it yields the required statement.
3. Since $A \odot_j B \in \mathrm{Tp}_k$, then $s(A) \ge 1$ and $s(A) - 1 + S(B) \le k$, which implies that $(A/J) \cdot B \in \mathrm{Tp}_k$. It suffices to prove the statement for $j = s(A) = s(A/J) + 1$. But $(A \odot_{(s(A/J)+1)} B) \sim ((A/J) \cdot J) \odot_{(s(A/J)+1)} B \sim (A/J) \cdot B$ by parts 1 and 6 of Lemma 5, which is what was required.

**Corollary 2.** *For any type $A \in \mathrm{Tp}_k$ there is a continuous $A' \in \mathrm{Tp}_k$, which is conjoinable with $A$.*

*Proof.* Induction on the number of discontinuous connectives in $A$. In every step we apply Lemma 9 to decrease the number of discontinuous connectives.

**Definition 2.** *A type $A \in \mathrm{Tp}_k$ is called reduced if it has no occurrences of the type $I$ and contains the basic type $J$ and primitive types $q_j$ of strictly positive sort only in subtypes of the form $q_j/J^{s(q_j)}$.*

Note that by definition every continuous reduced type is zero-sorted.

**Lemma 10.** *For any types $A, B, C$ from $\mathrm{Tp}_k$ and any positive numbers $i, j$, such that the considered types are correct, the following statements hold:*

1. $(A \cdot C^i) \cdot (B \cdot C^j) \sim (A \cdot B) \cdot C^{i+j}$,
2. $(A \cdot C^i)/(B \cdot C^j) \sim (A/B) \cdot C^{i-j}$,
3. $(B \cdot C^j)\backslash(A \cdot C^i) \sim (B\backslash A) \cdot C^{i-j}$.

*Proof.* The first statements easily follows from Lemma 7. The second statement is justified by the chain $(A \cdot C^i)/(B \cdot C^j) \sim (C^i \cdot A)/(C^j \cdot B) \sim ((C^i \cdot A)/B)/C^j = (C^i \cdot (A/B))/C^j \sim (A/B) \cdot (C^i/C^j) \sim (A/B) \cdot C^{i-j} \cdot (C^j/C^j) \sim (A/B) \cdot C^{i-j}$, which follows from Lemmas 7, 5 and 6. It is straightforward to verify that all the types in this chain belong to $\mathrm{Tp}_k$. The third statement is proved analogously.

**Lemma 11.** *For any type $A \in \mathrm{Tp}_k$ there exists a continuous reduced type $\widehat{A}$, such that the types $A$ and $\widehat{A} \cdot J^{s(A)}$ are conjoinable.*

*Proof.* Induction on the type $A$. We set $\widehat{p} = p$ for a primitive type $p$ of sort 0 and $\widehat{q} = q/J^s$ for a primitive type $q$ of sort $s > 0$, also $\widehat{I} = \widehat{J} = p_0/p_0$, where $p_0$ is a new primitive type of sort 0. Then the induction base follows from the part 1 of Lemma 5 and Lemma 6.

It suffices to prove the lemma for continuous types, so there are two cases to consider: $A = B \cdot C$ and $A = B/C$ (the case $A = C\backslash B$ follows trivially since $B/C \sim C\backslash B$). Then we set $\widehat{B \star C} = \widehat{B} \star \widehat{C}$ for any $\star \in \{\cdot, /\}$ and apply Lemma 10. The Lemma is proved.

*Example 4.* Let $s(p_1) = s(p_2) = 0$, $s(q_1) = s(q_2) = 1$, $s(r) = 2$. Consider the continuous type $(q_2/((q_1/(p_1 \cdot J)) \cdot p_2))\backslash r$ of sort 1: it is conjoinable with the type $(((q_2/J)/(((q_1/J)/(p_1 \cdot (p_0/p_0))) \cdot p_2))\ r) \cdot J$, which has the required form.

Let us rename all the primitive types of sort $i > 0$ as $q_{i,j}$, $j \in \mathbb{N}$. For any $q_{i,j}$ we introduce a corresponding primitive type $q'_{i,j}$ of sort 0. We call a $q$-image of the type $A \in \mathrm{Tp}_k$ the type $A'$, which is the result of replacing all the subtypes of the form $q_{i,j}/J^i$ by the primitive type $q'_{i,j}$.

*Example 5.* Let $A = (((p_1/(q_{2,1}/J^2))/(q_{1,1}/J))/(q_{1,2}/J))\backslash p_2$, then its 0-image equals $A' = (((p_1/q'_{2,1})/q'_{1,1})/q'_{1,2})\backslash p_2$.

**Lemma 12.** *Let $A', B'$ be 0-images of the continuous reduced types $A, B$. Then the equality $[\![A]\!] = [\![B]\!]$ implies that also $[\![A']\!] = [\![B']\!]$.*

*Proof.* Note that $A$ and $B$ do not contain discontinuous connectives, hence by Lemma 3 the equality $[\![A]\!] = [\![B]\!]$ is valid iff the condition $[\![A]\!]_p = [\![B]\!]_p$ holds for all primitive types. The counters of zero-sorted primitive types are preserved during the conversion of types to their $q$-images. If $q_{i,j}$ is a primitive type of sort $i > 0$ then it holds that $[\![A']\!]_{q'_{i,j}} = [\![A]\!]_{q_{i,j}} = [\![B]\!]_{q_{i,j}} = [\![B']\!]_{q'_{i,j}}$. Also $[\![A']\!]_J = [\![B']\!]_J = 0$ which implies that $[\![A']\!] = [\![B']\!]$ by Lemma 3. The lemma is proved.

**Lemma 13.** *Let $A', B'$ be 0-images of the reduced types $A, B$. Then the statement $A' \sim B'$ implies that also $A \sim B$.*

*Proof.* Let the types $A'$ and $B'$ be conjoinable and $C'$ be their join. Let $C$ be the type obtained from $C'$ by substituting all occurrences of $q'_{i,j}$ by the corresponding type $q'_{i,j}/J^i$. If the derivation of the sequent $A' \to C'$ is subjected to the same procedure, we obtain the derivation of the sequent $A \to C$. The derivation of the sequent $B \to C$ is analogous. The lemma is proved.

Let us prove the main theorem of the paper. The key construction is taken from theorem 1 in [7]. Let $\mathrm{Tp}^0_\sim$ denote the set containing equivalence classes of 0-images of continuous types with respect to $\sim$. We denote the equivalence class of the element $A$ (which is merely the set of types conjoinable with $A$) by $[A]$. Let us impose the group structure on the set $\mathrm{Tp}^0_\sim$, setting $[A]_\sim \circ [B]_\sim = [A{\cdot}B]_\sim$, $[A]^{-1}_\sim = [A\backslash A/A]_\sim$, $1 = [p/p]_\sim$, where $p$ is an arbitrarily chosen primitive type. The definition is correct since the conjoinability relation is a congruence on types.

**Lemma 14.** *The structure $\langle \mathrm{Tp}^0_\sim, \circ, ^{-1}, 1\rangle$ forms an abelian group.*

*Proof.* The accociativity of the group operation $\circ$ follows from the accociativity of $\cdot$ connective. The equality $[A]_\sim \circ [B]_\sim = [B]_\sim \circ [A]_\sim$ follows from the conjoinability of the types $A \cdot B$ and $B \cdot A$ in $\mathrm{HDL}_k$, thus $\circ$ is commutative. Also $[A]_\sim \circ [A]^{-1}_\sim = [A \cdot (A\backslash A/A)]_\sim = [A/A]_\sim = 1$, where the last equation follows from the conjoinability of the types $A/A$ and $p/p$. Finally, the relation $A \sim A \cdot (p/p)$ is valid for any $A$, which means the equality $[A]_\sim = [A]_\sim \circ 1$. So all the condition of the abelian group definition are satisfied and the lemma is proved.

**Theorem 1.** *It holds that $A \sim_k B$ if and only if $[\![A]\!] = [\![B]\!]$.*

*Proof.* The necessity of the condition $[\![A]\!] = [\![B]\!]$ is proved in Corollary 1, we are to prove the sufficiency, let $[\![A]\!] = [\![B]\!]$. Since the conjoinability relation is transitive by Corollary 2 it suffices to prove the theorem for continuous types, so we assume that $A$ and $B$ are continuous. Then by Lemma 11 there exist continuous reduced types $\widehat{A}$ and $\widehat{B}$, such that in case $s(A) = s(B) = 0$ the relations $A \sim \widehat{A}$ and $B \sim \widehat{B}$, and in the case $s(A) = s(B) = l > 0$ the relations $A \sim \widehat{A}{\cdot}J^l$ and $B \sim \widehat{B}{\cdot}J^l$ hold. Then from Corollary 1 we deduce that $[\![\widehat{A}]\!] = [\![\widehat{B}]\!]$. Since conjoinability is the congruence relation from the similarity $\widehat{A} \sim \widehat{B}$ follows the similarity $A \sim B$. Then its enough to prove the statement for continuous reduced types $A$ and $B$.

Let $A'$ and $B'$ be 0-images of the types $A$ and $B$; then by Lemma 13 it suffices to prove that $A'$ and $B'$ are conjoinable in $\mathrm{HDL}_k$. By Lemma 12 the equality $[\![A']\!] = [\![B']\!]$ holds so it suffices to prove the theorem for 0-images of reduced continuous types. Note that all the primitive subtypes of these types have sort 0. Let us denote by $\mathcal{G}_0$ the free abelian group generated by zero-sorted primitive types. The remaining part of the proof repeats the proof of theorem 1 in [7].

Consider the mapping $h\colon \mathcal{G}_0 \to \mathrm{Tp}^0_\sim$, defined as $h(p) = [p]_\sim$ and $h(p^{-1}) = [p\backslash p/p]_\sim$, and extend it to homomorphism from $\mathcal{G}_0$ to $\mathrm{Tp}^0_\sim$. Let us prove the equalities $h([\![E]\!]) = [E]_\sim$ and $h([\![E]\!]^{-1}) = [E\backslash E/E]_\sim$ by induction on the structure of 0-image $E$. The induction base follows from the definition of the mapping $h$.

Let us prove the induction step; there are two cases according to the main connective of the type $E = C \star D$, where $\star \in \{\cdot, /\}$. In the case of the $\cdot$ connective it holds that $h([\![C\cdot D]\!]) = h([\![C]\!][\![D]\!]) = [C]_\sim \circ [D]_\sim = [C\cdot D]_\sim$. Also $h([\![C\cdot D]\!]^{-1}) = h([C]^{-1}[D]^{-1}) = [C\backslash C/C]_\sim \circ [D\backslash D/D]_\sim \sim [(C\backslash C/C)\cdot(D\backslash D/D)]_\sim$. Then by the second part of Lemma 5 and Lemma 6 we obtain $(C\backslash C/C)\cdot(D\backslash D/D) \sim C\backslash((C/C)\cdot(D\backslash D/D)) \sim C\backslash(D\backslash D/D) \sim C\backslash((D\backslash D)\cdot(C/C))/D \sim C\backslash(D\backslash(D\cdot C)/C)/D \sim (D\cdot C)\backslash(D\cdot C)/(D\cdot C)$. Hence, $h([\![C\cdot D]\!]^{-1}) = [(D\cdot C)\backslash(D\cdot C)/(D\cdot C)]_\sim = [(C\cdot D)\backslash(C\cdot D)/(C\cdot D)]_\sim$, which is what was required.

Now consider the case $E = C/D$, then $h([\![C/D]\!]) = h([\![C]\!][\![D]\!]^{-1}) = [C]_\sim \circ [D\backslash D/D]_\sim = [C\cdot(D\backslash D/D)]_\sim = [(C\cdot(D\backslash D))/D]_\sim = [C/D]_\sim$. Also $h([\![C/D]\!]^{-1}) = h([\![D]\!][\![C]\!]^{-1}) = [D/C]_\sim$. By Lemmas 5 and 6 it holds that $D/C \sim D/((C/D)\cdot D) \sim (D/D)/(C/D) \sim (C/D)\backslash(C/D)/(C/D)$, which is what was required.

Hence, the condition $h([\![E]\!]) = [E]_\sim$ is valid for any 0-image $E$. Then the equality $[\![A']\!] = [\![B']\!]$ implies that $[A']_\sim = [B']_\sim$ which means that the types $A'$ and $B'$ are conjoinable. The theorem is proved.

We have proved that the equality $[\![A]\!] = [\![B]\!]$ is the criterion of conjoinability in the calculus $\mathrm{HDL}_k$. Let us prove that it also forms the criterion in the full calculus HDL. By Lemma 4 the equality of interpretation is a necessary condition. On the other hand the conjoinability of $A$ and $B$ in the fragment $\mathrm{HDL}_k$ a fortiori subsumes their conjoinability in HDL. Summarizing, the following criterion holds:

**Theorem 2.** *It holds that $A \sim B$ if and only if $[\![A]\!] = [\![B]\!]$.*

**Corollary 3.** *The types $A, B \in \mathrm{Tp}_k$ are conjoinable in the calculus* $\mathrm{HDL}_k$ *exactly when they are conjoinable in the calculus* HDL.

## 5 Conclusion

We have proved that the conjoinability relation in discontinuous Lambek calculus is characterized by equality of type interpretations in the free abelian group generated by primitive types. The structure of the proof might be useful to study how this calculus allows to simulate commutative and partially commutative language phenomena. Since the same criterion also holds for Lambek-Grishin calculus, it is also interesting to study, which requirements should satisfy a multimodal calculus to have an abelian characterization of conjoinability.

## References


1. Béchet, D.: Parsing pregroup grammars and Lambek grammars using partial composition. Studia Logica 87(2-3), 199–224 (2007)
2. Buszkowski, W., Moroz, K.: Pregroup grammars and context-free grammars. In: Computational Algebraic Approaches to Natural Language, pp. 1–21. Polimetrica (2008)
3. Lambek, J.: The mathematics of sentence structure. American Mathematical Monthly 65(3), 154–170 (1958)
4. Moortgat, M., Pentus, M.: Type similarity for the Lambek-Grishin calculus. In: Proceedings of the 12th Conference on Formal Grammar, Dublin (2007)
5. Morrill, G., Valentín, O.: On calculus of displacement. In: Proceedings of the 10th International Workshop on Tree Adjoining Grammars and Related Formalisms, pp. 45–52 (2010)
6. Morrill, G., Valentín, O., Fadda, M.: The displacement calculus. Journal of Logic, Language and Information 20(1), 1–48 (2011)
7. Pentus, M.: The conjoinability relation in Lambek calculus and linear logic. ILLC Prepublication Series ML–93–03, Institute for Logic, Language and Computation, University of Amsterdam (1993)
8. Pentus, M.: Lambek grammars are context-free. In: Logic in Computer Science, Proceedings of the LICS 1993, pp. 429–433 (1993)
9. Sorokin, A.: Conjoinability in 1-discontinuous Lambek calculus. In: Casadio, C., Coecke, B., Moortgat, M., Scott, P. (eds.) Lambek Festschrift. LNCS, vol. 8222, pp. 393–401. Springer, Heidelberg (2014)
10. Valentín, O.: Theory of discontinuous Lambek calculus: PhD Thesis. Universitat Autònoma de Barcelona, Barcelona (2012)

# The Cantor-Bendixson Analysis of Finite Trees

Christian Wurm

Universität Düsseldorf, Germany
cwurm@phil.uni-duesseldorf.de

**Abstract.** We present a measure on the structural complexity of finite and infinite trees and provide some first result on its relation to context-free grammars and context-free tree grammars. In particular this measure establishes a relation between the complexity of a *language as a set*, and the complexity of the *objects it contains*. We show its precise nature and prove its decidability for the formalisms we consider.

## 1 Introduction

We introduce a measure on the complexity of trees, working equally well on finite and infinite trees. This finitary Cantor-Bendixson rank is a slight adaptation from the concept of Cantor-Bendixson rank (CB-rank), which is however only meaningful on infinite trees. Our modification makes the concepts relevant to the theory of formal languages and grammars, a field where it seems to be unknown so far.

The finitary CB-rank (henceforth: FCB-rank) provides a measure of the complexity of trees, as they are for example used as representatives of derivations of sentences/words in formal language theory. Its main advantage is: it is based on discrete structural properties and thus highly informative on the structure of a tree; and at the same time, it is insensitive to things as unary branches, binary versus ternary branching etc. *Prima facie*, the FCB-rank does not say anything about the complexity of a certain grammar, only about certain objects it generates. We can refer to the latter as *syntagmatic complexity*, as a property of a single object of a language, to the former as *paradigmatic complexity*, as a property of the set as an entire collection. There is usually no immediate relation between the two – simple languages as $\Sigma^*$ can be generated by complicated grammars.[1] Using the FCB-rank, we can establish this sort of relation: we will assign an FCB-rank to context-free grammars, and show that this rank corresponds exactly to the hierarchy of $k$-linear languages. Moreover, we show that the rank of a grammar can be effectively computed.

As we said, we relate the theory of formal grammars/languages to the theory of relations/structures. This means we can talk about trees regardless of their mode of generation. So it does not matter whether we talk about generation trees of context-free grammars or trees generated by regular tree grammars, tree

[1] With the well-known consequence that the universality problem is undecidable for CFG.

G. Morrill et al. (Eds.): Formal Grammar 2014, LNCS 8612, pp. 185–200, 2014.

adjoining grammars, context-free tree grammars etc. We show how the FCB-rank can be computed for (simple) context-free tree grammars (CFTG). Whereas for CFG, bounded FCB-rank coincides with an established class of grammars, for CFTG this allows us to define new classes of tree- and string languages.

## 2 Definitions: Trees, Derivatives, Cantor-Bendixson Rank

We will present trees as structures in a well-known fashion; we take the concepts of linear order, partial order for granted.[2]

**Definition 1.** *A well-founded tree is a structure* $(T, \trianglelefteq)$, *where* $T$ *is a set (the set of nodes) and* $\trianglelefteq \subseteq T \times T$ *is a partial order with a smallest element* $r$ *and with the property that for each* $t \in T$, *the set* $\{s : s \trianglelefteq t\}$ *is 1. finite and 2. linearly ordered by* $\trianglelefteq$.

If $s \trianglelefteq t$, we say $s$ dominates $t$. We denote the non-reflexive restriction of $\trianglelefteq$ by $\triangleleft$, and the immediate dominance relation by $\triangleleft_i$, defined by $x \triangleleft_i y :\Longleftrightarrow x \triangleleft y$ and $x \trianglelefteq z \trianglelefteq y \rightarrow x = z$ or $y = z$. Note that a tree in this sense does not yet have a precedence order on the nodes which are not ordered by dominance; what we have specified is only the dominance order. We will throughout this paper assume that trees are well-founded.

A **path** in a tree $\mathcal{T}$ (or $\mathcal{T}$-path) is a set $P \subseteq T$ which is linearly ordered by $\trianglelefteq$ and convex.[3] A **complete** $\mathcal{T}$ **path** is a path which is maximal wrt. inclusion, that is, is not a proper subset of any other $\mathcal{T}$-path. The **depth** of a tree $depth(\mathcal{T})$ is the length of its longest path, where by length of a path $P$ - as paths are sets - we mean its cardinality $|P|$. If $P$ is a complete path in $\mathcal{T}$ with $t \in P$, $t$ is $\trianglelefteq$-maximal in $P$, then we also say that $|P|$ is the depth of $t$ ($depth(t)$), with $\mathcal{T}$ intended. By a **chain** we denote a tree $(T, \trianglelefteq)$ where $T$ is linearly ordered by $\trianglelefteq$. A **full** $n$**-ary tree** is a tree in which every node has zero or $n$ children. A **complete** $n$**-ary tree** is a full $n$-ary tree, such that if there is a leaf $t$ with $depth(t) = k$, then for all $t' \in T$, $depth(t') = k$ if and only if $t'$ is a leaf (this covers the case where there are no leaves).

**Definition 2.** *Let* $\mathcal{T} = (T, \trianglelefteq)$ *be a tree. We define the* **restriction** $\mathcal{T} \restriction S$ *of* $\mathcal{T}$ *to* $S \subseteq T$ *as* $(S, \trianglelefteq_{\restriction S})$, *where* $\trianglelefteq_{\restriction S} := \trianglelefteq \cap (S \times S)$.

We say that $\mathcal{S} = (S, \trianglelefteq)$ is a **subtree** of $\mathcal{T} = (T, \trianglelefteq)$, if $\mathcal{S}$ is a tree, and $\mathcal{S} = \mathcal{T} \restriction S$ for some $S \subseteq T$. We slightly abuse notation and write $\mathcal{S} \subseteq \mathcal{T}$ for the subtree relation; similarly, if $t \in T$, we also write $t \in \mathcal{T}$. [4]

---

[2] For a good general introduction into the theory of relations and structures, consider [3].

[3] A subset $P$ of a partially ordered set $(Q, \leq)$ is convex, if from $x, y \in P$, $x \leq z \leq y$ it follows that $z \in P$.

[4] Note that, if we think of trees as graphs, a subtree of $\mathcal{T}$ need not be a contiguous subgraph of the graph of $\mathcal{T}$. Our definition of subtree is, up to isomorphism, equivalent to an order theoretic definition, which says: $\mathcal{S} \subseteq \mathcal{T}$ if there is an order embedding $i : \mathcal{S} \rightarrow \mathcal{T}$. Then $\mathcal{S}$ is the isomorphic copy of a subtree of $\mathcal{T}$.

**Definition 3.** *The **core** $c(\mathcal{T})$ of a tree $\mathcal{T}$ is the set of nodes which lie on two distinct complete paths of the tree: $c(\mathcal{T}) := \{t \in T :$ there are complete $\mathcal{T}$-paths $P_1, P_2 \subseteq T$, $t \in P_1 \cap P_2$, and $P_1 \neq P_2\}$. Define the **derivative** of a tree as $d(\mathcal{T}) = \mathcal{T} \restriction c(\mathcal{T})$. Furthermore, for $n \in \mathbb{N}$, $d^n(\mathcal{T}) = d(d^{n-1}(\mathcal{T}))$, where $d^0(\mathcal{T}) = \mathcal{T}$.*

The derivative of a tree cuts away all nodes which lie only on a single complete path. $\mathcal{T}_\emptyset$ denotes the empty tree, i.e. the tree with empty domain.

**Definition 4.** *For $\mathcal{T}$ a tree, the finitary CB-rank $FCB(\mathcal{T})$ is defined as the least natural number $n$ such that $d^n(\mathcal{T}) = \mathcal{T}_\emptyset$. We put $FCB(\mathcal{T}) = \omega$, if there is no $n \in \mathbb{N}$ such that $d^n(\mathcal{T}) = \mathcal{T}_\emptyset$.*

Some observations: every finite tree has a finite FCB-rank, and every finite tree will eventually converge to $\mathcal{T}_\emptyset$. If $\mathcal{S} \subseteq \mathcal{T}$, then $d(\mathcal{S}) \subseteq d(\mathcal{T})$, and so $FCB(\mathcal{S}) \leq FCB(\mathcal{T})$. Furthermore, if $s, t \in \mathcal{T}$, $s \trianglelefteq t$ and $t \in d(\mathcal{T})$, then $s \in d(\mathcal{T})$. There are infinite trees $\mathcal{T}$ for which $FCB(\mathcal{T})$ is finite. $FCB(\mathcal{T})$ can be infinite in two distinct cases: either there is an $n \in \mathbb{N}$ such that $d^n(\mathcal{T}) = d^{n+1}(\mathcal{T}) \neq \mathcal{T}_\emptyset$; that is, our derivatives reach a fixed point. The other possibility is that for all $n \in \mathbb{N}$, $d^n(\mathcal{T}) \supsetneq d^{n+1}(\mathcal{T})$, which entails that for all $n \in \mathbb{N}$, $d^n(\mathcal{T}) \neq \mathcal{T}_\emptyset$. If $FCB(\mathcal{T}) = n$, then for any $m \leq n$, there is $\mathcal{S} \subseteq \mathcal{T}$ with $FCB(\mathcal{S}) = m$.

The Cantor-Bendixson rank is originally used with infinite trees, where the *core* is defined as in definition 3 with the additional requirement that paths be infinite (see for example [10],[12]); and $CB(\mathcal{T})$ is the least ordinal $\alpha$ such that $d^\alpha(\mathcal{T}) = d^{\alpha+1}(\mathcal{T})$. Skipping the infinity condition makes this notion applicable in the finite; the other modification allows us to avoid some unwanted conclusions in the infinite.[5] Note that however our concept works equally well with infinite trees. Consider some examples:

1. For the countably infinite complete binary tree $\mathcal{T}$, we have $FCB(\mathcal{T}) = \omega$, because $d(\mathcal{T}) = \mathcal{T}$.
2. Let $\mathcal{T}$ be the finite or infinite derivation tree of a regular string grammar. Then regardless of $depth(\mathcal{T})$, $FCB(\mathcal{T}) = 2$.
3. Let $\mathcal{T}$ be the complete, at least binary branching tree of depth $k$. Then $FCB(\mathcal{T}) = k$.
4. Let $\mathcal{T}$ be a tree of the form $\mathcal{T}_1[\mathcal{T}_2[\mathcal{T}_3[...]]]$, where $\mathcal{T}_i$ is the complete binary tree of depth $i$, and the root of $\mathcal{T}_{i+1}$ is immediately dominated by an arbitrary leaf of $\mathcal{T}_i$. Then we have for all $n \in \mathbb{N}$, $d^n(\mathcal{T}) \neq d^{n+1}(\mathcal{T})$, and so we have $FCB(\mathcal{T}) = \omega$.
5. If $FCB(\mathcal{T})$ is finite, then $d^{FCB(\mathcal{T})}(\mathcal{T}) = \mathcal{T}_\emptyset$.

The following lemma will help in understanding the concept, and in some sense generalizes the above examples:

[5] Under the definition of CB-rank, the countable infinite complete binary has CB-rank 0, contrary to finite trees. So in the infinite, we lose the property of monotonicity: $\mathcal{S} \subseteq \mathcal{T} \not\Rightarrow FCB(\mathcal{S}) \leq FCB(\mathcal{T})$.

**Lemma 5.** *For $k \in \mathbb{N}$, $FCB(\mathcal{T}) \geq k$, if and only if there is a complete binary branching subtree of $\mathcal{T}$ with depth $k$.*

**Proof.** *If*: Easy exercise.

*Only if*: by induction on the (inverse) derivation. Assume that $FCB(\mathcal{T}) \geq k$. We have $\mathcal{T}_\emptyset \subseteq d^k(\mathcal{T}) \subsetneq d^{k-1}(\mathcal{T})$. So $d^{k-1}(\mathcal{T})$ must contain a complete binary tree of depth $\geq 1$ (complete binary is trivially satisfied for depth 1, that is, the singleton tree). Next we make the induction step: assume that for $1 \leq i < k$, $d^{k-i}(\mathcal{T})$ contains a full binary tree of depth $i$. As $\trianglelefteq$ is acyclic and we are in the finite, we have a set $L$ of leaves, which are maximal with respect to $\trianglelefteq$ in $d^{k-i}(\mathcal{T})$. As $d^{k-i}(\mathcal{T}) \neq d^{k-(i+1)}(\mathcal{T})$, these nodes must have been in the core of $d^{k-(i+1)}(\mathcal{T})$. Consequently, for each $l \in L$, there must have been two nodes $m, n$, such that $l \trianglelefteq m$, $l \trianglelefteq n$, and $m, n$ are on distinct paths. Given these nodes, we know that $d^{k-(i+1)}(\mathcal{T})$ contains the full binary tree of depth $i+1$. □

Another observation we now can make is the following: if $d(\mathcal{T}) = \mathcal{T}$, then either $\mathcal{T} = \mathcal{T}_\emptyset$, or $\mathcal{T}$ is infinite, more precisely: every node in $\mathcal{T}$ dominates a complete infinite binary subtree.

We will not ponder on questions of psycholinguistic nature, but for completeness of exposition, we will very shortly illustrate what in our view are the major advantages of the FCB-rank with respect to other current (linguistic) methods of measuring the complexity of (finite) trees (see for example [5]). (1) FCB-rank has values in $\mathbb{N} \cup \{\omega\}$. (2) FCB-rank is computed locally from the most complex subtree (recall Lemma 5). (3) Given the FCB-rank of a tree $\mathcal{T}$, we can precisely say what kind of subtrees do *not* occur in $\mathcal{T}$, and which ones *do*, and which ones *can*; the measure is very informative about structure. (4) The FCB-rank is monotonous over the subtree relation: if $\mathcal{S} \subseteq \mathcal{T}$, then $FCB(\mathcal{S}) \leq FCB(\mathcal{T})$. This does not hold for many other measures, as can be easily shown. (5) FCB rank is insensitive to unary branching; moreover a tree of the form [[..][..][..]] [6] and a tree of the form [[..][[..][..]]] get the same FCB rank. In our view, this point speaks in favor of FCB-rank, as these are so to speak variants encoding the same dependencies.

## 3 Ordered Trees and Labelled Trees

The trees we presented in the previous section were unordered; there was no precedence specified between elements not dominating each other. Trees generated by phrase structure grammars usually have a precedence order specified as an intrinsic feature.

**Definition 6.** *A well-founded ordered tree is a structure $(T, \trianglelefteq, \preceq)$, where $(T, \trianglelefteq)$ is a well-founded tree, $\preceq \subseteq T \times T$ is a partial order, for each $t \in T$, $t \trianglelefteq t'$, $\{s : t \lhd_i s\} \cap \{s : s \preceq t'\}$ is finite, and where*

1. *for arbitrary elements $s, t$, we have $s \preceq t$ or $t \preceq s$ if and only if $s \not\trianglelefteq t$ and $t \not\trianglelefteq s$;*
2. *if $s \preceq t$, then for all $u, v$, if $s \trianglelefteq u, t \trianglelefteq v$, then $u \preceq v$.*

[6] For example linguists sometimes dislike such subtrees, for rather ideological reasons as X-Bar theory or assumption of binary branching

We write an ordered tree as $(\mathcal{T}, \preceq)$, but in the sequel also simply as $\mathcal{T}$, as long as misunderstandings can be excluded. The reason is that our definition of a FCB-rank can be equally well applied to ordered trees as to trees. Define $d(\mathcal{T}, \preceq) = (d(\mathcal{T}), \preceq_{\restriction d(\mathcal{T})})$. We can easily check that the derivative of an ordered tree $(\mathcal{T}, \preceq)$ is an ordered tree, and $FCB(\mathcal{T}, \preceq) = FCB(\mathcal{T})$. So the order $\preceq$, to which we also refer as precedence, does not play any role for the derivative.

An (ordered) **labelled tree** is a structure $(\mathcal{T}, X_1, ..., X_n)$, where $\mathcal{T}$ is an (ordered) tree, and $X_1, ..., X_n \subseteq T$, such that for $i \neq j$, $X_i \cap X_j = \emptyset$, and $\bigcup_{1 \leq i \leq n} X_i = T$.[7] $X_1, ..., X_n$ are the labels of $\mathcal{T}$. Given a labelled tree $\mathcal{T} = (T, \trianglelefteq, (\preceq), X_1, ..., X_n)$, $U \subseteq T$, we define $\mathcal{T} \restriction U := (U, \trianglelefteq_{\restriction U}, (\preceq_{\restriction U}), X_1 \cap U, ..., X_n \cap U)$. This is again an (ordered) labelled tree, so labels have no influence on core, derivatives etc., and so we can define the FCB-rank of labelled trees in the usual fashion.

## 4 Context-Free Grammars and Derivation Trees

Context-free grammars are very well-known; so we just fix our conventions. A context-free grammar (CFG) is a tuple $(\mathtt{S}, \mathcal{N}, A, R)$ with $\mathtt{S} \in \mathcal{N}$ (note the font to avoid confusion), $\mathcal{N}$ the non-terminals, $A$ the set of terminals, and $R \subseteq \mathcal{N} \times (\mathcal{N} \cup A)^*$ the set of production rules. To stick with conventional notation, we write $N \to \alpha$, if $(N, \alpha) \in R$; we assume lowercase Greek letters to be ranging over $(\mathcal{N} \cup A)^*$. The sets $A, \mathcal{N}, R$ are supposed to be finite. Derivability is defined as follows: let $\vdash_G \subseteq (\mathcal{N} \cup A)^* \times (\mathcal{N} \cup A)^*$ be defined by: if $N \to \alpha \in R$, then for all $\beta_1, \beta_2 \in (\mathcal{N} \cup A)^*$, $(\beta_1 N \beta_2, \beta_1 \alpha \beta_2) \in \vdash_G$, and nothing else. Let $\vdash_G^*$ be the reflexive and transitive closure of $\vdash_G$. We stick to common usage and write $\alpha \vdash_G^* \beta$ instead of $(\alpha, \beta) \in \vdash_G^*$.

We now define how what is usually called the "derivation tree" of a CFG relates to the relational notion of trees. Given a tree $\mathcal{T}$, its **elementary subtrees** are its subtrees $\mathcal{S}$ of depth 2, where for $t$ the root of $\mathcal{S}$, $t' \neq t$, $t' \in S$ if and only if $t \triangleleft_i t'$ in $\mathcal{T}$. Given a CFG $G = (\mathtt{S}, \mathcal{N}, A, R)$, we say a well-founded, ordered labelled tree $\mathcal{T}_L = (T, \trianglelefteq, \preceq, (N)_{N \in \mathcal{N}}, (a)_{a \in A})$ is a **derivation tree** of $G$, if for $r$ the unique $\trianglelefteq$-minimal element of the tree, we have $r \in \mathtt{S}$ (henceforth, we will stick to the convention that $r$ denotes the root), and for each elementary subtree $\mathcal{T} \subseteq \mathcal{T}_L$, $T = \{s_1, ..., s_i\}$, $s_1 \trianglelefteq s_2, ..., s_i$ and $s_2 \prec ... \prec s_i$, we have $s_1 \in X_1, ..., s_i \in X_i$ only if there is a rule $X_1 \to X_2 ... X_i \in R$. We denote the set of derivation trees of a grammar $G$ by $D(G)$. If we omit the condition that $r \in \mathtt{S}$, we say the tree is $G$-**conform** (note that we omit the condition that leaves have terminal labels; this will have some importance in the sequel). What is crucial for connecting results from formal language theory to structure theory is the connecting lemma, for which we omit the proof.

**Lemma 7.** *(Connecting Lemma) Let $G = (S, \mathcal{N}, A, R)$ be a CFG, $X_0, ..., X_i \in \mathcal{N}$. We have $\mathcal{T} \in D(G)$ if and only if for all subtrees $\mathcal{S} \subseteq \mathcal{T}$ the following*

[7] In words: every $t \in T$ has exactly one label. One sometimes assumes (equivalently) a labelling function; we choose another way to stay within the boundaries of strict relation theory.

*holds: for all* $s, t_1, ..., t_i \in S$, *where* $s$ *is the root of* $\mathcal{S}$ *and* $t_1 \prec ... \prec t_i$, *we have* $s \in X_0, t_1 \in X_1, ..., t_i \in X_i$ *if and only if* $X_0 \vdash_G^* \alpha_1 X_1 \alpha_2 ... \alpha_i X_i \alpha_{i+1}$.

The proof is by induction on the number of derivation steps. This notion thus captures our intuition on derivation trees, except for one important detail: we also allow for infinite derivation trees. This will be important for what is to follow, because we are interested in grammars generating trees with unbounded FCB-rank.[8]

## 5 Decidability Results

We now define what it means for a CFG to have a certain FCB-rank; given a set $M$ of ordinals, by the supremum $M$, in symbols $sup(M)$, we denote the smallest ordinal which is larger than (or equal to) all elements of $M$.

**Definition 8.** *Given a CFG G, define the FCB rank of G by* $FCB(G) := sup\{FCB(\mathcal{T}) : \mathcal{T} \in D(G)\}$.

In our case, the supremum actually coincides with the maximum: if a CFG has derivation trees of arbitrarily large FCB-rank, it has a derivation tree of infinite rank (we prove this later on, lemma 11). Given a grammar $G$, its FCB-rank can be finite or infinite, but it is well-defined in any case. The first main result shows that we can effectively compute the FCB-rank of a CFG. But first we need to introduce a construction which is essential for the proof of the theorem.

Let $I$ be an arbitrary index set, $\{\mathcal{T}_i : i \in I\}$ be a set of pairwise disjoint trees.[9] By $\otimes\{\mathcal{T}_i : i \in I\}$ we denote the tree $(\bigcup_{i \in I} T_i \cup \{r\}, \trianglelefteq)$, where $r \notin \bigcup_{i \in I} T_i$ and for $s \neq r \neq t$, $s \trianglelefteq t$ if and only if $s \trianglelefteq_{T_i} t$ for some $i \in I$, and $r \trianglelefteq s$ for all $s \in (\bigcup_{i \in I} T_i \cup \{r\})$. By convention, we put $\otimes\emptyset = \otimes\{\mathcal{T}_\emptyset\} = (\{r\}, \{\langle r, r \rangle\})$, the singleton tree. The proof of the first main theorem relies on the following simple, beautiful property, which also illustrates the "structural locality" of the notion of core and derivative.

**Lemma 9.** *(Locality Lemma) Assume* $\{\mathcal{T}_i : i \in I\}$ *contains at least 2 non-empty trees. Then* $d(\otimes\{\mathcal{T}_i : i \in I\}) = \otimes\{d[\{\mathcal{T}_i : i \in I\}]\} := \otimes\{\{d(\mathcal{T}_i) : i \in I\}\}$

**Proof.** 1. $d(\otimes\{\mathcal{T}_i : i \in I\}) \subseteq \otimes\{d[\{\mathcal{T}_i : i \in I\}]\}$. Assume $t \in d(\otimes\{\mathcal{T}_i : i \in I\})$. Assume $t \in \mathcal{T}_i$ for some $i \in I$. Then there are two distinct paths $P_1, P_2$ starting in $t$ and each containing a leaf, where all elements in this paths are in $\mathcal{T}_i$; therefore,

[8] For the reader who distrusts in infinite trees (as does the author) I should add: we could also restrict ourselves to finite trees, and then the statement: "there is a $\mathcal{T} \in D(G)$ with infinite FCB-rank" translates: "there is no finite upper bound to the FCB-rank of finite trees $\mathcal{T} \in D(G)$", and the two statements can be shown to be equivalent. But the latter usually requires an additional step, so using infinite trees is a matter of convenience.

[9] We use this as a shorthand for: trees with disjoint domains. In the sequel, for this and similar constructions we will always assume that trees are disjoint.

$t \in d(\mathcal{T}_i)$, and so $t \in \otimes\{d[\{\mathcal{T}_i : i \in I\}]\}$. Otherwise, assume we have $t = r$. Then by definition, $r \in \otimes\{d[\{\mathcal{T}_i : i \in I\}]\}$.

2. $\otimes\{d[\{\mathcal{T}_i : i \in I\}]\} \subseteq d(\otimes\{\mathcal{T}_i : i \in I\})$. Assume $t \in \otimes\{d[\{\mathcal{T}_i : i \in I\}]\}$. Then either $t \in c(\mathcal{T}_i)$ for some $i \in I$, and so $t \in c(\otimes\{\mathcal{T}_i : i \in I\})$, because the core is monotonous over the subtree relation. Or $t = r$. As $\{\mathcal{T}_i : i \in I\}$ contains two non-empty trees, there are at least two complete paths in $\otimes\{\mathcal{T}_i : i \in I\}$, and so $r \in d(\otimes\{\mathcal{T}_i : i \in I\})$.

This shows equality of the domains; we skip the proof of equality of relations, which follows in a straightforward fashion.[10] □

**Corollary 10.** *Let $\{\mathcal{T}_i : i \in I\}$ be a set of trees. Put $k = max\{FCB(\mathcal{T}_i) : i \in I\}$.*

1. *Assume there are $\mathcal{T}_i, \mathcal{T}_j : i \neq j, i, j \in I$, such that $FCB(\mathcal{T}_i) = FCB(\mathcal{T}_j) = k$. Then $FCB(\otimes\{\mathcal{T}_i : i \in I\}) = k + 1$.*
2. *Assume there are no $\mathcal{T}_i, \mathcal{T}_j : i \neq j, i, j \in I$, such that $FCB(\mathcal{T}_i) = FCB(\mathcal{T}_j) = k$. Then $FCB(\otimes\{\mathcal{T}_i : i \in I\}) = k$.*

**Proof.** 1. Assume we have $FCB(\mathcal{T}_i) = FCB(\mathcal{T}_j) = k$. Then $d^{k-1}(\mathcal{T}_i), d^{k-1}(\mathcal{T}_j)$ are two non-empty chains. As they are disjoint, there are two complete paths through $r$ in $d^{k-1}(\otimes\{\mathcal{T}_i : i \in I\}) = \otimes(d^{k-1}[\{\mathcal{T}_i : i \in I\}]$. Therefore, $d^k(\otimes\{\mathcal{T}_i : i \in I\}) = (\{r\}, \trianglelefteq_{\restriction\{r\}})$, and so, $FCB(\otimes\{\mathcal{T}_i : i \in I\}) = k + 1$.

2. Assume there is only one $\mathcal{T} \in \{\mathcal{T}_i : i \in I\}$ such that $FCB(\mathcal{T}) = k$, and all other trees have strictly smaller FCB-rank. It follows that for $s \notin \mathcal{T} \cup \{r\}$, we have $s \notin \otimes(d^{k-1}[\{\mathcal{T}_i : i \in I\}]) = d^{k-1}(\otimes\{\mathcal{T}_i : i \in I\})$. Moreover, $d^{k-1}(\mathcal{T})$ is a chain; so $d^{k-1}(\otimes\{\mathcal{T}_i : i \in I\}) = \otimes\{d^{k-1}(\mathcal{T})\}$, which is also a chain; so $d^k(\otimes\{\mathcal{T}_i : i \in I\}) = \mathcal{T}_\emptyset$. □

**Lemma 11.** *(1) If for a CFG $G = (\mathtt{S}, \mathcal{N}, A, R)$, we have $FCB(G) = \omega$, then there is $\mathcal{T} \in D(G)$ such that the complete infinite binary tree is a subtree of $\mathcal{T}$; moreover, (2) there is a reachable $N \in \mathcal{N}$ such that $N \vdash^*_G \alpha N \beta N \gamma$ for some $\alpha, \beta, \gamma \in (\mathcal{N} \cup A)^*$.*

**Proof.** We first prove (2) by contradiction. Assume $FCB(G) = \omega$. Then for any $k \in \mathbb{N}$, there is $\mathcal{T} \in D(G)$ with $FCB(\mathcal{T}) \geq k$. Choose $\mathcal{T}$ with $FCB(\mathcal{T}) \geq |\mathcal{N}|+1$. Then take the root $r$ with label $\mathtt{S}$. $r$ dominates two subtrees $\mathcal{S}_1, \mathcal{S}_1'$ of rank $\geq |\mathcal{N}|$ (see lemma 13). If nodes with label $\mathtt{S}$ occur in both $\mathcal{S}_1, \mathcal{S}_2$, then (2) follows; if not, there is a subtree of rank $\geq |\mathcal{N}|$, in which only $|\mathcal{N}| - 1$ labels occur ($\mathtt{S}$ does not). Take its root $t$ with label $X$. It dominates two subtrees $\mathcal{S}_2, \mathcal{S}_2'$ of rank $\geq |\mathcal{N}| - 1$. If $X$ occurs in both $\mathcal{S}_2, \mathcal{S}_2'$, then (2) follows; if not, there is a subtree of rank $\geq |\mathcal{N}| - 1$, in which only $|\mathcal{N}| - 2$ labels occur ($\mathtt{S}, X$ do not). We iterate this, until we are left with a subtree of rank $\geq 1$, in which 0 labels occur - contradiction, as our labelling is exhaustive. Thus there are $t, t', t'' \in \mathcal{T}$ with $t, t', t'' \in X$ for some label $X$, $t \trianglelefteq t', t''$ and $t' \prec t''$. By the connecting lemma, this holds if and only if $X \vdash^*_G \alpha X \beta X \gamma$.

[10] Note that if $\{\mathcal{T}_i : i \in I\}$ does not contain two non-empty trees, the result does not obtain: let $\mathcal{C}$ be a chain. $d(\otimes\{\mathcal{C}\}) = \mathcal{T}_\emptyset$, and $\otimes(d(\mathcal{C})) \neq \mathcal{T}_\emptyset$.

Now (1) follows easily: by $\mathtt{S} \vdash_G^* \alpha N\beta$, $N \vdash_G^* \alpha N\beta N\gamma$ and the connecting lemma we construct a tree having the complete infinite binary tree as a subtree. Take $\mathcal{T}$, in which there are infinitely many distinct $t, t', t'' \in \mathcal{T}$ such that $t, t', t'' \in N$, and $t \lhd t', t''$ and $t' \prec t''$ (this exists by the connecting lemma). Then there is a subtree $\mathcal{S} \subseteq \mathcal{T}$ which consists only of the nodes $s \in N$; this is a complete infinite binary tree. □

**Theorem 12.** *Given a context-free grammar $G$, there is an algorithm which computes $FCB(G)$.*

**Proof.** Part 1: the algorithm

Assume without loss of generality that every non-terminal is reachable; all those who are not can be thrown out anyway.

a) The infinite case: we check whether $FCB(G) = \omega$.

It is well-known that we can compute for any $N \in \mathcal{N}$ the set $der_G^1(N) := \{M : N \vdash_G^* \alpha M\beta\}$.[11] From this it follows that we can also compute $der_G^2(N) := \{(N_1, N_2) : N \vdash_G^* \alpha N_1\beta N_2\gamma\}$; we quickly explain the proof: we transform a grammar $G$ to $G'$ by adding non-terminals in $\mathcal{N} \times \mathcal{N}$, and rules $N \to (N_1, N_2)$ if $N \vdash_G \alpha N_1\beta N_2\gamma$ (NB: $\vdash_G$ is not the reflexive and transitive!); furthermore, if $N_1 \vdash_G^* \alpha N_3\beta$, $N_2 \vdash_G^* \alpha' N_4\beta'$, then we add rules $(N_1, N_2) \to \{(X, Y) : X \in \{N_1, N_3\}, Y \in \{N_2, N_4\}\}$. This procedure terminates, because the set of possible pairs is finite, and we can compute $der_G^1(N)$ for each $N \in \mathcal{N}$; the only point where we introduce new pairs is for the immediate derivability $\vdash_G$, which is trivially decidable. Call the resulting grammar $G'$. It can be shown that $(N_1, N_2) \in der_G^2(N)$ if and only if $(N_1, N_2) \in der_{G'}^1(N)$. Now if in a grammar $G = (\mathtt{S}, \mathcal{N}, A, R)$ we have an $N \in \mathcal{N}$, such that $(N, N) \in der_G^2(N)$, then we put $FCB(G) = \omega$.

b) The finite case. If $FCB(G) \neq \omega$, proceed as follows:

1. Pick out all $G$-rules of the form $N \to \alpha a\beta$ where $a \in A$. If $\alpha\beta \neq \epsilon$, put $val(N) := 2$, otherwise, $val(N) := 1$.
2. If there is a non-terminal $N$ such that $N \to M \in R$ for some $M$ such that $val(M) := n$, $val(N) \leq n$, we put $val(N) := n$.
3. If there is an $N$ such that $N \to \alpha M\beta O\gamma \in R$, where $min\{val(M), val(O)\} \geq n$, and $val(N) \leq n$, we put $val(N) := n + 1$.
4. Iterate steps 2 and 3 until we cannot assign any higher value to any non-terminal.
5. Put $FCB(G) = val(\mathtt{S})$.

Importantly, this procedure will converge only if there is no $N \in \mathcal{N}$ such that $N \vdash_G^* \alpha N\beta N\gamma$; so we first have to check whether $FCB(G) = \omega$.

Part 2: Correctness and completeness

a) The infinite case

*Completeness*: It follows from lemma 11 (2) that if $FCB(G) = \omega$, we have $N \in \mathcal{N}$ with $N \vdash_G^* \alpha N\beta N\gamma$, which is our algorithm can decide.

[11] For a more recent reference on decision problems of CFG, consider [8].

*Correctness*: If for $G = (\mathtt{S}, \mathcal{N}, A, R)$, $N \in \mathcal{N}$, we have $N \vdash_G^* \alpha N \beta N \gamma$, such that our algorithm assigns the grammar the FCB-rank $\omega$, then there is $\mathcal{T} \in D(G)$ with $FCB(\mathcal{T}) = \omega$; see the construction in proof of lemma 11 (1).

b) The finite case
Assume $FCB(G) \neq \omega$. Then we can establish by induction that we get the correct rank: we assign the correct rank to the elementary trees which are conform to a grammar $G$ (that is, 1 or 2). And if we assign the correct values to the elementary trees, then we assign the correct values to larger $G$-conform trees when we construct $G$-conform trees with the $\otimes$-method, as follows from corollary 10. Moreover, as the number of non-terminals is finite, the possible trees of relevant (binary) structure become less and less, otherwise we have $FCB(G) = \omega$; so the procedure terminates. □

# 6 Expressive Power

The next question is: what is the class of languages for which there is a CFG with bounded FCB-rank?[12] Recall that a context-free grammar is **linear**, if every rule in $R$ has the form: $N \to \alpha N' \beta$ with $\alpha, \beta \in A^*$. A language is linear if it is generated by some linear CFG. Obviously, linear grammars are related to bounded FCB-rank: $L$ is linear if and only if there is a grammar $G$ such that $L(G) = L$ and $FCB(G) \leq 2$. This is fairly straightforward. But note that there are grammars $G$ with $FCB(G) \leq 2$ which are not linear: we might have a rule $\mathtt{S} \to N_1 N_2 N_3$, where $N_1, N_3$ only generate unary branches. These can of course be brought into linear form; but we have to be careful in distinguishing properties of grammars and languages. For simplicity, we will mostly assume that our CFGs do not have unary rules; call such grammars *neat*. Then we can easily check that: $G$ is linear if and only if $G$ is neat and $FCB(G) \leq 2$. We now generalize this result. Let $A$ be an alphabet. A **(linear) substitution** is a map $\sigma : A \to \{L_a : a \in A\}$, where each $L_a$ is a (linear) language; this map is extended to strings and sets in the usual fashion, and can be applied to arbitrary languages $L \subseteq A^*$ (we use square brackets $f[-]$ to indicate the pointwise extension of a function to subsets of the domain. In general, if the $L_a : a \in A$ are all contained in a class of languages $\mathcal{C}$, then we say $\sigma$ is a substitution into $\mathcal{C}$. A **linear grammar substitution** $G = (\mathtt{S}, \mathcal{N}, A, R)$ by a set $\{G_a : a \in A\}$ of linear grammars is obtained by substituting all occurrences of a terminal $a$ in $R$ by $\mathtt{S}_{G_a}$, and then taking the union of the rules of $\{G_a : a \in A\}$ with the modified rules of $G$. Obviously, there is a close correspondence of linear substitutions and linear grammar substitutions. As this is obvious and well-known, we will be sometimes a bit sloppy when moving from grammars to languages and vice versa, avoiding some tedious details.

---

[12] In this section, we presuppose familiarity of the reader with standard techniques in formal language theory. For those who want some background in this topic, we refer to [1],[8].

Call a language of the form $\sigma_{k-1}[...\sigma_1[L]]$, where $L$ is linear and $\sigma_1, ..., \sigma_{k-1}$ are linear substitutions, $k$-**linear**; we denote the class of $k$-linear languages by $k$-**LIN**. The hierarchy of $k$-linear languages is well-known (see [1],p.209), and it has been shown in [4] that these classes form a proper infinite hierarchy.[13] Before we can present the main theorem connecting the FCB-rank with $k$-**LIN**, we need to establish some additional results.

Given a tree $\mathcal{T}$, $t \in \mathcal{T}$ a node, we define $rank(t, \mathcal{T})$ as follows: if there is $n \in \mathbb{N}$ such that $t \notin d^n(\mathcal{T})$, then $rank(t, \mathcal{T})$ is the smallest $n'$ such that $t \notin d^{n'}(\mathcal{T})$; and $\omega$ otherwise. It is obvious that $FCB(\mathcal{T}) \geq k$ if and only if there is $t \in \mathcal{T}$ such that $rank(t, \mathcal{T}) = k$, and for $r$ the root of $\mathcal{T}$, $FCB(\mathcal{T}) = rank(r, \mathcal{T})$. The first simple lemma is the following:

**Lemma 13.** *Given a tree $\mathcal{T}$, we have $FCB(\mathcal{T}) \geq k + l - 1$, if and only if there is a subtree $\mathcal{S} \subseteq \mathcal{T}$, such that 1. for all leaves $s$ of $\mathcal{S}$ , we have $rank(s, \mathcal{T}) = k$, and $FCB(\mathcal{S}) = l$*

**Proof.** *If*: As the rank of the leaves of $\mathcal{S}$ is $k$, we have $\mathcal{S} \subseteq d^{k-1}(\mathcal{T})$; as $FCB(\mathcal{S}) = l$, we have $d^{l-1}(d^{k-1}(\mathcal{T})) \neq \mathcal{T}_\emptyset$.

*Only if*: Assume $FCB(\mathcal{T}) \geq k + l - 1$. Put $\mathcal{S}' = d^{k-1}(\mathcal{T})$. For all leaves $s$ of $\mathcal{S}'$, we have $rank(s, \mathcal{T}) = k$. Moreover, $FCB(\mathcal{S}) \geq l$, because $FCB(\mathcal{T}) \geq k + l - 1$. We thus have to choose an appropriate subtree of $\mathcal{S} \subseteq \mathcal{S}'$. By the subtree properties, this tree exists. □

The next result concerns a property of substitutions. For two (linear) substitutions $\sigma_1$, $\sigma_2$, define $\sigma_1 \circ \sigma_2$ by $\sigma_1 \circ \sigma_2(a) = \sigma_1(\sigma_2(a))$. $\sigma_1 \circ \sigma_2$ is again a substitution (though not necessarily a linear one, if both $\sigma_1, \sigma_2$ are linear). It is easy to see that for any language $L$, $\sigma_1[\sigma_2[L]] = \sigma_1 \circ \sigma_2[L]$, for if $w \in \sigma_1 \circ \sigma_2[L]$, then there is $a_1...a_i \in L$, and $w \in \sigma_1 \circ \sigma_2(a_1)...\sigma_1 \circ \sigma_2(a_i)$; consequently, $w \in \sigma_1(\sigma_2(a_1))...\sigma_1(\sigma_2(a))$, and thus $w \in \sigma_1[\sigma_2[L]]$. Conversely, if $w \in \sigma_1[\sigma_2[L]]$, then there is $a_1...a_i \in L$, and $w \in \sigma_1(\sigma_2(a_1))...\sigma_1(\sigma_2(a_i))$. Consequently, $w \in \sigma_1 \circ \sigma_2(a_1)...\sigma_1 \circ \sigma_2(a_i)$, and $w \in \sigma_1 \circ \sigma_2(a_1...a_n) \subseteq \sigma_1 \circ \sigma_2[L]$. From this, we immediately obtain the following:

**Corollary 14.** *1. $\sigma_1 \circ (\sigma_2 \circ \sigma_3)[L] = (\sigma_1 \circ \sigma_2) \circ \sigma_3(L)$.*
*2. $\sigma_1[..[\sigma_k[L]]...] = \sigma_1 \circ ... \circ \sigma_k[L]$.*
*3. If $\sigma_1$ is a substitution into k-**LIN**, $\sigma_2$ into l-**LIN**, then $\sigma_1 \circ \sigma_2$ is a substitution into $k + l$-**LIN**.*

1. and 2. are immediate, 3. follows the easily from 1. and 2. We are now ready for the second main theorem.

**Theorem 15.** *There is a CFG $G$ with $L(G) = L$ and $FCB(G) \leq k + 2$, if and only if $L \in k$-**LIN**.*

[13] Note that 2-**LIN** is already a superclass of the class of languages recognized by $k$-turn pushdown automata, that is, by PDA which can change from pushing to popping and vice versa at most $k$ times (see [1] for reference). This is because each such language is obtained by a linear substitution on a finite language, and every finite language is linear.

**Proof.** *If*: Induction. For $k = 0$ this means: if $L$ is linear, then there is a $G$ with $L(G) = L$ and $FCB(G) \leq 2$. We leave this to the reader. Now assume (induction hypothesis) that for some $k$, if $L \in k$-**LIN**, then there is a $G$ with $L(G) = L$ and $FCB(G) \leq k + 2$. Assume (induction step) $L' \in k + 1$-**LIN**. Then there is $L \in k$-**LIN** and a linear substitution $\sigma$ such that $L' = \sigma[L]$. Furthermore, there is $G$ with $L(G) = L$ and $FCB(G) \leq k + 2$. We can now effect a linear grammar substitution of $G$ simulating $\sigma$, thereby obtaining $G'$ with $L(G') = L'$. Now take an arbitrary $\mathcal{T} \in D(G')$. It consists of a tree $\mathcal{S}$ in $D(G)$, with (possibly empty) linear grammar derivation trees departing from its leaves. If we now take $d(\mathcal{T})$, all these linear trees become chains, and as they are attached to leaves of $\mathcal{S}$, we have $FCB(d(\mathcal{T})) = FCB(\mathcal{S})$, and so $FCB(\mathcal{T}) \leq FCB(\mathcal{S}) + 1$, and so $FCB(G') \leq FCB(G) + 1$.

*Only if*: Induction. The base case is simple: if $FCB(G) \leq 2$, then $G$ is linear (modulo unary branching, which can be eliminated anyway). Now assume (induction hypothesis) that for all grammars $G$ with $FCB(G) \leq k+2$, $L(G) \in k$-**LIN**. Induction step: assume $FCB(G) \leq k+3$. Then by lemma 13, for each tree $\mathcal{T} \in D(G)$, we have a subtree $\mathcal{S} \subseteq \mathcal{T}$, with $FCB(\mathcal{S}) = 2$, and for each leaf $s$ of $\mathcal{S}$, we have $rank(s, \mathcal{T}) \leq k + 2$. Now for each of these $s$, the non-terminal corresponding to its label derives a set of trees of FCB-rank $\leq k+2$, as otherwise it contradicts, with lemma 13, our induction hypothesis. Moreover, by induction hypothesis, the languages derivable from these non-terminals (for $N$, this is defined as $\{w \in A^* : N \vdash^*_G w\}$) are in $k$-**LIN**. Call the grammars which result each from making one of these non-terminals the start-symbol of the grammar $G_N : N \in \mathcal{N}$ (however, that does not concern all non-terminals, only those which are only labels of nodes of rank $\leq k+2$!). This for the "outer trees". The "inner trees" $\mathcal{S}$ of FCB-rank 2 (modulo unary branching) are all linear grammar derivation trees, and if we consider the language formed by their leaves, we get a linear language. We can now construct a grammar which generates exactly the "inner trees" of FCB-rank 2 of the $G$-derivation trees, such that the leaves, instead of being non-terminals, are terminals being in one-to-one correspondence with non-terminals of $G$ (we write $a_N : N \in \mathcal{N}$). Call this grammar $G^i$; it is clear that $L(G^i)$ is linear. We can now obtain $L(G)$ as follows: define $\sigma$ by $\sigma(a_N) = L(G_N)$ (on other letters, let it compute the identity). We now have $\sigma[L(G^i)] = L(G)$, and by construction and corollary 14, $\sigma[L(G^i)]$ is a $k + 1$-linear language, so $L(G) \in k + 1$-**LIN**. □

We have already mentioned the following result of [4]:

**Theorem 16.** *(Greibach) For each $k \in \mathbb{N}$, $k$-**LIN** $\subsetneq k + 1$-**LIN**.*

As an immediate consequence, it follows that:

**Corollary 17.** *$CFL \supsetneq \bigcup_{k\in\mathbb{N}} k$-**LIN**; there are context-free languages $L$ such that $L \notin k$-**LIN** for all $k \in \mathbb{N}$.*

**Proof.** Assume the contrary: for some $k \in \mathbb{N}$, $k$-**LIN** $\supseteq CFL$. As for all $n \in \mathbb{N}$, $n$-**LIN** $\subseteq CFL$, that means that $k$-**LIN** $\supseteq (k + 1)$-**LIN**, contradiction. □

So the classes of languages generated by CFG of FCB-rank $k$ also form a proper, infinite hierarchy; furthermore, there are CFLs for which no grammar of bounded FCB rank exists:

**Corollary 18.** *For every $k \in \mathbb{N}$, there is a language $L$ such that there is a CFG $G$ with $FCB(G) = k+1$ and $L(G) = L$, but no CFG $G'$ with $FCB(G') \leq k$ and $L(G') = L$. Moreover, there is a CFL $L$, such that if $L(G) = L$, it follows that $FCB(G) = \omega$.*

The typical candidate for CFLs which are not in $k$-**LIN** is the family of Dyk-languages over alphabets of arbitrary cardinality. We do not have to explain that inverse implications of the form: "if $FCB(G) > k$ or $FCB(G) = \omega$, then $L(G)$ is X" are not legitimate: as is well-known, it is undecidable whether a CFG generates a regular or linear language. So knowing the FCB rank of a CFG does not help us in establishing a *lower bound* of the complexity of its language, it only might give an upper bound.

## 7 Beyond Context-Free Grammars

We have seen that a bounded FCB-rank for CFG corresponds to a well-established class of languages. The interesting thing is that our treatment is by no means restricted to CFG, as trees arise in many ways in formal language theory: in particular, we can look at formalisms beyond CFG. There are two primary ways to go: firstly, there are formalisms which keep the rule format of context-free grammars, while allowing operations on strings which are more complex than just concatenation. The most prominent example for this are multiple context-free grammars (MCFG,[11]). On the other side, there are formalisms extending rule-schemes directly to trees, such as regular tree grammars (RTG), tree-adjoining grammars (TAG), and more generally, context-free tree grammars (CFTG) (see [9], [2]). Regarding formalisms as MCFG, there is little of interest we can say: derivation trees are usually defined in exactly the same manner as in CFG. In this sense all tree-based results for CFG transfer to MCFG. What changes of course are the language-theoretic properties, but those are not the results we are interested in in the first place. So we will use the rest of this paper to show how the FCB-rank works for tree grammars.

We have mentioned RTG, TAG and CFTG. For reasons of space, we will only define and look at (a particular class of) CFTG, because the two former can be seen as special cases of the latter, and so they are *a fortiori* covered by our treatment.[14] The foundation for tree grammars is the representation of finite, ordered, labelled trees as terms. Let $\Sigma$ be an alphabet. *term*($\Sigma$) is defined as the smallest set such that 1. if $a \in \Sigma$, then $a \in$ *term*($\Sigma$), and if $\mathtt{t}_1, ..., \mathtt{t}_i \in$ *term*($\Sigma$), $a \in \Sigma$, then $a(\mathtt{t}_1, ..., \mathtt{t}_i) \in$ *term*($\Sigma$). Let $X$ be a countable set of variables. *term*$_X$($\Sigma$) is defined as follows: if $a \in \Sigma$, then $a \in$ *term*$_X$($\Sigma$); if $x \in X$, then

[14] More precisely, an RTG is a CFTG where all non-terminals have arity 0, and TAG correspond to simple monadic CFTG, see definitions below.

$x \in term_X(\Sigma)$; if $\mathtt{t}_1, ..., \mathtt{t}_i \in term_X(\Sigma)$, $a \in \Sigma$, then $a(\mathtt{t}_1, ..., \mathtt{t}_i) \in term_X(\Sigma)$. We generally use $\mathtt{t}$ as meta-variable for terms. By $\mathtt{t}[x_1, ..., x_i]$ we intend a term, such that the variables occurring therein are among $\{x_1, ...x_i\}$. It is easy to see how to translate terms into finite, ordered, labelled trees (up to isomorphism). A problem is that a term does not specify the names of nodes, yet we need some address to identify nodes. We therefore assume that in trees corresponding to terms, nodes are named as in a **tree domain**. A tree domain is a set $T \subseteq \mathbb{N}^*$ (we think here of numbers as abstract entities, not in binary, ...,decimal etc. representation), such that for $n, m \in \mathbb{N}$, $\overline{n}, \overline{m} \in \mathbb{N}^*$, if $\overline{nm} \in T$, then $\overline{n} \in T$, and if $\overline{n}n\overline{m} \in T$, $m < n$, then $\overline{n}m \in T$. Given a tree domain, we define the precedence order $\preceq$ as the (reflexive) lexicographic order *lex*, and $\trianglelefteq$ as the prefix relation *pref*. Tree domains give us a "canonical" node labels for any (ordered) tree, and so for $\mathtt{t} \in term_X(\Sigma)$, we denote by $\hat{\mathtt{t}}$ its associated canonical tree. In the sequel, we will sometimes mix notions of trees and notions of terms; this will usually not lead to confusion and keep the treatment to a manageable size. Contrary to subtrees, subterms are generally thought to be contingent: if $\mathtt{t} = a \in \Sigma$, then $subterm(\mathtt{t}) = \{a\}$; if $\mathtt{t} = a(\mathtt{t}_1, \ldots, \mathtt{t}_i)$, then $subterm(\mathtt{t}) = \{\mathtt{t}\} \cup \bigcup_{j=1}^{i} subterm(\mathtt{t}_i)$. We write $\mathtt{t}[\mathtt{t}_1, \ldots, \mathtt{t}_i]$ if $\mathtt{t}_1, \ldots, \mathtt{t}_i \in subterm(\mathtt{t})$. For $\sigma_1, \ldots, \sigma_i \in \Sigma$, we write $\mathtt{t}\langle\sigma_1, ..., \sigma_i\rangle$, if there are nodes $t_1, ..., t_i$ in $\hat{\mathtt{t}}$ labelled with $\sigma_1, ..., \sigma_i$, respectively, and for no two distinct $t_n, t_m \in \{t_1, ..., t_i\}$, we have $t_n \trianglelefteq t_m$. Let $\mathtt{t}[x_1, \ldots, x_i] \in term_X(\Sigma)$ with $x_1, \ldots, x_i \in X$. By $\mathtt{t}[\mathtt{t}_1/x_1, \ldots, \mathtt{t}_i/x_i]$ we denote the term which results by substituting $\mathtt{t}_1$ for $x_1, \ldots, \mathtt{t}_1$ for $x_1$.

A **CFTG** is a tuple $(\mathtt{S}, \mathcal{N}, \Sigma, R)$, where $\mathcal{N}$ and $\Sigma$ are disjoint, finite sets (of terminals and non-terminals), $\mathtt{S} \in \mathcal{N}$, and $R \subseteq (term_X(\Sigma \cup \mathcal{N}))^2$, where all rules in $R$ have the form $(N(x_1, ..., x_i), \mathtt{t}[x_1, ..., x_i])$, by which we hereby include the possibility that $x = 0$, and thus $N(x_1, ..., x_i) = N$. In general, we assume that non-terminals have a fixed arity, that is, they always occur with the same number of variables/subterms in $R$. The derivability relation $\vdash_G \subseteq (term_x(\Sigma \cup \mathcal{N}))^2$, for $G$ a CFTG, is defined as follows. We have $\mathtt{t}[N(\mathtt{t}_1, ..., \mathtt{t}_i)] \vdash_G \mathtt{t}[\mathtt{t}'[\mathtt{t}_1/x_1, ..., \mathtt{t}_i/x_i]]$, iff $(N(x_1, ..., x_i), \mathtt{t}'[x_1, ..., x_i]) \in R$; $\vdash_G^*$ is the reflexive and transitive closure, and $L(G) = \{\mathtt{t} \in term(\Sigma) : \mathtt{S} \vdash_G^* t\}$. We now define OI-derivations: We write $\mathtt{t}[N(\mathtt{t}_1, ..., \mathtt{t}_i)] \vdash_{GOI} \mathtt{t}[\mathtt{t}'[\mathtt{t}_1/x_1, ..., \mathtt{t}_i/x_i]]$, if in $\hat{\mathtt{t}}[N(\mathtt{t}_1, ..., \mathtt{t}_i)]$, there is no node labelled by a non-terminal which dominates (a node labelled by) the non-terminal we expand in the derivation. $\vdash_{GOI}^*$ is the reflexive and transitive closure of $\vdash_{GOI}$. For $\mathtt{t} \in term_X(\Sigma)$, we have $\mathtt{S} \vdash_G^* \mathtt{t}$ if and only if $\mathtt{S} \vdash_{GOI}^* \mathtt{t}$; this is an important fact of which we will make use in the sequel.

A CFTG is said to be **linear**, if for $(N(x_1, ..., x_i), \mathtt{t}[x_1, ..., x_i]) \in R$, each $x_1, ..., x_i$ occurs at most once in $\mathtt{t}[x_1, ..., x_i]$; it is **non-deleting**, if each variable occurs at least once; grammars which are both linear and non-deleting are said to be **simple**. To simplify our treatment, we will assume that our grammars are simple. Note that this restriction does not come without loss of generality (see [6]). But it is also well-known that linear and non-deleting CFTG still have considerable expressive power: for example, TAG are equivalent to CFTG which are simple and allow at most one variable to occur on each side of a rule (modulo some details, see [7]). Simple CFTG are particularly interesting, because the languages they generate are semilinear. Note that for $\mathtt{t} \vdash_G^* \mathtt{t}'$, we also allow variables to occur in $\mathtt{t}, \mathtt{t}'$, but by definition of $\vdash_G^*$ and assumption of non-deleting rules, a variable $x$ then occurs in $\mathtt{t}$ if and only if it occurs in $\mathtt{t}'$. Furthermore, we will assume (this time without loss of generality) that all nonterminals of our CFTG are both reachable and groundable. FCB-rank of a CFTG is defined in a straightforward fashion by $FCB(G) := sup\{FCB(\hat{\mathtt{t}}) : \mathtt{t} \in L(G)\}$. Note that we do not use derivation trees, but the trees generated by the grammar. As we do not generate infinite trees, we cannot use maximum instead of supremum, so we have to consider this in our constructions.

By $yield : term(\Sigma) \to \Sigma^*$ we denote the function mapping trees to words by forming the concatenation of their leaves; it is extended to sets in the canonical fashion. The above mentioned correlation of TAG and simple, monadic CFTG entails that we get a CFTG $G$ with $FCB(G) = 2$, where $yield(L(G))$ is not context-free. Consider the rules:

$$\mathtt{S} \to e(a, N(e(b, c)), d);\; N(x) \to e(a, N(e(b, x, c)), d);\; N(x) \to e(x).$$

This yields $\{a^n b^n c^n d^n : n \in \mathbb{N}\}$; yet, all derived trees have rank 2.

## 8 FCB-Rank of CFTG

We will now show that it is decidable whether $FCB(G) = \omega$ for a simple CFTG $G$. The procedure is more complicated than for CFG, so we will show its steps separately. A particularly useful normal-form for CFTG would be one where all trees on the right-hand side of rules have depth 2; this however is not available for the general case we consider. We will however assume without loss of generality that CFTG-rules are in **non-terminal normal form** (NTNF): all right-hand sides of rules have the form $\tau(\mathtt{t}_1, ..., \mathtt{t}_i)$, where $\tau \in \mathcal{N} \cup \Sigma$, and $\mathtt{t}_1, .., \mathtt{t}_i \in term_X(\mathcal{N})$. We thus do not allow terminals on the right hand side, except for the root of term-trees. It is easy to see that any CFTG can be brought into nonterminal normal form without substantial modification. Recall that we assume that all non-terminals are reachable and groundable.

**Lemma 19.** *Let $G$ be a simple CFTG. $FCB(G) = \omega$, if and only if there is an $N \in \mathcal{N}$, such that $N(x_1, ..., x_i) \vdash_G^* \mathtt{t}\langle N, N\rangle$.*

**Proof.** *If*: Assume we have such an $N \in \mathcal{N}$. Then $\mathtt{S} \vdash_{GOI}^* \mathtt{t}[N(\mathtt{t}_1, ..., \mathtt{t}_i)]$, where the node labelled by $N$ is not dominated by any nonterminal. As its position will not change in the final derived tree, we can already give him its address in the tree domain, say $\alpha$. By assumption, $N \vdash_{GOI}^* t\langle N, N\rangle$, where none of the two nodes labelled $N$ is dominated by any non-terminal; so having the address $\alpha_1, \alpha_2$ in $t\langle N, N\rangle$, we have $\mathtt{t}[N(\mathtt{t}_1, ..., \mathtt{t}_i)] \vdash_{GOI}^* \mathtt{t}'\langle N, N\rangle$, where the two nodes labelled $N$ have address $\alpha\alpha_1, \alpha\alpha_2$. Iterating this argument, we can generate a tree with an arbitrarily large finite complete binary subtree.

*Only if*: Assume there is no such $N \in \mathcal{N}$. The argument we use is similar to the CFG case, though slightly more delicate. If $S$ derives a tree of arbitrarily large FCB-rank, then it follows that there must be $N_1, N_2$, such that $S \vdash_G^* t\langle N_1, N_2\rangle$, none of the two is dominated by another non-terminal, and both $N_1, N_2$ derive trees of arbitrarily large FCB-rank (this uses OI-derivation and NTNF); otherwise, we run into a contradiction.[15] Now by assumption, only one of $N_1, N_2$ can derive $S$; assume without loss of generality it is $N_1$. For $N_2$ it also holds that there must be $N_{21}, N_{22}$, such that $N_2 \vdash_G^* t\langle N_{21}, N_{22}\rangle$, none of the two is dominated by another non-terminal, and both $N_1, N_2$ derive trees of arbitrarily large FCB-rank; none of them can derive $S$, and only one of them $N_2$. Pick the other one etc. Iterating this, we finish up with a non-terminal $M$, such that $M$ can derive trees of arbitrarily large FCB-rank, yet it cannot introduce any non-terminals - contradiction. □

**Lemma 20.** *For a CFTG $G$, $N \in \mathcal{N}$, it is decidable whether $N \vdash_G^* \mathtt{t}\langle N, N\rangle$ for some $\mathtt{t}$.*

**Proof.** We start by an algorithm checking whether $N \vdash_G^* \mathtt{t}\langle M\rangle$. That is easily decidable: just check whether $N \vdash_G \mathtt{t}\langle M\rangle$; and next check for all $N' : N \vdash_G \mathtt{t}\langle N'\rangle$ (except $N$) whether $N' \vdash_G \mathtt{t}\langle M\rangle$ etc. The procedure for checking $N \vdash_G^* \mathtt{t}\langle N, N\rangle$ is based on this: we check whether $N \vdash_G \mathtt{t}\langle N, N\rangle$, and then check for all $M, M'$ such that $N \vdash_G \mathtt{t}\langle M, M'\rangle$ whether $M, M' \vdash_G^* \mathtt{t}\langle N\rangle$, $M \vdash_G^* \mathtt{t}\langle N, N\rangle$ or $M' \vdash_G^* \mathtt{t}\langle N, N\rangle$ and iterate this until we have checked immediate derivability for all non-terminals. □

**Corollary 21.** *Given a simple CFTG $G$, we can decide whether $FCB(G) = \omega$ or $FCB(G) < \omega$.*

So we have decidability for the infinite case. How about the finite case? Here are several problems to overcome; for reasons of space, we do not present the algorithm we think does the job, but only put forward the following conjecture:

**Conjecture 22.** *Given a simple CFT $G$, $FCB(G) < \omega$, there is a terminating algorithm which computes $FCB(G)$.*

The proof of this is based on $n$-cuts of trees: intuitively, an $n$-cut of a tree is a subset of its domain, such that its $\trianglelefteq$-predecessors form a subtree with FCB-rank $n$ (thus an $n$-cut must contain at least $2^{n-1}$ elements). We can use these $n$-cuts to determine how CFTG-rules increase the rank of derived trees. Important open questions are the following: what classes of (string) languages correspond with CFTG with (a particular) bounded FCB-rank? It is clear that they are not contained in the context-free languages; do they conversely contain the context-free languages for a certain rank?

[15] Actually, this uses still another fact: if we substitute trees of finite FCB-rank for the nodes of a tree of finite FCB-rank, we get a tree of finite FCB-rank. This argument is rather simple, though, and we do not have the space to introduce the relevant notions.

## 9 Conclusions

The main goal of this paper was to introduce the notion of finitary Cantor-Bendixson rank of trees, and establish its major properties. The first main result was that the rank of a context-free grammar is decidable. This is surprising, because many similar properties of CFGs are undecidable. The explanation is that it is a property concerning the strong generative capacity of a CFG, rather than the language generated. Our second main result was that there is a correspondence of two proper infinite hierarchies, the hierarchy of $k$-linear languages, and the class of grammars with FCB-rank $k+2$. This is an interesting result, because the former notion is languages-theoretic, whereas the latter comes from relation-theory, and when applied to grammars, refers to their strong generative capacity. We thus have an interesting relation between notions from very diverse fields. The third main result concerned the decidability of the FCB-rank of simple context-free tree grammars. We have shown that it is decidable whether a simple CFTG has infinite rank and put forward the conjecture that its precise rank is computable; for reasons of space, we could not present the algorithm for its computation and its correctness proof.

## References

1. Berstel, J.: Transductions and Context-free Languages. Teubner, Stuttgart (1979)
2. Engelfriet, J., Schmidt, E.M.: Io and oi. i. J. Comput. Syst. Sci. 15(3), 328–353 (1977)
3. Fraïssé, R.: Theory of Relations. Studies in logic and the foundations of mathematics, vol 118. North-Holland (1986)
4. Greibach, S.A.: Chains of full AFL's. Mathematical Systems Theory 4, 231–242 (1970)
5. Hawkins, J.A.: Efficiency and Complexity in Grammars. Oxford Univ. Pr., Oxford (2004)
6. Hofbauer, D., Huber, M., Kucherov, G.: Some results on top-context-free tree languages. In: Tison, S. (ed.) CAAP 1994. LNCS, vol. 787, pp. 157–171. Springer, Heidelberg (1994)
7. Kepser, S., Rogers, J.: The equivalence of tree adjoining grammars and monadic linear context-free tree grammars. Journal of Logic, Language and Information 20(3), 361–384 (2011)
8. Kracht, M.: The Mathematics of Language. Studies in Generative Grammar, vol. 63. Mouton de Gruyter, Berlin (2003)
9. Rounds, William C.: Mappings and grammars on trees. Mathematical Systems Theory 4(3), 257–287 (1970)
10. Rubin, S.: Automata presenting structures: A survey of the finite string case. Bulletin of Symbolic Logic 14(2), 169–209 (2008)
11. Seki, H., Matsumura, T., Fujii, M., Kasami, T.: On multiple context-free grammars. Theor. Comp. Sci. 88, 191–229 (1991)
12. Simmons, H.: The extendend Cantor-Bendixson analysis of trees. Algebra Universalis 52, 439–468 (2005)

# Author Index